Structural Integrity

Volume 5

The *Structural Integrity* book series is a high level academic and professional series publishing research on all areas of Structural Integrity. It promotes and expedites the dissemination of new research results and tutorial views in the structural integrity field.

The Series publishes research monographs, professional books, handbooks, edited volumes and textbooks with worldwide distribution to engineers, researchers, educators, professionals and libraries.

Topics of interested include but are not limited to:

- Structural integrity
- Structural durability
- Degradation and conservation of materials and structures
- Dynamic and seismic structural analysis
- Fatigue and fracture of materials and structures
- Risk analysis and safety of materials and structural mechanics
- Fracture Mechanics
- Damage mechanics
- Analytical and numerical simulation of materials and structures
- Computational mechanics
- Structural design methodology
- Experimental methods applied to structural integrity
- Multiaxial fatigue and complex loading effects of materials and structures
- Fatigue corrosion analysis
- Scale effects in the fatigue analysis of materials and structures
- Fatigue structural integrity
- Structural integrity in railway and highway systems
- Sustainable structural design
- Structural loads characterization
- Structural health monitoring
- Adhesives connections integrity
- Rock and soil structural integrity

More information about this series at http://www.springer.com/series/15775

Emmanuel E. Gdoutos

Editor

Proceedings of the First International Conference on Theoretical, Applied and Experimental Mechanics

 Springer

Editor
Emmanuel E. Gdoutos
Office of Theoretical and Applied
  Mechanics
Academy of Athens
Athens
Greece

ISSN 2522-560X   ISSN 2522-5618   (electronic)
Structural Integrity
ISBN 978-3-319-91988-1   ISBN 978-3-319-91989-8   (eBook)
https://doi.org/10.1007/978-3-319-91989-8

Library of Congress Control Number: 2018944306

Printed on acid-free paper

This Springer imprint is published by the registered company Springer International Publishing AG
part of Springer Nature
The registered company address is: Gewerbestrasse 11, 6330 Cham, Switzerland

Preface

This volume contains 42 six-page papers and 59 two-page abstracts presented at the "First International Conference on Theoretical, Applied and Experimental Mechanics," (ICTAEM_1) held in Paphos, Cyprus, June 17–20, 2018. The papers/abstracts are arranged in four topics and two special symposia with 69 and 32 papers/abstracts, respectively. The papers of the tracks have been contributed from open call, while the papers of the symposia have been solicited by the respective organizers.

ICTAEM_1 will focus in all aspects of theoretical, applied, and experimental mechanics including biomechanics, composite materials, computational mechanics, constitutive modeling of materials, dynamics, elasticity, experimental mechanics, fracture, mechanical properties of materials, micromechanics, nanomechanics, plasticity, stress analysis, structures, wave propagation. During the conference, special symposia covering major areas of research activity organized by members of the Scientific Advisory Board will take place.

The attendees of ICTAEM_1 will have the opportunity to interact with the most outstanding world leaders and get acquainted with the latest developments in the area of mechanics. ICTAEM_1 will be a forum of university, industry, and government interaction and exchange of ideas in an area of utmost scientific and technological importance.

I am sure that besides the superb technical program, you will enjoy the majestic town of Paphos with its unique beaches and scenic beauty, many areas of historical interest and archeological importance, the delicious local cuisine, and the traditional Cypriot hospitality.

More than a hundred participants attended ICTAEM_1. The participants of ICTAEM_1 came from 25 countries. Roughly speaking, 36% came from Europe, 25% from the Far East, 7% from the Americas, and 32% from other countries. I am happy and proud to have welcomed in Paphos well-known experts who came to discuss problems related to the analysis and prevention of failure in structures. The tranquility and peacefulness of this small town provided an ideal environment for a group of scientists and engineers to gather and interact on a personal basis. Presentation of technical papers alone is not enough for effective scientific communication. It is the healthy exchange of ideas and scientific knowledge, formal and informal discussions, together with the plenary and contributed papers that make a fruitful and successful meeting. Informal discussions, personal acquaintance, and friendship play an important role.

I am proud to have hosted ICTAEM_1 in the beautiful town of Paphos, and I am pleased to have welcomed colleagues, friends, old and new acquaintances.

I very sincerely thank the authors who have contributed to this volume, the symposia/sessions organizers for their hard work and dedication, and the referees who reviewed the quality of the submitted contributions. The tireless effort of the members of the Organizing Committee as well as of other numerous individuals and people behind the scenes is appreciated. I am deeply indebted to Dr. Stavros Shiaeles for his hard work and dedication in the organization of the conference. Finally, a special word of thanks goes to Dr. Maria Shiaele for her continuous collaboration and support.

March 2018   Emmanuel E. Gdoutos

Contents

Contents

Miscellaneous (Biomechanics, Compu-tational mechanics, Dynamics, Nano-mechanics, Plasticity, Structures, Wave propagation)

Materials: Properties, Manufacturing, Modelling

Influence of Sample Preparation on Determined Nanomechanical Properties of Metastable Calcium Carbonate Polymorphs

Radek Ševčík[1(✉)] and Vladimír Hrbek[2]

[1] Centre of Excellence Telč, Institute of Theoretical and Applied Mechanics
of the Czech Academy of Sciences,
Batelovská 485-486, 588 56 Telč, Czech Republic
sevcik@itam.cas.cz
[2] Faculty of Civil Engineering, Czech Technical University in Prague,
Thakurova 7/2077, Prague 6, Czech Republic

Abstract. Synthetically prepared metastable calcium carbonates ($CaCO_3$) polymorphs, vaterite and aragonite, were used for nanoindentation testing. Nanomechanical measurements were done on samples embedded in epoxy resins. Hardness and reduced modulus were determined and both values were found to be higher for vaterite. Hardness was determined to be in the ranges 1.1 to 0.2 and 0.5 to 0.1 GPa for vaterite and aragonite, respectively. Reduced modulus was found to be in the ranges 32 to 5 and 18 to 4 GPa, for vaterite and aragonite, respectively. Reduced modulus of vaterite was found to be approximately three times lower in comparison with vaterite sample prepared as pressed pellets. These results may be helpful for designing new products containing $CaCO_3$ for applications.

Keywords: $CaCO_3$ · Hardness · Reduced modulus

1 Introduction

Calcium carbonate ($CaCO_3$) is one of the most widely studied compounds due to high importance in many industrial fields as well as in natural processes [1, 2]. It can be present as three anhydrous polymorphs, two hydrated phases and as amorphous $CaCO_3$ [3]. The most stable anhydrous polymorph, under Earth surface conditions, is calcite. Other two metastable polymorphs – vaterite and aragonite – can be converted to calcite under specific conditions following the Wolfgang Ostwald's law of stages.

Due to different chemical-physical properties of these anhydrous polymorphs, the measurements of their mechanical properties are necessary for the critical assessment of their usage in various applications. In literature, there is lack of such information, only few studies dealt with determination of hardness (H) and reduced modulus (E_r) of $CaCO_3$ from different origin, mostly biogenic [4], not directly comparable with synthetic one.

In this contribution, nanoindentation testing was used to obtain information about H and reduced modulus E_r of synthetically prepared vaterite and aragonite. The aim of

© Springer International Publishing AG, part of Springer Nature 2019
E. E. Gdoutos (Ed.): ICTAEM 2018, SI 5, pp. 3–7, 2019.
https://doi.org/10.1007/978-3-319-91989-8_1

this investigation was to compare two different approaches of sample preparations for nanoindentation testing and their possible influence on determined H and E_r values. In our previous study [5], samples were prepared as pressed pellets whereas in this work samples were cast in epoxy resin and carefully ground and polished (grinding method).

2 Experimental Part

2.1 Sample Preparation

Vaterite and aragonite were synthesized as previously described without use of additives [6, 7]. The composition of obtained powders was examined using quantitative phase analysis of collected XRPD patterns (for more details see ref. [6]).

Each powder $CaCO_3$ polymorph was mixed with the epoxy resin and left over 24 h to harden in cylindrical mold. Diamond cut-off wheel was used to separate approximately 15 mm thick part of each sample. MD-Piano equipped with 1200, 2000 and 4000 grain cm^{-2} grids was used for grinding. Fine polishing was done using alcohol based DP-Suspension containing 3 and 1 µm polycrystalline diamonds on MD-Dur and MD-Nap cloth, respectively. MD-Chem cloth combined with alcohol-based DP-Spray containing 0.25 µm diamond particles was selected for final polishing.

2.2 Nanodentation Testing

Nanoindentor TI 750 L Ubi (Hysitron) was used for testing. The indentation was done on the total area of 190 × 190 µm on the grid consists of 20 × 20 indents with 10 µm spacing between them using three sided pyramidal diamond Berkovich indenter. Prior to the testing, area of 2 500 µm^2 (50 by 50 µm) in center of each indentation position was scanned using in-situ scanning probe microscopy to determine surface roughness of samples. Measured root mean square roughness (R_q) of specimens containing vaterite and aragonite was found to be 243 and 250 nm, respectively. The load-controlled quasi-static indentation was performed with maximum applied force 500 µN. The loading/unloading time was set to 2 s with a 5 s peak hold time.

3 Results and Discussion

Quantitative phase analysis performed using the Rietveld method revealed that pure vaterite (100 ± 2 wt%) and aragonite with small amount of calcite (4.7 ± 0.2 wt%) were synthetized. The prepared testing vaterite specimen is depicted on Fig. 1 together with the vaterite specimen prepared directly from powder as pressed pellet using hydraulic press [5]. On Fig. 2 are shown mean P-h curves of testing specimens of vaterite and aragonite prepared with these two different methods. The same trend is clearly visible. Maximum load was reached at lower displacement for samples of vaterite. On the contrary, positions of P-h curves for vaterite and aragonite are much closer to each other in case of the usage of grinding method.

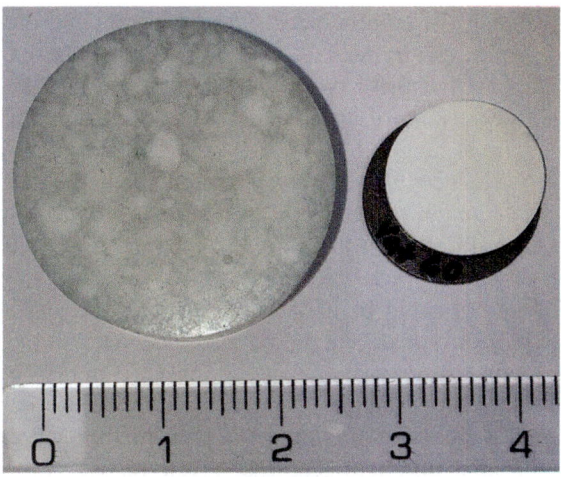

Fig. 1. Photographic image of the prepared specimens, in both cases containing synthetized vaterite, used for nanoindentation analysis. On the left, specimen prepared using grinding method. On the right, specimen prepared as pressed pellet.

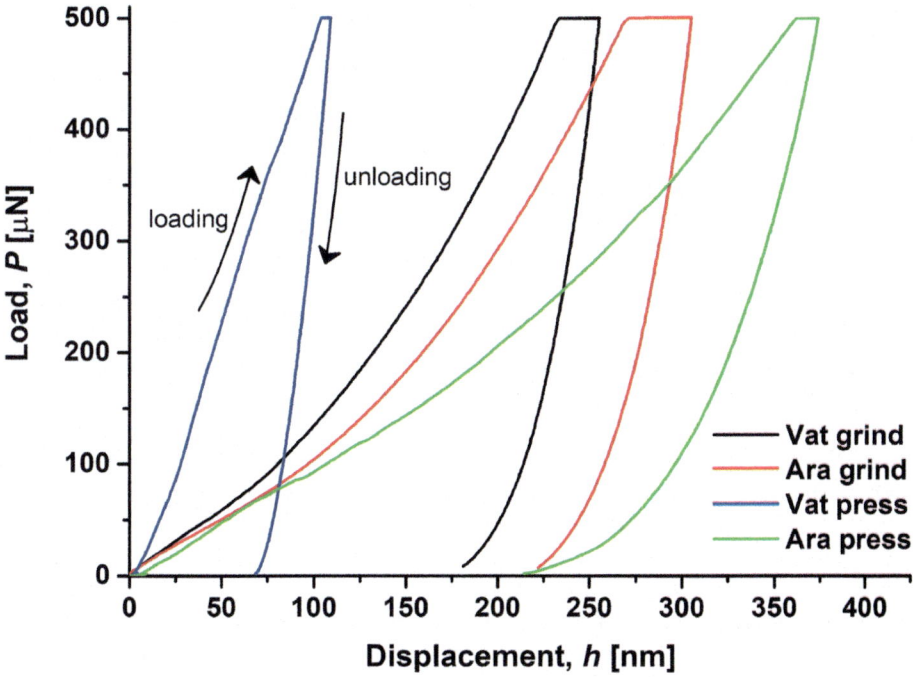

Fig. 2. Comparison of the mean load-displacement (P-h) curves of synthetized metastable $CaCO_3$ polymorphs. Testing specimens were prepared with two different methods: grind = grinding method, press = pressing method (Vat – vaterite, Ara – aragonite).

As reported in Table 1, the highest mean E_r and H were determined for vaterite. E_r and H determined for aragonite were found to be lower about 23% and 27%, respectively. One can see that these mean values introduce high standard deviations due to the significant decreasing of these values with increasing contact depth. Hardness was found to be in the ranges 1.10 to 0.18 and 0.51 to 0.13 GPa for vaterite and aragonite, respectively. Reduced modules were found to be in the ranges 32 to 5 and 18 to 4 for vaterite and aragonite, respectively. In comparison with samples prepared as pressed pellets, the main difference was detected for Er and H of vaterite. These values are about three times higher. Aragonite's H and E_r values is only slightly higher (Table 1). Such differences could be ascribed to the compactness of the samples as consequence of the applied force during pressing and the fact that the upper layer, used for nanoindentation testing, was mostly affected as reported in [5]. Higher variations of E_r and H values detected for vaterite specimens could be ascribed to the vaterite morphology. Synthetized vaterite was composed of microsized spherules built up from nanosized particles. For this morphological reason, vaterite was probably more affected by the force application during specimens' preparation rather than needle-like shaped aragonite particles with lengths up to 60 μm [5].

Table 1. Summary of mean values of reduced modulus (E_r) and hardness (H) as determined using nanoindentation instrument for specimens prepared with two different methods (grind – grinding method, press – pressing method).

Sample	E_r [GPa]	H [GPa]
Vaterite (grind)	9.5 ± 5.3	0.33 ± 0.16
Vaterite (press) [5]	31 ± 8	0.9 ± 0.6
Aragonite (grind)	7.3 ± 2.6	0.24 ± 0.07
Aragonite (press) [5]	10 ± 5	0.3 ± 0.2

Comparison with other literature data is complicated due to lack of information about mechanical properties of synthetic vaterite and aragonite. E_r and H were detected for biogenic vaterite originated from fresh water carp astericus and were found to be 57 and 3.2 GPa [8]. In case of aragonite originated from fresh water carp lapillus, E_r and H were found to be 67 and 4.9 GPa [8]. It is well known that $CaCO_3$ originated from biomineralization processes are formed with highly organized structures and exhibit better mechanical properties [9]. Recently, E_r and H of a biomimetic prismatic-type vateritic thin film were reported to be 32.5 ± 2.4 and 2.2 ± 0.2 GPa, respectively, values comparable with those of vaterite prepared as pressed pellets.

The pressing method was applied because of the significant reduced time (from hours to minutes) needed for preparation of specimens, and also because contamination with other chemicals, like epoxy resins, that could interfere the measurements [4], is avoided. However, as it was shown above, disadvantage may be the increased compactness of samples, probably influencing the measured nanomechanical properties. Investigation of the effect of different forces applied during the pressing procedure on values of E_r and H is the subject of the future research.

4   Conclusion

In this contribution, nanomechanical properties, reduced modulus and hardness, of synthetized metastable anhydrous $CaCO_3$ polymorphs were investigated using nanoindentation instrument. The testing specimens of vaterite and aragonite were prepared using epoxy resin with subsequent grinding and polishing. E_r were found to be in the ranges 32 to 5 and 18 to 4 GPa for vaterite and aragonite, respectively. H were found to be in the ranges 1.10 to 0.18 and 0.51 to 0.13 GPa. In comparison with the previously obtained results of the same synthetized $CaCO_3$ polymorphs prepared as pressed pellets, E_r and H for vaterite were found always to be higher than for aragonite. In case of pressed samples, determined values were found to be higher for both $CaCO_3$ polymorphs, especially for vaterite. The possible explanation could be found in the effect of samples compactness that is more affecting smaller vaterite particles rather than larger particles of aragonite. The obtained results may have implication to producing $CaCO_3$ containing materials with improved properties for applications.

Acknowledgement. The authors gratefully acknowledge support from the Czech Science Foundation GA ČR grant 17-05030S and project LO1219 under the Ministry of Education, Youth and Sports National sustainability program I of Czech Republic. We thank Eva Pažourková for helping with preparation of testing specimens.

References

1. Dobrev, J., Markovic, P.: Calcite: Formation, Properties, and Applications. Nova Science Publishers, New York (2012)
2. Dhami, N.K., Reddy, M.S., Mukherjee, A.: Biomineralization of calcium carbonates and their engineered applications: a review. Front. Microbiol. **4**, 1–13 (2013)
3. Meldrum, F.C., Cölfen, H.: Controlling mineral morphologies and structures in biological and synthetic systems. Chem. Rev. **108**, 4332–4432 (2008)
4. Dhami, N.K., Mukherjee, A., Reddy, M.S.: Micrographical, minerological and nano-mechanical characterisation of microbial carbonates from urease and carbonic anhydrase producing bacteria. Ecol. Eng. **94**, 443–454 (2016)
5. Ševčík, R., Šašek, P., Viani, A.: Physical and nanomechanical properties of the synthetic anhydrous crystalline CaCO3 polymorphs: vaterite, aragonite and calcite. J. Mater. Sci. **53**, 4022–4033 (2018)
6. Ševčík, R., Pérez-Estébanez, M., Viani, A., Šašek, P., Mácová, P.: Characterization of vaterite synthesized at various temperatures and stirring velocities without use of additives. Powder Technol. **284**, 265–271 (2015)
7. Sarkar, A., Mahapatra, S.: Synthesis of all crystalline phases of anhydrous calcium carbonate. Cryst. Growth Des. **10**, 2129–2135 (2010)
8. Ren, D., Meyers, M.A., Zhou, B., Feng, Q.: Comparative study of carp otolith hardness: lapillus and asteriscus. Mater. Sci. Eng. C **33**, 1876–1881 (2013)
9. Xiao, C., Li, M., Wang, B., Liu, M.-F., Shao, C., Pan, H., Lu, Y., Xu, B.-B., Li, S., Zhan, D., Jiang, Y., Tang, R., Liu, X.Y., Cölfen, H.: Total morphosynthesis of biomimetic prismatic-type CaCO3 thin films. Nat. Commun. **8**, 1398–1406 (2017)

Crystallization and Dissolution of Common Salts - Damage Potential to Porous Media

Veronika Koudelková[(⊠)] and Benjamin Wolf

Institute of Theoretical and Applied Mechanics CAS, v.v.i., Prosecká, 809/76,
19000 Prague, Czech Republic
koudelkova@itam.cas.cz

Abstract. Growing crystals of soluble salts could cause degradation of porous building materials due to generation of crystallization pressure inducing tensile stress inside porous system. Considerable damage potential has been observed in case of sodium sulfate through phase change and rapid formation of hydrated phase mirabilite from highly supersaturated solution rising from dissolution of anhydrous phase thenardite after changing of surrounding conditions. Crystallization of sodium chloride can also lead to damage but the intensity is not as evident in comparison with sodium sulfate. The extent of salt attack strongly depends particularly on the environmental conditions and salt content in the material. The morphology of crystals (NaCl, Na_2SO_4 and mixture of both in ratio 1:1) and phenomena related to dissolution were studied with optical microscope. Conclusions from microscopic observation were applied to real porous system - sandstone subjected to salinization and wetting-drying cycles. The massive damage (>50%) showed the specimen containing single sodium sulfate crystals which are during wetting subjected to phase transition accompanied by volume change. The damage caused by sodium chloride and by mixture was much lower - 1% and 3% respectively. Such low mass change could be explained by greater amount of efflorescence and also by lower damage potential of NaCl and the mixture.

Keywords: Soluble salts · Porous media · Salt deterioration

1 Introduction

Crystallization of salts is one of the main harmful phenomenon causing deterioration of porous building materials. Salts are mostly transported in the dissolved form with water that can rise in to the masonry from ground and thus the most endangered are historical buildings with no waterproofing. Growing crystals of salts in the contact with pore walls are able to generate tensile stress in the material and possibly evoke damage, i.e. creation of cracks and loss of material. Sodium chloride and sodium sulfate can either generate high crystallization pressures whose magnitude is dependent on the supersaturation of salt solution and rate of crystallization. In other words supersaturation is a driving force for the crystallization against obstacle [1]. Sodium chloride is most common salt crystallizing in the porous network due to its using as deicing agent in the winter period. Sodium sulfate is used in standard tests for testing the resistivity of

© Springer International Publishing AG, part of Springer Nature 2019
E. E. Gdoutos (Ed.): ICTAEM 2018, SI 5, pp. 8–13, 2019.
https://doi.org/10.1007/978-3-319-91989-8_2

stones against salt crystallization. Sodium sulfate has two stable phases – anhydrous thenardite (Na_2SO_4) and decahydrate mirabilite ($Na_2SO_4 \cdot 10H_2O$). The phase transition between both phases can occur in the common environmental conditions. The anhydrous phase is stable at lower relative humidity and higher temperatures whereas decahydrate is stable at higher values of relative humidity and lower temperatures [2]. Nevertheless, porous system of real stone rather than single salt type contains mixture of salts and their hydrates which are characterized by broad diversity and distribution in the space. Characterization of behavior such mixture is highly complicated and depends among others on the temperature, relative humidity and composition of pore solution [3, 4].

2   Materials and Methods

The dynamics of salt crystallization was observed firstly under optical microscope to reveal how the single salts and mixture of them behave. Small volume (approx. 1 μm) of saturated salt solutions of NaCl (6.16 mol $\cdot$ kg^{-1} at 20 °C), Na_2SO_4 (4.56 mol $\cdot$ kg^{-1} at 20 °C) and its mixture 1:1 was dropped on the microscopic slides. The salts were left to crystallize and its morphology was investigated. The crystals were later wetted with demineralized water (Carl Roth, Germany) and the dissolution of salts was observed. For the experiment with sodium sulfate it is necessary to reveal how the anhydrous phase behave after contact with water. The second part of experiment was based on the observation of degradation of sandstone during wetting-drying cycles. Feldspathic coarse grained sandstone from the locality near Kamenné Žehrovice in the Czech Republic is characterized by its good quality. This sandstone was frequently used in medieval architecture. From the sandstone were cut testing specimens with dimensions $2 \times 1 \times 0.5$ cm. The specimens were weighted and then immersed into saturated solutions and left to dry in the climatic chamber at 20 °C and 40% of relative humidity. Specimens containing salts were thereafter wetted with 100 μl of demineralized water and left to dry in climatic chamber at 20 °C and 40% RH. The wetting-drying cycles were repeated 2 times. The specimens were after each cycle weighted and the loss of salts from the porous space, i.e. the amount of efflorescence, was monitored. To prevent specimens from undesirable loss of material during manipulation, both largest planes on the specimens were fixed with tape.

3   Results

3.1   Microscopic Observation

The morphology of crystals of NaCl, Na_2SO_4 and their mixture is depicted in the Fig. 1. The shape of crystals strongly depends on the rate of evaporation of salt solution. Evaporation causes supersaturation close to liquid-vapor interface, i.e. at the boundary of sessile droplet, followed by formation of equilibrium crystal shapes known as subflorescence. Sodium chloride creates typical cubic-shape crystals (see Fig. 1a) whereas sodium sulfate creates dendritic and octahedral-shape crystals (see Fig. 1b).

Moreover, rapid evaporation mostly induces forming of considerable amount of microcrystals known as efflorescence which can be typically found on the surface of stones referred to as salt crust. Group of well-developed crystals is surrounded with this "crust" along the external part (see Fig. 1a). Efflorescence can thus reduce the evaporation from the subsurface zone in the porous media and cause crystallization of subflorescence which is able to generate crystallization pressures. The morphology of crystals formed from the mixture of NaCl and Na_2SO_4 solution in ratio 1:1 was identical with shapes formed from both single NaCl and Na_2SO_4 saturated solutions respectively. Octahedral-shaped crystals of Na_2SO_4 tended to grow predominantly inside the droplet whereas cubic crystals and microcrystals of NaCl created a rim around the sodium sulfate phases (see Fig. 1c).

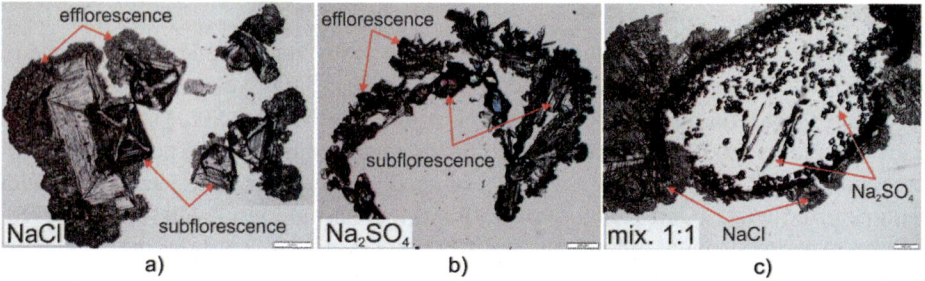

Fig. 1. Morphology of crystals: (a) sodium chloride, (b) sodium sulfate, (c) mixture of NaCl and Na_2SO_4 in ratio 1:1.

Whenever crystals are wetted with water they rapidly start to dissolve until the solution become saturated since the system requires to be equilibrated, i.e. to have the smallest possible energy. The distribution of saturation in the droplet is not homogenous. The highest concentration of ions appears near the sessile droplet boundary whereas the lowest concentration is in the center of the droplet. This phenomenon was observed on the base of dissolution of crystals close to the center of droplet and subsequent crystallization at the interface (see Fig. 2a). Sufficient concentration of ions for development of supersaturation followed by crystallization is sustained by evaporation as well as by migration of solute due to diffusion in liquid or by advection with liquid [5]. Hence, salt solution in the porous system tends to migrate towards the zone with maximum evaporation, i.e. to the surface and near subsurface space of porous specimen. Formation of crystals in the subsurface zone causes delamination and disintegration of stone. Sodium sulfate has two stable phases – anhydrous (thenardite) and decahydrate (mirabilite) whose transition is dependent on environmental conditions. When thenardite comes to contact with water, it starts to dissolve and the surrounding solution becomes highly supersaturated with respect to mirabilite. It was observed that mirabilite crystallizes very rapidly on the surface of thenardite crystal which acts as nucleation area. The phase transition is accompanied by extensive volume change (see Fig. 2b). This phenomenon is the reason of extreme damage to which are exposed porous specimens during testing of resistivity against salt attack.

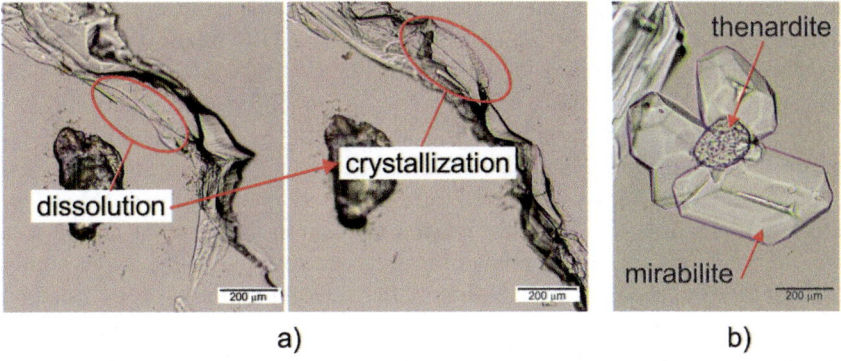

Fig. 2. Sodium sulfate: (a) rapid dissolution of crystals provides ions for crystallization at the liquid/air interface, (b) phase change of thenardite to mirabilite during wetting.

3.2   Sandstone Degradation

Assessment of variability of the moisture content in the wet specimens of sandstone as well as the salt content in dried specimens was performed by gravimetric analysis. Figure 1 shows dependence of the dry specimen weights on the number of wetting-drying cycles after removing of efflorescence (Fig. 3).

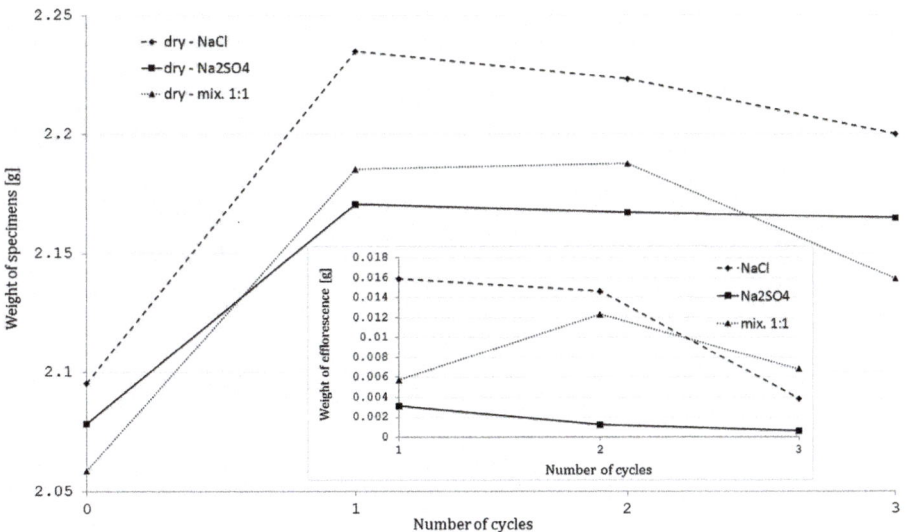

Fig. 3. Diagram showing dependence of mass change on the number of performed cycles.

The small diagram embedded in the large one represents the amount of efflores-cence crystallized during cycles 1 to 3 on the surface of specimens. Solution containing sodium chloride tended to migrate towards the specimen surface in contrast to sodium sulfate that mostly remained in the porous system of the stone. In form of efflorescence crystallized 27% of NaCl, 9% of Na_2SO_4 and 23% of the mixture. The amount of efflorescence in case of mixture of both salts suggests that the addition of NaCl to Na_2SO_4 solution changes among others properties mobility of solution (surface tension) and thus it can more easily migrate through the porous system [6]. Weights referred to the cycle 0 represent the initial reference state at the beginning of experiment for the dried specimens without salts. The specimens were immersed in the saturated solutions of NaCl, Na_2SO_4 and mix of them in the ratio 1:1 and then dried. The weights of dried specimens containing salts are referred to the cycle number 1. Loss of material was assessed by weighting after removing of salts from tested bodies by means of leaching with pure demineralized water. The specimen containing sodium sulfate was the most damaged one, overall loss of material reached 58% and the specimen broke into two pieces. The damage caused by NaCl and mixture 1:1 was insignificant in comparison with attack of sodium sulfate (NaCl – 1% and mixture – 3%). Low damage of specimens caused by NaCl and mixture could be most likely explained by (i) flow of solution towards the surface of specimen where salt crystallized as efflorescence and (ii) lower damage potential of NaCl and mixture of salts. Even if NaCl and Na_2SO_4 crystallized from the mixed solution as two separate phases (see Fig. 1c), it is obvious that properties of mixed solution are quite different in comparison to single salt solutions. Hence, for the generation of more damage in the sandstone should be performed more dissolution-wetting cycles. Photographs of the specimens before and after salt attack are shown in the Fig. 4.

reference   mix. 1:1 Na_2SO_4

Fig. 4. Specimens before (marked as "reference") and after salt attack (marked as "mix. 1:1" and "Na_2SO_4").

4   Conclusion

Microscopic investigation of crystallization and dissolution dynamic of sodium chloride, sodium sulfate and mixture of them confirmed strong dependence of crystal morphologies on the evaporation rate of solution. High evaporation of solution in dry conditions induces developing of microcrystalline forms know as efflorescence whereas in the zone with reduced evaporation crystallizes well-defined crystals able to create damage known as subflorescence. Sodium chloride tends to crystallize rather on the surface as efflorescence in contrast to sodium sulfate which tends to remain in the pore system of material. Sodium sulfate has extreme damage potential due to volume change during phase transition from anhydrous thenardite to decahydrate mirabilite after wetting. The results obtained from the microscopy observation as well as from the experiment with sandstone confirmed this phenomena. Nevertheless, development of damage is considerably restrained in real porous systems that hardly ever contain salts in single form. Characterization of properties of salt mixtures is very complicated and is primarily related to the temperature, relative humidity and composition of pore solution. Hence, in this experiment more realistic is behavior and subsequent damage generated by mixture of $NaCl$ and Na_2SO_4. Characterization of distribution of salts and salt mixtures in the porous sandstone and their damage potential during humidity cycling will be subject of future research.

References

1. Scherer, G.W.: Crystallization in pores. Cem. Concr. Res. **29**(8), 1347–1358 (1999)
2. Flatt, R.J.: Salt damage in porous materials: how high supersaturations are generated. J. Cryst. Growth **242**(4), 435–454 (2002)
3. Steiger, M., Heritage, A.: Modelling the crystallization behaviour of mixed salt systems: input data requirements. In: 12th International Congress on Deterioration and Conservation of Stone Proceedings, pp. 1–13. ICOMOS – ISCS, New York (2012). http://iscs.icomos.org/cong-12.html
4. Lindström, N., Heitmann, N., Linnow, K., Steiger, M.: Crystallization behavior of $NaNO_3$-Na_2SO_4 salt mixtures in sandstone and comparison to single salt behavior. Appl. Geochem. **63**, 116–132 (2015)
5. Pel, L., Huinink, H.: Ion transport and crystallization in inorganic building materials as studied by nuclear magnetic resonance. Appl. Phys. Lett. **81**(15), 2893 (2002)
6. Ruiz-Agudo, E., Mees, F., Jacobs, P., Rodriguez-Navarro, C.: The role of saline solution properties on porous limestone salt weathering by magnesium and sodium sulfates. Environ. Geol. **52**(2), 269–281 (2007)

Torsional Shear Testing of Mortar

Miloš Drdácký$^{(\boxtimes)}$, Michal Hlobil, Jiří Kunecký, Miloš Černý,
and Benjamin Wolf

Institute of Theoretical and Applied Mechanics of the Czech
Academy of Sciences, Prosecká 76, 190 00 Praha 9, Czech Republic
{drdacky,hlobil,kunecky}@itam.cas.cz

Abstract. This paper presents a complex multidisciplinary approach to characterize the torsional shear on thin-walled cocciopesto lime mortar tubes. The approach employs (i) an in-house developed biaxial electro-mechanical loading frame designed for torsional loading, (ii) an optical measurement setup based on the digital image correlation technique to monitor and record the deformation during loading and (iii) a numerical model using finite element simulation to assess localization of shear failure. Prior to torsional loading, the mortar specimen needs to be firmly secured into the loading frame without exposing it to unwanted tensile or bending loading and possible crushing at fixture points. During the torsional loading, the mortar tube must be able to freely move axially. An optical measurement setup based on the digital image correlation technique allowed to quantify deformation measurements during torsional testing and provided the strain field on the observed surface. The mode of torsional failure was finally assessed from finite element simulations and compared to experimental measurements.

Keywords: Torsional shear test · Thin-walled cocciopesto mortar tube
Digital image correlation

1 Introduction

Torsional shear of a circular thin-walled tube is considered an ideal test method to determine the shear modulus and the shear strength of any material. The method is rather complex and not so widely used because it requires preparation of specific test specimens as well as restrictive loading conditions. Starting with the experimental setup, the specimen needs to be firmly secured into the loading frame without exposing it to unwanted tensile or bending loading and possible crushing at fixture points. During the torsional loading, the specimen must be able to freely move axially.

Regardless of the difficulties involved, torsional testing is a promising test method since it allows for repeated or cyclic loading of quasibrittle composites. In this paper we focus a pilot study of shear loading of cocciopesto lime mortar tube.

© Springer International Publishing AG, part of Springer Nature 2019
E. E. Gdoutos (Ed.): ICTAEM 2018, SI 5, pp. 14–20, 2019.
https://doi.org/10.1007/978-3-319-91989-8_3

2   Materials

The studied cocciopesto lime mortar composition was optimized to suit the production of small laboratory testing specimens. The constituent dosage was based on volumetric ratios of raw materials. The binder consisted of a natural hydraulic lime Calcidur 3.5 intermixed with white Portland cement HET in a mass ratio 1:2. Aggregates consisting of river sand Borek together with crushed bricks were then added to the mixture in 5 and 2 volumetric parts, respectively. Lastly, sufficient amount of water was added to create a workable fresh mortar with a resulting initial water-to-binder mass ratio of fresh paste equal to 0.714.

2.1   Test Specimens

Former experimental programs for testing efficiency of various consolidation and strengthening agents on mechanical characteristics of lime mortars successfully used thin-walled mortar specimens [1]. The developed manufacturing procedure was applied for preparing cocciopesto mortar tubes of the following nominal dimensions - tubes of 40 mm in outer diameter, of the height of 40 mm, and a wall thickness of 4 mm. The tubes were formed by casting into a stainless steel formwork which was removed shortly after compacting the fresh mortar mixture. The mortar tubes were cured by a gentle spraying with water for two weeks, and then let to mature for more than two years.

Additionally, mortar prisms were fabricated from the same batch of fresh mortar to determine flexural and compressive strength. Prisms were 105 mm long with a square cross-section of 20 × 20 mm. After subjecting the prisms to a three-point bending test, cubes with side length of 20 mm were dry-cut out using a diamond saw from the remains of the broken prisms.

3   Experimental Testing

Mortar prisms were first subjected to a three-point bending test using an electro-mechanical loading machine Wolpert equipped with Lukas 2 kN load cell. The testing protocol followed ČSN 12372 (721145). The mortar prism was supported over a span of 80 mm and centrally loaded with a constant speed of 0.15 mm/min until failure. The average bending strength measured on 6 prisms amounted to 11.92 ± 0.65 MPa.

The broken remains of the mortar prisms were subsequently dry-cut to create cubes, which were then tested for uniaxial compression using the same machine equipped now with an MTS 25 kN load cell. A constant loading speed of 0.45 mm/min was used on all cubic specimens in accordance with a ČSN 1926 test protocol. The resulting measured compressive strength from 6 measurements amounted to 50.38 ± 11.70 MPa.

3.1   Loading Frame for Torsional Testing

A biaxial electro-mechanical loading frame of our own construction (see Fig. 1) was constructed to carry out the torsional testing. Two stepper motors with lead screws perpendicularly arranged each other ensure axial and torsional loading of a tested specimen.

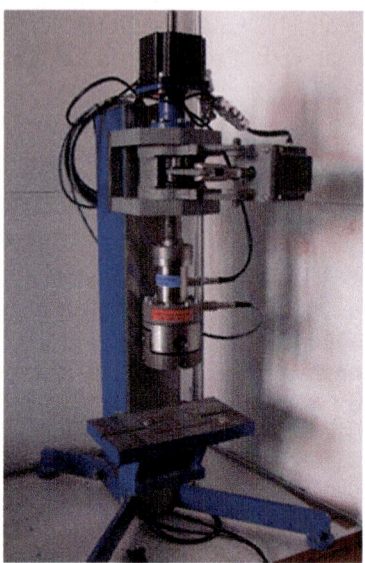

Fig. 1. Loading frame developed for torsional testing.

The forces are measured with two independent loading cells. The loading cell for the axial force has a range of ±10 kN while the torsional one features the range of ±50Nm. The loading head is controlled via a regulation loop with a selectable precision of a step size of the smallest motor movement – in order to ensure the smallest regulation error and a smooth force increase on hard and fragile specimens. A force measurement provided by two loading cells processed by two high-precision 24-bits ADCs ensures enough resolution to regulate the force and position of the head during the whole measurement. This arrangement allows to compensate the axial force generated due to a firm fixing of the specimen and thus can simulate its necessary free axial movement. The advantage of this arrangement is that it also provides the ability to perform measurements under defined and controlled axial force.

3.2   Test Setup

Prior to torsional testing, two supporting metal plates were firmly attached to the top and bottom surfaces of the mortar tube using polyurethane glue. Both plates contained a shallow rugged groove used to seat the tube, centralize its position during hardening of the glue, and provide increased resistance against rotation. The specimens were then fixed in a loading frame designed specifically for torsional testing.

Initially, the mortar tube was preloaded by a constant axial load equal to 30 N to facilitate the installation of the specimen into the loading frame and to avoid tensile loading that could originate from tightening the specimen into the frame. Once the tube was fixed into place, a constant horizontal displacement was applied until a maximum horizontal angle 3° reached or failure of tube occurred.

3.3   Shear Deformation Measurements

The deformation of the mortar tube during torsional testing was measured using an optical measurement setup based on the digital image correlation technique. A 5MPix digital camera Basler ac2440-20gm equipped with a fixed focus 25 mm lens with less than 1% distortion was used for the capturing pictures. The camera setup was calibrated before the image sequence was taken. Images were captured every 3 s. The influence of the curved surface was studied on a wooden dummy specimen [2]. An array of tracking characteristics was defined on the testing scene and the tracking was performed using Digital Image Correlation (DIC) algorithm, see Fig. 2. The algorithm uses IC-GN method [3] to correlate the points and calculate the displacements. Subsequently, a MATLAB script was used to compute the shear strain.

Fig. 2. An array of tracking points defined on mortar tube specimen used to monitor displacements during torsional loading.

3.4   Numerical Modeling

A 3D numerical analysis using finite elements was carried out to provide an additional insight into the torsional failure mode of mortar tubes. The full geometry of the tube including top and bottom loading plates was discretized by a finite element mesh consisting of 11298 tetrahedrons with linear displacement interpolations. The displacements in all directions on all nodes located at the bottom loading plate were fixed. The top plate was fixed in the axial direction of the mortar tube but free to move in a plane perpendicular to the axial tube direction. An incremental displacement was applied on the corner points of the top plate as loading. A linear elastic material (E = 70 GPa, nu = 0.2) was assigned to both top and bottom loading plates, while an isotropic damage material was used for the mortar tube, with the following material characteristics E = 11.3 GPa, nu = 0.3, fc = 50.38 MPa, Gf = 20 J/m2. The preprocessing of the simulation was carried out using Salome software, the numerical computation was solved by the OOFEM software package [4] and post-processing of results provided by Paraview.

4   Results

The mortar tube was loaded by a constant horizontal displacement until failure. Peak torsion moment recorded at failure was 22.3 kN and the horizontal twist angle was 2°, see Fig. 3 for full loading diagram.

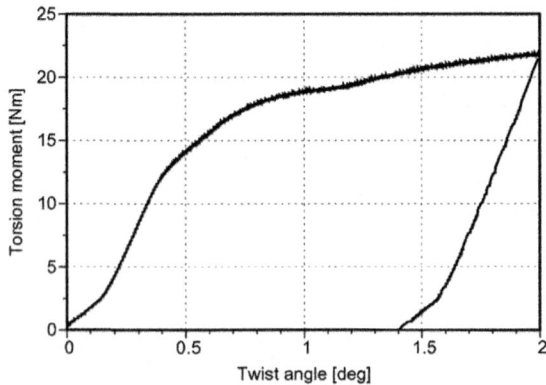

Fig. 3. Experimentally measured torsion moment and horizontal twist angle.

Failure of the mortar tube localizes into only a single shear band inclined by 45°. Results from finite element simulations indicate an identical failure but multiple shear bands are present, compare both modes of failure in Fig. 4.

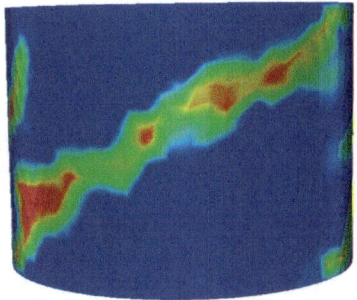

Fig. 4. Mortar sample after torsional failure (left); finite element model showing a localized shear crack.

Digital Image Correlation technique accurately captured the initiation and development of the shear crack the even though the fracture occurred on a rounded surface of the tube, see Fig. 5.

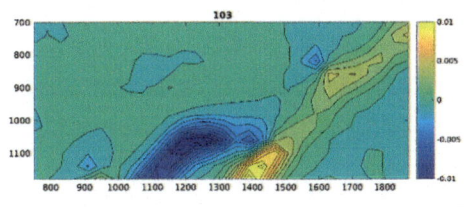

Fig. 5. Shear failure after loading (left); strain field obtained from digital image correlation (right).

5 Conclusions

Torsional shear of a circular thin-walled cocciopesto mortar tube was analyzed using a complex multidisciplinary approach involving (i) an in-house developed biaxial electro-mechanical loading frame designed for torsional loading, (ii) an optical measurement setup based on the digital image correlation technique to monitor and record the deformation during loading and (iii) a numerical model using finite element simulation to assess localization of shear failure. The developed testing methodology will be further refined and extended to repeated (cyclic) loading in the near future.

Acknowledgement. This paper is based on results and experience acquired with support from the MEYS grant LO1219 and the Czech Science Foundation Project No. P105/12/G059.

References

1. Drdácký, M., Slížková, Z.: Lime-water consolidation effects on poor lime mortars. APT Bull. J. Preserv. Technol. 1(43), 31–36 (2012)
2. Doktor, T., Koudelka, P., Černý, M., Wolf, B., Drdácký, M.: Strain measurements in tubular specimens under combined tensile-torsional loading. In: Gdoutos, E. (ed.) 17th International Conference on Experimental Mechanics (ICEM 17), Rhodes, Greece (2016)
3. Pan, B., Li, K., Tong, W.: Fast, robust and accurate digital image correlation calculation without redundant computations. Exp. Mech. 7(53), 1277–1289 (2013)
4. Patzák, B.: OOFEM - an object-oriented simulation tool for advanced modeling of materials and structures. Acta Polytech. 52(6), 59–66 (2012)

Comparative Tests of Strengthening Effects on Weak Mortars Consolidated with Various Agents

Zuzana Slížková$^{(\boxtimes)}$, Dita Frankeová, and Miloš Drdácký

Institute of Theoretical and Applied Mechanics of the Czech Academy of
Sciences, Prosecká 76, 190 00 Praha 9, Czech Republic
slizkova@itam.cas.cz
http://itam.cas.cz, http://cet.arcchip.cz

Abstract. The capacity of various consolidating liquids to strengthen weak
mortars is reported in this paper. Commercial products (CaLoSiL IP 15, KSE
100) and liquids prepared in laboratory by dissolving chemicals in water (barium hydroxide, ammonium oxalate, ammonium phosphate) or by dilution more
concentrated products (Syton X 30) were used for the experiment. In order to
study the influence of the mineralogical composition of the mortar, various sorts
of sand were alternatively used in the experiment for the mortar preparation. The
efficiency of the treatment was evaluated as mechanical strength change measured on different shape specimens - on tubes for compression strength, on thin
plates for tension test and on beams for flexural strength. All the tested consolidation agents showed positive strengthening effect when they were used for
consolidation of the poor lime mortar, with the exception of the influence of
lime water on the flexural strength. Very different results were achieved on
soil-based mortar.

Keywords: Lime mortar · Consolidating treatment · Non-standard testing
Strengthening

1 Introduction

Consolidation of degraded mortar or render represents one of the most complicated
restoration technology as well as an intervention which to a large extent influences
material qualities of a historical object. Conservation science streams constantly to
produce new agents for stone and mortar conservation which are demanded on the
market. However, their introduction in conservation practice requires a careful investigation of their impact on the historical material from the sustainability and conservation policy of views.

An assessment and evaluation of impact of new agents on cultural heritage is not an
easy task which usually has no ambition to study all possible effects or to attain
absolute standard values because the relative data are mostly sufficient for decision
making. Ideally, the effects should be examined on real materials in real conditions of
their application and future action. In reality, the investigations are mostly carried out in
laboratories with some complementary limited checks on real objects in situ. In fact,

© Springer International Publishing AG, part of Springer Nature 2019
E. E. Gdoutos (Ed.): ICTAEM 2018, SI 5, pp. 21–26, 2019.
https://doi.org/10.1007/978-3-319-91989-8_4

there are not available appropriate testing facilities and procedures for exact determination of some characteristics in situ. Therefore, an optimum extent of tests is usually set up based on study of performance of the most critical effects which might compromise the required compatibility of intervention with the historic material and its condition and operational environment. They mostly involve testing of mechanical and other physical and chemical characteristics in the case of consolidation [1].

2 Experimental

2.1 Consolidation treatment

The intent of the study was to compare the capacity of various consolidating liquids to strengthen weak mortars. Lime water investigated in previous experiment on weak lime mortar [2] proved to be effective after a large number of applications (160 saturations). Less labour-intensive treatments were tested in the present experiment.

Commercial products (CaLoSiL IP 15, KSE 100) and liquids prepared in laboratory by dissolving chemicals in water (barium hydroxide, ammonium oxalate, ammonium phosphate) or by dilution more concentrated products (Syton X 30) were used for the experiment. Also some combinations of lime based and silica based consolidating liquids were tested to study synergetic effects of both products. Consolidating liquids were applied on specimen surface using an automatic manual pipette. The volume of the applied liquid in one action (usually in one day) was chosen individually for each specimen in order to reach full saturation of mortar. If the liquid need to be applied repeatedly on specimen surface, the following dose was applied after the previous one had dried. The condition of the experiment was that so much consolidating liquid would be applied on mortar surface (by its repeated impregnation), to deliver 30 mg of the binder, contained in the liquid, into 1 cm^3 of the mortar specimen. Table 1 gives data of used binders contained in various consolidating liquids. As the binder concentration in various consolidation liquids was different, also the quantity of these liquids to be applied needed be various, as well as the time required for specific treatments.

Table 1. Binder forming components in tested consolidation liquids

Binder forming solid component	Consolidation liquid	Symb.	Solvent
Calcium Hydroxide	Lime water, 0.16 w%	L	Water
Calcium Hydroxide	Nanolime CaLOSiL IP 15, 15 g/L	CL	Isopropyl alcohol
Calcium Hydroxide	Nanolime ZFB, 15 g/L	ZFB	Isopropyl alcohol
Barium Hydroxide	BH-water solution, 4 w%	B	Water
Silica	Syton X 30, dil. to 10 w%	S	Water
Silica	KSE 100	KSE	Organic solvent
Ammonium Oxalate	AO water solution, 2.5 w%	AO	Water
Dihydrogen Ammonium Phosphate	DAP water solution, 5 w%	AP	Water

After completion of impregnation, the mortar specimens were left for at least 90 days at laboratory conditions and then subjected to mechanical tests. In addition to the mechanical properties that were the main subject of study, also microstructure and water transport properties (the water absorption rate, vapor permeability) of mortars were observed.

2.2   Testing Specimens

Former experimental studies for testing efficiency of lime water on mechanical characteristics of lime mortars successfully used thin-walled mortar specimens in the shape of tubes for compression strength and thin plates for tension test [2, 3]. Thin specimens enable even distribution of the consolidation agent across the entire volume of the test specimen, and shorten the time for maturing. The same specimen shapes were therefore used also in the present study. In addition to the thin specimens, beams for measuring the bending strength of mortars were prepared and tested. Figure 1 illustrates shapes of specimens for different strength testing, the specimens parameters were following: tube 29 × 39 × 5.5 mm, plates 60 × 40 × 6 mm, beams 101 × 20 × 20 mm.

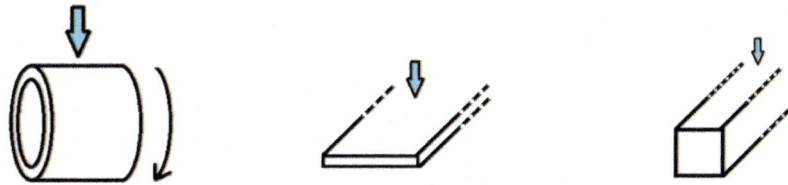

Fig. 1. Specimens shape: tube for compression, plate for tension, beam for flexural strength.

The specimens were fabricated from a weak lime mortar which was previously prepared using sand (9 volume parts) and small quantity of lime hydrate (1 volume part) with water. Six specimens of the same shape and same mortar composition were tested as reference samples to found out mechanical characteristics of the specific mortar and then sets of six specimens were tested for each consolidation liquid after the consolidation treatment of the mortar. In order to study the influence of the mineralogical composition of the mortar, various sorts of sand were alternatively used in the experiment for the mortar preparation: siliceous (mainly quartz) sand with semiangular to semirounded grains, quartzite crashed sand and limestone crashed sand. Possible chemical reaction of used sands with the lime binder was studied in special research by conducting of pozzolanic activity test [4]. All mortar sorts were prepared using the same part of sand of the same granularity and the same part of lime hydrate and water. Besides four sorts of lime mortar with different type of sand, one sort of earth mortar was prepared and tested in the study. The consolidation treatment of prepared mortar specimens was conducted after all lime mortar specimens were fully carbonated. The carbonation rate of lime mortars was checked by means of thermal analysis. The illustration of the consolidated tubes and the following compressive test are documented in Fig. 2.

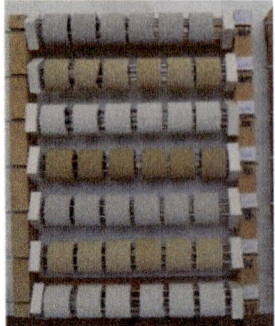

Fig. 2. Mortar tubes after the consolidation treatment (left), compression test (right).

3   Results and Discussion

The experiments conducted on various types of mortar composition brought the knowledge that the strengthening efficiency due to consolidation treatment is influenced significantly by the characteristics of consolidated mortar. In terms of lime mortars and especially the mortar with limestone aggregate, the "nano-lime" (dispersion of calcium hydroxide nanoparticles in alcohol), marketed under the trade name CaLoSiL IP 15, proved to be very effective strengthening agent. Also the combined applications of lime-based consolidants (lime water or CaLoSiL IP 15) with silica-based consolidants (diluted Syton X30 or KSE 100) have shown very good consolidation effects on lime mortars. Results are given in Figs. 3, 4 and 5.

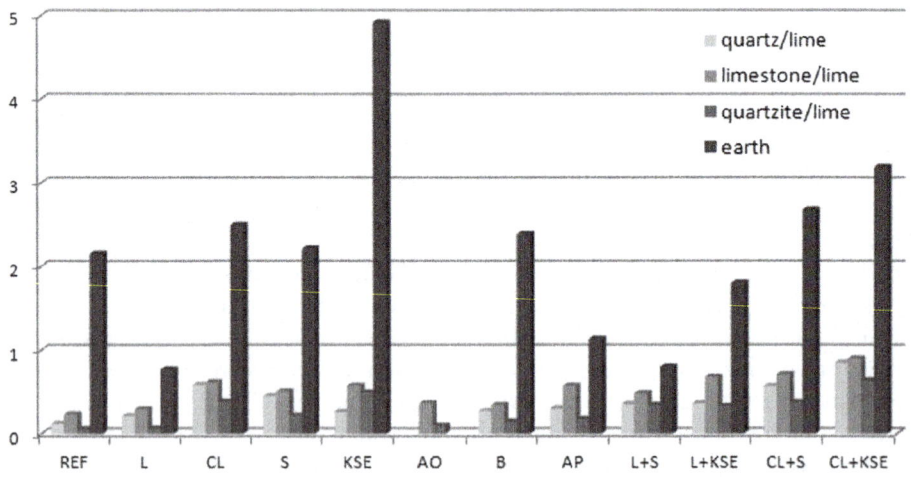

Fig. 3. Compressive strength (MPa) of mortar tubes: non-treated (REF) and treated with different consolidation treatment

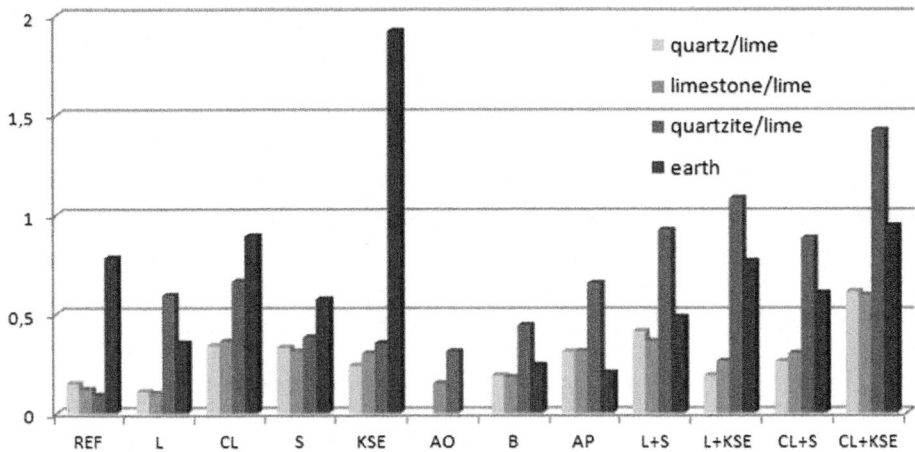

Fig. 4. Flexural strength (MPa) of mortar beams: non-treated (REF) and treated with different consolidation treatment

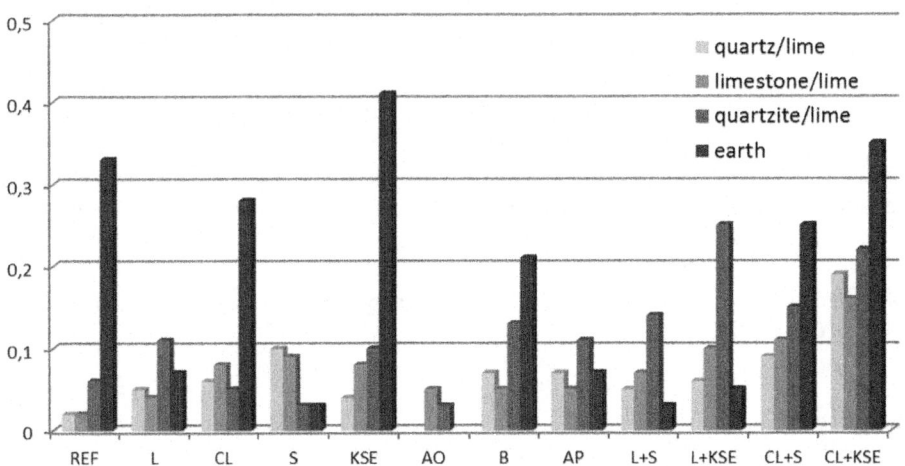

Fig. 5. Tension strength (MPa) of mortar plates: non-treated (REF) and treated with different consolidation treatment

Entirely different results were achieved when the same consolidants were used for strengthening of earth mortar. For that mortar, KSE 100, product based on TEOS, was the most efficient consolidant in relation to strength improvement. Unfortunately, KSE 100 treatment caused significant inhibition of the water drop absorption into the treated mortar due to the change of the surface tension. However, water vapour permeability of tested mortars was not significantly changed after the consolidation treatments.

Contrary to the lime mortars results, only slight effect was found in the case of the earth mortar for CaLoSiL IP 15, probably due to low penetration of the consolidant into the material.

For the practical use, the financial costs of consolidation intervention represent one of the most fundamental aspects. Table 2 shows that treatments including lime water are several times time demanding than other.

Table 2. Total period of treatment (in days) of beam samples

Sand/binder mortar composition	L	CL	S	KSE	AO	B	AP	L+S	L+KSE	CL+S	CL+KSE
Quartz/lime	108	18	2	2		3	3	57	57	16	16
Limestone/lime	108	10	2	2	8	3	3	57	57	9	9
Quartzite/lime	108	13	2	2	13	4	3	57	57	9	9
Earth	64	23	2	2		5	4	66	66	15	15

4 Conclusions

All the tested consolidation agents showed positive strengthening effect when they were used for consolidation of the poor lime mortar, with the exception of the influence of lime water on the flexural strength. Study of the pozzolanic activity of various types of sand in mortars showed that the grains of sand may be reactive with calcium hydroxide and can participate in the formation of hydraulic compounds when they are subjected to calcium hydroxide for a long time (several months at room temperature). In treated lime mortar samples, however, the products of this reaction were not detected, probably because of the relatively small amount of these substances in mortar and their X-ray diffraction amorphous character.

Acknowledgement. The authors acknowledge kind support from the Czech Science Foundation project GAČR P105/12/G059.

References

1. van Hees, R., Veiga, R., Slížková, Z.: Consolidation of renders and plasters. Mater. Struct. **50** (1), 65 (2017)
2. Drdácký, M., Slížková, Z.: Lime-water consolidation effects on poor lime mortars. APT Bull. **43**(1), 31–36 (2012)
3. Moreau, C., Slížková, Z., Drdácký, M.: Consolidation of pure lime mortars with nanoparticles of calcium hydroxide. In: Proceeding of the 2nd Historic Mortars Conference HMC 2010 and RILEM TC 203-RHM 'Repair Mortars for Historic Masonry' Final Workshop, Prague, CR, pp. 1113–1121 (2010)
4. Frankeová, D., Slížková, Z.: Determination of the pozzolanic activity of mortar's components by thermal analysis. J. Therm. Anal. Calorim. **125**(3), 1115–1123 (2016)

Novel Device for 4-Point Flexural Testing of Quasi-Brittle Materials During 4D Computed Tomography

Petr Koudelka[1]($\boxtimes$), Tomáš Fíla[1], Daniel Kytýř[1], Leona Vavro[2],
Martin Vavro[2], Kamil Souček[2], Daniel Vavřík[1], and Miloš Drdácký[1]

[1] Czech Academy of Sciences, Institute of Theoretical and Applied Mechanics,
v. v. i, Prosecká 809/76, 190 00 Praha 9, Czech Republic
koudelkap@itam.cas.cz
[2] Czech Academy of Sciences, Institute of Geonics, v. v. i,
Studentská 1768, 708 00 Ostrava-Poruba, Czech Republic

Abstract. This paper deals with development and validation of a novel device for investigation of the crack behavior in quasi-brittle materials using radiographically observed flexural testing. Instead of standard horizontal arrangement of three-point bending devices, a novel approach consisting in a vertically oriented four-point bending setup is proposed. In the paper, technical description of the proposed device together with its advantages over existing methods and results of validation experiments with natural rocks are presented. The validation experiments were concentrated on the ability of the device to capture the fracture behavior of the samples using X-ray transmission radiography and 4D X-ray micro-tomography. Particular attention was paid to evaluate the ability to perform tomographic scans during post-peak softening, i.e. on intermittent loading of samples after formation of the crack. The acquired results showed very good performance in terms of both the mechanical characteristics of the device (stiffness and loading precision) and the X-ray imaging properties.

Keywords: Four-point bending · Fracture process zone · 4D micro-CT

1 Introduction

Fracture propagation characteristics of quasi-brittle materials describing their ability to resist the fracture evolution have been studied for several decades [1]. Several methods of testing based on various experimental arrangements and specimen geometry have been established, but intensive research is still devoted to evaluation of effects influencing reliability and reproducibility of acquired results. Among these effects, particularly the size effect (i.e. representative volume element problem), problem of contact at interface between the specimen and the loading device and influence of the specimen free surfaces have been investigated [2, 3].

To improve understanding of effects determining the process of macroscopic crack evolution, the method based on parameters of the zone of material failure at the crack tip, the fracture process zone (FPZ), has been developed and already employed [4, 5]. In addition to analytical models and numerical approach to investigation of FPZ

© Springer International Publishing AG, part of Springer Nature 2019
E. E. Gdoutos (Ed.): ICTAEM 2018, SI 5, pp. 27–32, 2019.
https://doi.org/10.1007/978-3-319-91989-8_5

characteristics, experimental techniques in different loading modes utilizing acoustic emission measurement, holographic interferometry, infrared thermography and radiographic imaging have been reported. We have already published studies, where we have investigated FPZ of silicate composites and natural rocks during radiographically observed three-point bending loading [6–8]. Here, the radiographic imaging involved both the X-ray digital transmission radiography and 4D X-ray computed micro-tomography (micro-CT).

It had been found out that radiographic imaging provides more reliable results than acoustic emission measurement, but several problems have also emerged from the experiments [6]. Although the continuous transmission radiography coupled with digital image correlation (DIC) provides a useful insight into crack formation process and evolution of FPZ, it was shown that micro-CT and the resulting 3D images are substantially more powerful in characterization of fracture properties of particularly the cementitious composites [6] and the natural rocks. Based on the results of aforementioned studies and by appropriate application of 4D micro-CT together with carefully optimized in-situ loading, it is possible to perform several tomographic scans on the descending part of the material response, and thus to obtain a highly detailed models of the evolving crack and FPZ. However, such a method is highly demanding in terms of both performance of the used radiographic imaging setup and the loading device itself. Among the encountered difficulties during such three-point bending experiments coupled with time-lapse micro-CT belong problems with mechanical contact between the specimens and the supports. The three-point bending also requires very high stiffness of the loading setup as its deformation may cause sudden rupture of the specimens at each loading step, where the loading has to be stopped until tomographic scan is finished. Another problem is directly related to resulting quality of the tomographic reconstruction and its resolution. Standard three-point bending assemblies have the specimen oriented horizontally on the two supports, while loading is induced into the specimen using vertical movement of the third support, which is identical with axis of rotation of the CT device. When specimen loaded using such a setup is subjected to a tomographic scan, revolution of the whole assembly causes significant change in volume of the material, through which the X-rays have to travel. The resulting variation of attenuation in individual radiograms (reconstructed to the final 3D image) make the process of tomographic reconstruction difficult.

In this paper, we propose a novel method for 4D CT measurements of FPZ and crack propagation in quasi-brittle materials subjected to bending loading model. The approach is based on utilization of a novel four-point bending device specifically designed for high-resolution micro-CT scans, which is made possible by vertical orientation of the investigated specimen identical with rotational axis of the CT devices. This not only eliminates problems with variation of attenuation, but entirely different geometries of the imaging chain (source-detector and source-object distances) can be used as particularly the source-object distance is limited only by dimensions of the loading frame, rather than length of the specimen itself. Aside from technical description of the loading device, we show particularly the preliminary results of validation experiments.

2   Loading Device

The modular device for four-point bending during 4D micro-CT measurements is composed of three main components: a pair of a motorized loading units with an integrated outer supports of the four-point bending arrangement, a pair of a stationary inner supports of the four-point bending arrangement, and the cylindrical load bearing frame housing the loaded specimen together with the loading units and all the supports. The device is during the radiographically observed loading sequence oriented vertically. The middle part of the loading frame is manufactured from a carbon fiber composite with low attenuation of X-rays (MTM57 series epoxy resin, T700S carbon fibers, shell nominal thickness 1.95 mm) to ensure high contrast in the acquired radiograms. Loading of the specimens in four-point bending arrangement is performed by two sets of supports. The inner supports are in a fixed position during loading, while the outer supports are movable and integrated with driving units with precision captive stepper linear actuators (23-2210, Koco Motion DINGS, USA) with precision linear guideways (MGW12, Hiwin, Japan). The longitudinal axis of the specimen, emplaced in the test-ready position, is identical with the rotational axis of the device. As such, both axes are simultaneously coincident with the rotational axis of the CT scanner. Both of the two independent driving units are equipped with linear encoders (LM10, resolution 1 μm, Renishaw Inc., United Kingdom) for highly precise positioning and with individual load cells (LCM300, Futek, USA) for force measurements on both the outer supports. The maximum load capacity of the device is 1500 N per single support. Position accuracy and repeatability in single-direction loading is better than 10 μm with sensitivity better than 1 μm.

Such a design of the loading device has many advantages over the standard arrangements of the horizontally oriented three/four-point bending machines. It enables to use optimum dimensions of the specimens with respect to their representative volume elements and guarantees uniform attenuation of X-rays during rotation of the samples. The length of samples can be also increased without the need to increase also the diameter of the loading device, which would necessarily lead to lower resolution achievable in the acquired radiograms. The supports can be moved independently, which enables to select optimum ratio of distances between the inner and outer supports, depending on geometry and material of the specimen. Furthermore, input and output wiring is equipped with a pair of slip rings that enables infinite number of revolutions of the whole device, which significantly facilitates measurements with multiple tomographic scans during loading of a single specimen. The device and its scheme in partial section is shown in Fig. 1.

3   Results

The proposed device for four-point bending experiments during 4D micro-CT described in the previous section was subjected to design evaluation in a series of pilot experiments. In the first stage of the evaluation, the device was subjected to tests concentrated to prove its ability of intermittent displacement during post-peak softening without sudden crack of the specimen. For this reason, transmission continuous

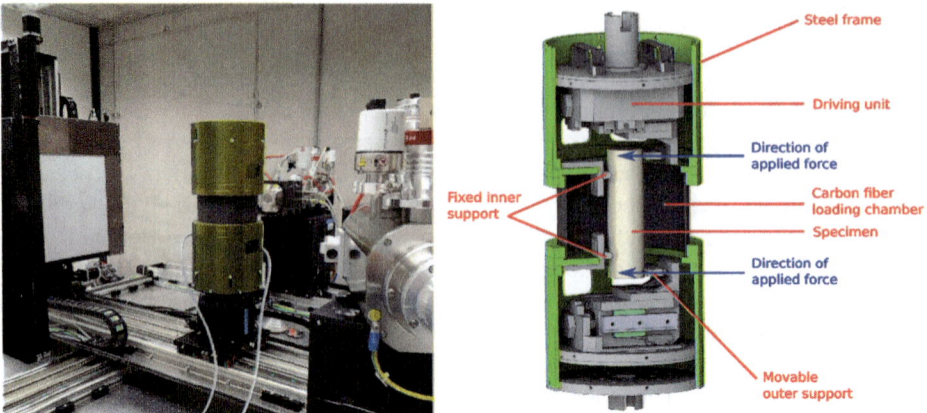

Fig. 1. Loading device attached to rotational stage of the TORATOM DSCT/DECT device (Centre of Excellence Telč) during validation measurement (left) and device configuration (right).

radiographic measurement with five breaks in displacement for simulation of duration of the tomographic scans was performed. The specimen used in this case was a cylindrical specimen (36 mm in diameter, 195 mm in length) drilled from a large Godul sandstone block. In the middle of the specimen, the chevron edge notch with thickness 1.4 mm was cut using water jet cutting. The acquired loading curve can be seen in the Fig. 2.

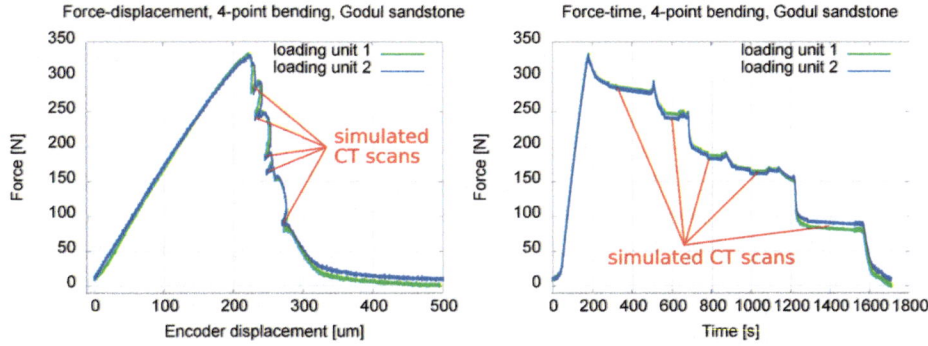

Fig. 2. Loading curves of the simulated CT of a cylindrical Godul specimen.

In the Fig. 2, force measured using both load cells is plotted against displacement captured using both encoders and time respectively, while positions of the simulated CT scans are indicated by text labels. This measurement showed that even such a large specimen, where the peak force per single support before the crack formation reached approx. 325 N, can be successfully subjected to loading with discrete steps during post-peak softening.

In the following study, the Godul specimen with the same geometry and dimensions was subjected to actual radiographic measurement to evaluate particularly the imaging properties of the device with such a large specimen. The Fig. 3 shows the radiograms acquired during loading of the specimen. In every case, the macroscopic view on the crack is coupled with detailed image of the area near the tip of the notch. The left-most image shows the intact specimen before start of the loading procedure. In the middle image, the detailed view clearly depicts the crack evolving from the tip of the notch and the right images shows crack passing through the whole specimen. This demonstrates that the carbon fiber composite material used for the imaging chamber exhibits very low attenuation of X-rays in comparison with specimen and thus, the damage of the specimen can be radiographically captured in all its stages including the possibility to quantitatively track the crack geometry thanks to the sufficient resolution.

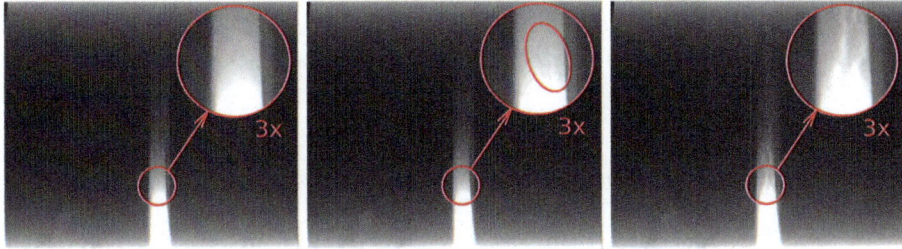

Fig. 3. Radiograms acquired during loading of the specimen showing evolution of damage under the tip of chevron edge notch.

In the last measurement, actual micro-CT measurement with cylindrical sample of Mšené sandstone with nominal diameter 29 mm and length 195 mm was performed to evaluate quality of the reconstructed tomographic image and simultaneously shape of the measured material response during post-peak softening. The reconstructed images under the tip of the notch with resolution approx. 20 µm showing evolution of the crack and the related changes in microstructure are depicted in the Fig. 4.

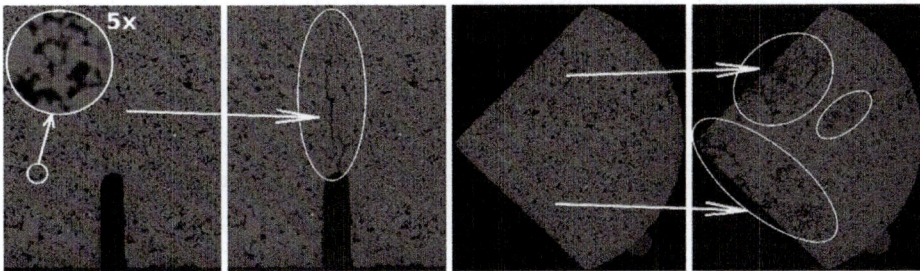

Fig. 4. Visualization of the reconstructed micro-CT measurement showing the apparent crack under the top of the notch (left two images) and voids forming in the microstructure around the crack (right two images).

4 Conclusion

The proposed four-point bending device designed for 4D micro-CT measurements of FPZ and fracture characteristics of quasi-brittle materials was subjected to a series of tests to proof the design and reliability of the results. In these studies, it was validated that the device is successfully able to capture post-peak softening during response to bending moment and that the synchronized movement of both loading units does not influence the measured results. In case of imaging performance, it was shown that it is possible to perform both the continuous X-ray transmission radiography and X-ray 4D micro-CT measurement during post-peak softening. During both of these experiments, cylindrical specimens of natural rocks with chevron edge notch and diameter of 29 mm or 36 mm were used. It was possible to capture the evolving damaged zone in the material during flexural loading proving that the carbon fiber loading chamber has very low attenuation of X-rays. To conclude, the measurements showed that the concept of vertically oriented four-point bending device for 4D micro-CT provides considerable advantages over the standard horizontally oriented three-point or four-point bending setups.

Acknowledgement. This work was supported by the Czech Science Foundation (project No. P105/12/G059) and by institutional support of RVO: 68378297.

References

1. Karihaloo, B.L.: Fracture Mechanics and Structural Concrete. Longman, New York (1995)
2. Yu, Q., Le, J.-L., Hoover, C.G., Bažant, Z.P.: Problems with Hu-Duan boundary effect model and its comparison to size-shape effect law for quasi-brittle fracture. J. Eng. Mech. **136**, 40–50 (2010)
3. Duan, K., Hu, X.-Z., Wittmann, F.H.: Size effect on specific fracture energy of concrete. Eng. Fract. Mech. **74**, 87–96 (2007)
4. Veselý, V., Frantík, P., Keršner, Z.: Cracked volume specified work of fracture. In: Proceedings of the 12th International Conference on Civil, Structural and Environmental Engineering Computing, Civil-Comp Press, Stirlingshire (2009)
5. Veselý, V., Frantík, P.: Reconstruction of a fracture process zone during tensile failure of quasi-brittle materials. Appl. Comput. Mech. **4**, 237–250 (2010)
6. Kumpová, I., Fíla, T., Vavřík, D., Keršner, Z.: X-ray dynamic observation of the evolution of the fracture process zone in a quasi-brittle specimen. J. Instrum. **10**, C08004 (2014)
7. Veselý, V., Keršner, Z., Merta, I.: Quasi-brittle behaviour of composites as a key to generalized understanding of material structure. Procedia Eng. **190**, 126–133 (2017)
8. Vavro, L., Souček, K., Kytýř, D., Fíla, T., Keršner, Z., Vavro, M.: Visualization of the evolution of the fracture process zone in sandstone by transmission computed radiography. Procedia Eng. **191**, 689–696 (2017)

Effects of Long-Term Aging on the Low-Cycle Fatigue Behavior of Inconel 718 Superalloy

Lei Wang$^{(\boxtimes)}$, Jin-lan An, Yang Liu, and Xiu Song

Key Lab for Anisotropy and Texture of Materials, Ministry of Education,
Northeastern University, Shenyang 110819, People's Republic of China
wanglei@mail.neu.edu.cn

Abstract. Inconel 718 superalloy is one kind of important metallic materials used for manufacturing turbine discs in aero-engine. Since the turbine disc usually bears overloading which will lead to the low cycle fatigue (LCF) damage in real working, and because it is added with high ratio alloying elements, it is meaningful to decide the relationship between the LCF degradation and microstructure evolution. In the present study, LCF behavior of Inconel 718 alloy during long-term aging was investigated. The microstructure evolutions of Inconel 718 alloy during long-term aging at 750 °C for 500, 100, 1500 and 2000 h, respectively and the influences of long-term aging on the LCF behavior were investigated. The results show that the size of γ'' phases increases and the volume fraction decreases with the increase of aging time, compared with the increase of both size and volume fraction of δ phases. Both the fatigue strength and fatigue life of the alloy decrease with the increase of aging time.

Keywords: Inconel 718 · Long-term aging · Low-cycle fatigue
Microstructure

Precipitation strengthening type nickel base superalloy Inconel 718 is main strengthened by solid solution strengthening and second phase precipitation strengthening [1], and because of its excellent comprehensive performance, it has been widely used for manufacturing hot structure parts for air engine, gas turbine engine [2–5]. However, due to the high degree alloying of Inconel 718 alloy, complicated composition and long-term servicing at high temperature [6], the microstructure stability is very important for the safety servicing [7, 8]. During the actual servicing, especially for high thrust-weight-ratio engine, in response to some sudden load demand, overload or over-temperature will be bound to accelerate the microstructure evolution of the turbine disk, thus property stability and service security will be endangered. Therefore, the low cycle fatigue behavior of superalloy is widely concerned. The parameters such as temperature, load time, waveform and strain amplitude are external factors which influence the behavior of fatigue behavior of alloy [9–15], as well as the microstructure evolution characteristics of the alloy [16]. Therefore, the present study focuses on the Inconel 718 alloy, to investigate the effects of long-term ageing on the low cycle fatigue behavior of the alloy and the mechanism, the quantitative description of relationship the fatigue behavior and microstructure characteristics, in order to provide a basis prediction of fatigue behavior based on microstructure evolution.

© Springer International Publishing AG, part of Springer Nature 2019
E. E. Gdoutos (Ed.): ICTAEM 2018, SI 5, pp. 33–37, 2019.
https://doi.org/10.1007/978-3-319-91989-8_6

1 Experimental

A commercial Inconel 718 alloy was used for the present research with the chemical composition: C 0.023, 17.30, Fe Cr 17.92, 2.98, Mo Al 0.52, 1.00, Ti, Nb, 5.40 Mn 0.08, P = 0.009, 0.004 B, Ni balance. Alloy was prepared by vacuum induction and vacuum consumable smelting, cast, hot-forged, hot-rolled, re-hot-forged into a diameter of 230 mm, 95 mm plate, then homogenization treated. Testing pieces were cut from disc center along the radial in 11 mm × 11 mm × 115 mm square billet. Alloy standard heat (1050 °C, 1 h, air cooling, 720 °C, 8 h, furnace cooling, 50 °C/h rate after cooling to 620 °C, 620 °C, 8 h, air cooling), marked with SHT. The SHT samples were long-term aged at 750 °C for 500 h, 1000 h, 1500 h and 2000 h, respectively. After long-term aging, the samples were machined into rod-type LCF specimen with a gage range of 6 mm diameter, 30 mm long.

LCF testing was carried out by MTS-810 fatigue testing machine, with reverse axial total strain control at room temperature. The strain waveform is sine wave, total strain $\Delta\varepsilon_t/2 = \pm0.5\%$, strain ratio R = 1, cyclic frequency f = 0.3 Hz. A GX71 type optical microscope (OM) was used to observe microstructure. Both the size and volume fraction of precipitated phase were calculated by quantitative metallographic analysis method. The microstructure and fracture morphology were examined by JMS-6510 scanning electron microscope (SEM), a JEOL-2100F transmission electron microscope (TEM) were used for the observing precipitation phase and dislocation configuration.

2 Results and Discussion

LCF Strain-Stress Characteristics. The LCF strain-stress characteristics of the alloy is shown in the Fig. 1. It is clear that under a certain total strain, with the increasing of cyclic number, the stress response of alloy is in cyclic hardening, cyclic stability, then cyclic softening. The maximum cyclic stress response decreased with the increasing of aging time, as well as the LCF, the detail results are shown in Table 1.

Table 1. The max cyclic stress and LCF life of the alloy aged at 750 °C for different times

Aging time h	The max cyclic stress MPa	Low cycle fatigue life Cyc
SHT	927	8504
500	850	7247
1000	763	6848
1500	705	5816
2000	627	5533

Cyclic Stress Response Characteristics of LCF. At beginning, the cyclic hardening is depended on cutting of smaller γ'' phases with dislocation. Figure 2 shows the dislocation configuration of the alloy aged at 750 °C for 1500 h. As shown in Fig. 2b, after 50 cycles, the dislocation formed network around the δ phase. It means that the dislocation is early interacted with δ phase during cyclic deformation, dislocation has to around the δ phases, so that the motion resistance increased and presented cyclic hardening characteristics. Compared with that in SHT state, the maximum stress response of alloy after long-term aging needs cost more cycles. Because in SHT state, there is no δ phase, the cyclic hardening is only caused by the cutting of γ″ with dislocation. However, after certain cycles, dislocation multiplication and annihilating reached a stable, strengthening and weakening offset each other, so that the stress response presented saturated.

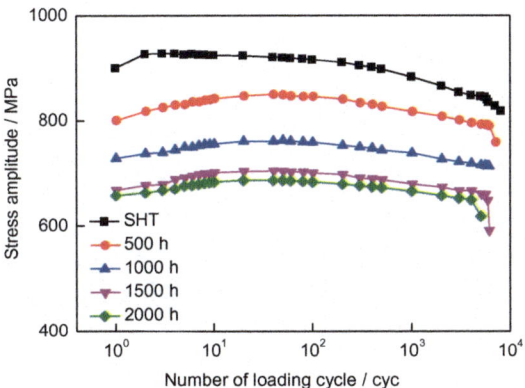

Fig. 1. Cyclic stress amplitude curves of the alloy aged at 750 °C for different times

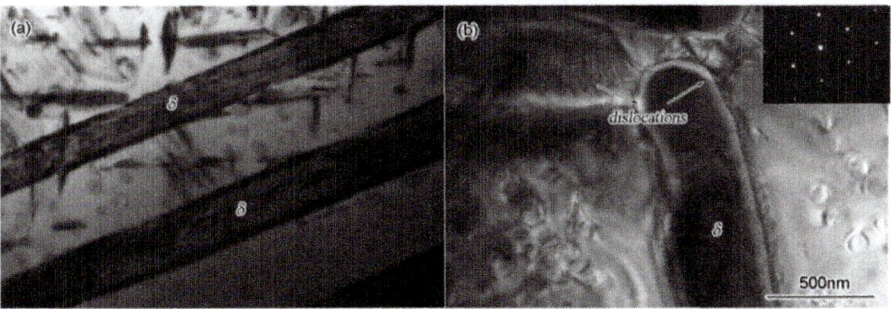

Fig. 2. Dislocation configurations in the alloy aged at 750 °C for 1500 h after 0 cycles (a) and 50 cycles (b) (The inset in Fig. 2b shows the SAED pattern)

Effect of δ Phase on LCF Behavior. Figure 3 shows the LCF fracture morphology near the fracture surface. It can be seen that crack easily along the long needle δ phase or extend the precipitate free zone. It is suggested that the long needle δ phase and precipitate free zone around δ phase become the cracking preferred path. In the rapidly crack propagation stage, large number of dimples can be found on the fracture surface of the alloy aged for 500 h, while such dimples could not be found for the alloy aged for 1500 and 2000 h, as shown in the Fig. 4. After aged for 500 h, there are a small amount of granule, short rod-shaped δ phase, disc γ″ phase in the matrix of the alloy. Therefore, under the cyclic loading, the dimple easy occur around the γ″ phase, granule or short δ phase (Fig. 4a). While, after long-term aging, part of γ″ phase transforms into long needle δ phase. Under cyclic loading, the crack propagate to be along the long needle δ phase and the precipitate free zone, to form a long strip characteristic morphology (Fig. 4b and c). With the increasing of aging time, the fatigue fracture transient breaking district strip characteristic morphology increase, this is mainly due to the increasing of δ phase and precipitate free zone in the alloy.

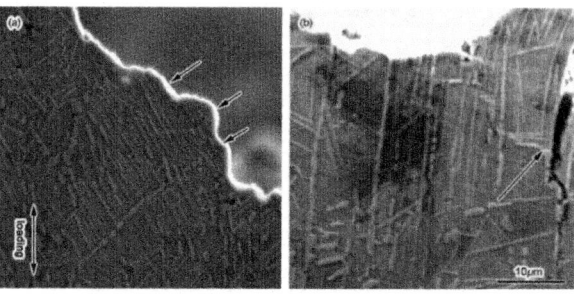

Fig. 3. Morphologies of cross-sectional near LCF fracture surface of the alloy aged at 750 °C for 2000 h, main crack along δ phases (a) and secondary cracks along δ phases (b)

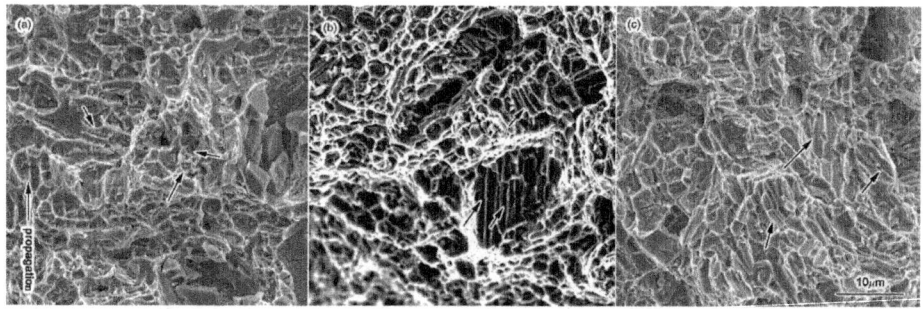

Fig. 4. Morphologies of fracture crack fast propagation zone of the alloy after LCF aged at 750 °C for 500 h (a), 1500 h (b) and 2000 h (c) (arrows indicate δ phases)

3   Conclusions

The characteristics of LCF deformation of the Inconel 718 alloy are in the cyclic hardening, cyclic stability and cyclic softening. But with the increasing of aging time, both the maximum cyclic stress response and LCF life are reduced.

The LCF behavior of the alloy is mainly depended on the evolution of γ'' phase and δ phase, as well as the changing of precipitate free zone during the long-term aging of the Inconel 718 alloy.

References

1. Huang, Q.Y., Li, H.K.: Superalloy, pp. 1–10. Metallurgy Industry Press, Beijing (2000)
2. Medeiros, S.C., Prasad, Y.V.R.K., Frazier, W.G., Srinivasan, R.: Microstructural modeling of metadynamic recrystallization in hot working of IN 718 superalloy. Mater. Sci. Eng. **A193**, 198–207 (2003)
3. Tian, S.G., Wang, X., Xie, J., Liu, C., Guo, Z.G., Liu, J., Sun, W.R.: Characteristic and mechanism of phase transformation of GH4169G alloy during heat treatment. Acta Metall. Sin. **49**, 845–852 (2013)
4. Zhang, H.Y., Zhang, S.H., Zhang, W.H., Cheng, M., Wang, Z.T.: The research on the optimization of the hot die forging of GH4169 turbine discs. J. Plast. Eng. **14**, 69–75 (2007)
5. Yang, Y.R., Liang, X.F., Cai, B.C., Huang, F.X.: Effect of δ phase on stress-rupture properties of GH4169 alloy. J. Aeronaut. Mater. **16**, 38–43 (1996)
6. Jiang, H.P.: Requirements and forecast of turbine disk materials. Gas Turbine Exp. Res. **15**, 1–6 (2002)
7. Liu, F., Sun, W.R., Yang, S.L., Li, Z., Guo, S.R., Yang, H.C., Hu, Z.Q.: Effects of Al content on microstructure and stability of GH4169 nickle base alloy. Acta Metall. Sin. **44**, 791–798 (2008)
8. Dong, J.X., Bai, Y.Q., Xu, Z.C., Xie, X.S., Zhang, S.H.: Microstructure of long time aging in GH4169 alloys. J. Univ. Sci. Tech. Beijing **15**, 567–571 (1993)
9. Coffin, L.F.: Effects of frequency and environment on fatigue crack growth in A286 at 1100 F. In: Carden, A.E., McEvily, A.J., Wells, C.H. (eds.) Fatigue at Elevated Temperatures, pp. 112–122. ASTM, Baltimore (1973)
10. Gell, M., Leverantm, G.R.: Mechanisms of High Temperature Fatigue. In: Carden, A.E., McEvily, A.J., Wells, C.H. (eds.) Fatigue at Elevated Temperatures, pp. 37–67. ASTM, Baltimore (1973)
11. Coffin, L.F.: Some physical aspects of high-temperature, low-cycle fatigue. Soc. Mater. Sci. **21**, 186–191 (1971)
12. Merrick, H.K.: The low cycle fatigue of three wrought nickel-base alloys. Metall. Trans. **5A**, 891–897 (1974)
13. Fournier, D., Pineau, A.: Low cycle fatigue behavior of Inconel 718 at 298 K and 823 K. Metall. Trans. **8A**, 1095–1105 (1977)
14. Day, M.F., Thyomas, G.B.: Microstructural assessment of fractional life approach to low-cycle fatigue at high temperatures. Met. Sci. **13**, 25–33 (1979)
15. Antolovich, S.D., Liu, S., Baur, R.: Low cycle fatigue behavior of René 80 at elevated temperature. Metall. Trans. **12A**, 476–481 (1981)
16. Yao, J., Guo, J.T., Yuan, C., Li, Z.J.: Low cycle fatigue behavior of cast nickel base superalloy K52. Acta Metall. Sin. **41**, 357–362 (2005)

Enhanced Young's Modulus in Percolative Cementitious Composites Reinforced with Carbon Nanotubes

Maria S. Konsta-Gdoutos[1], Panagiotis A. Danoglidis[1(✉)],
Maria G. Falara[1], Myrsini E. Maglogianni[1],
and Emmanuel E. Gdoutos[2]

[1] School of Engineering, Democritus University of Thrace,
671 00 Xanthi, Greece
pdanogli@civil.duth.gr
[2] Academy of Athens, 106 79 Athens, Greece

Abstract. The modulus of elasticity, resistivity and capacitive reactance were determined for carbon nanotube reinforced mortars, near percolation. It is shown that the abrupt decrease of the resistivity values observed at the CNT content of 0.1 wt% is associated with the onset of percolation. Mortars reinforced with 0.1 wt% CNTs exhibit an 89% increase in Young's modulus. Values of resistivity and capacitance are 27% and 90% lower than that of the plain mortar, respectively. After the conductive network is formed, resistivity values show a little dependence on the CNT content, reaching a plateau. Capacitance on the other hand was increased by an order of magnitude, showing an amplified energy storage ability, probably due to the existence of small CNT agglomerates. The observed relationship between capacitance values and modulus of elasticity may provide valuable information on the actual CNT dispersion state in the matrix.

Keywords: Young's modulus · Mortar · Carbon nanotubes
Percolation threshold

1 Introduction

Carbon nanotubes (CNTs) are excellent reinforcing materials, offering unprecedented improvements in Young's modulus of cement based materials [1–3]. The addition of CNTs also determines the electrical properties of the nanocomposites [2, 3]. Two electrical phases, resistive and capacitive, are observed in any CNT reinforced cement based material. The resistive phase arises from the intrinsic resistance of nanotubes [4], while the capacitive phase is due to the localized nanotube/mortar/nanotube capacitors, and is dependent to the formation of CNT agglomerates in the matrix [5]. In this study, Electrochemical Impedance Spectroscopy (EIS) measurements were conducted to evaluate both capacitance and resistivity of mortars reinforced with CNTs near percolation. Modulus of elasticity was assessed through three-point bending test. Resistivity values can determine the critical amount of CNTs that define the percolative behavior of the nanocomposites. Capacitance values indicate the state of dispersion and

© Springer International Publishing AG, part of Springer Nature 2019
E. E. Gdoutos (Ed.): ICTAEM 2018, SI 5, pp. 38–43, 2019.
https://doi.org/10.1007/978-3-319-91989-8_7

the existence of agglomerations in the matrix. It was shown that CNTs at an amount of 0.1 wt% result in the formation of a complete network of individual nanotubes, leading to the highest improvements in Young's modulus and simultaneously to the lowest energy storage ability of the nanoreinforced mortars.

2 Experimental Work

2.1 Materials and Specimens

Carbon nanotubes were added in mortar mixtures at amounts of 0.05%, 0.1%, 0.2%, 0.3% and 0.5 wt% of cement. CNT dispersion was achieved by applying ultrasonic energy with the use of a surfactant [6, 7]. Two types of specimens were prepared: 40 × 40 × 160 mm prisms for the three-point bending tests; and 20 × 20 × 80 mm beams for the determination of the electrical properties. Following demolding, the samples were cured in lime-saturated water for 28 days. For the EIS measurements, titanium grids with large opening (3 × 3 mm) were embedded into the specimens immediately after casting.

2.2 Experimental Evaluation of Young's Modulus

Three-point bending tests were conducted to assess the Young's modulus of the nanocomposites. 40 × 40 × 160 mm prismatic specimens were prepared and tested at the age of 28d based on the ASTM C348-14. The test was performed using a 25 kN MTS servo-hydraulic, closed-loop testing machine under displacement control. The rate of displacement was kept as 0.1 mm/min.

2.3 Assessment of Resistive and Capacitive Phases

The electrical resistivity and capacitance were measured using an Agilent E4980 LCR meter at 100 kHz frequency [8]. Amplitude of the sinusoidal voltage was chosen to be 2 V.

The capacitance C of the nanocomposites can be evaluated by the following equation:

$$C = A \cdot k \frac{\varepsilon}{d} \qquad (1)$$

Where A is the area of an individual CNT, k the permittivity of dielectric mortar, ε the permittivity of space and d the distance between two nanotubes (Fig. 1). The k and ε factors are constant. Assuming perfect dispersion, the area A of individual nanotubes is constant. Then, the distance d between two nanotubes is the region where electrical energy can be stored, therefore becomes a crucial factor for the investigation of nanocomposite's capacitance. If a continuous network of localized capacitors, in the form of individual nanotube/mortar/nanotube, is created, then the electrical charge can easily pass and cannot be stored. At the same time the capacitance of the nanocomposites is also decreased.

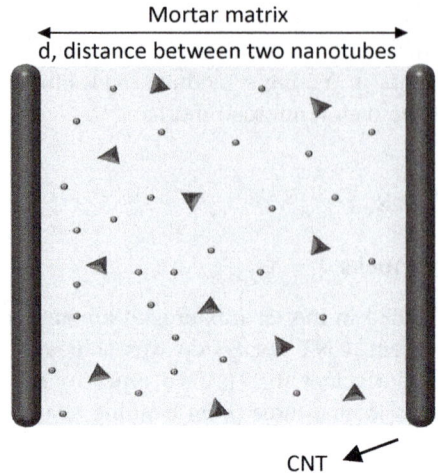

Fig. 1. Nanotube/mortar/nanotube capacitor structure

3   Results and Discussion

The resistivity values of 28d mortars reinforced with CNTs at amounts of 0.05%, 0.1%, 0.2%, 0.3% and 0.5 wt% are presented in Fig. 2. Resistivity gradually decreases with the CNT incorporation for amounts up to 0.2 wt%, where it reaches the lowest value of 4.0 kΩ·cm. The nanotube content of 0.1 wt% is associated with the onset of percolation. For higher CNT amounts the conductive nanotube pathways are already established and the resistive phase reaches a plateau demonstrating an independence of the CNT content for higher concentrations, 0.3% and 0.5 wt%.

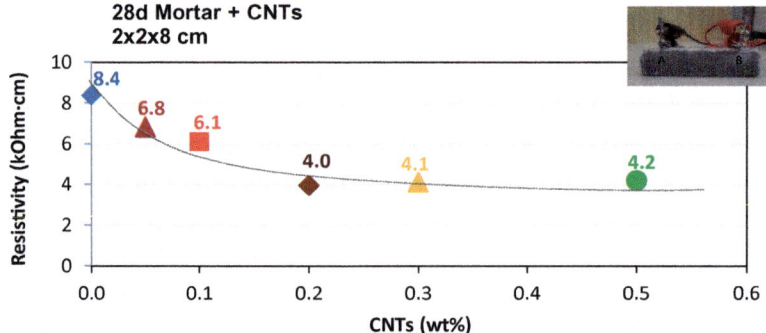

Fig. 2. Resistivity of plain mortar and mortars reinforced with CNTs, at amounts of 0.05%, 0.1%, 0.2%, 0.3% and 0.5 wt%

The resistivity and capacitance results of 0.1%, 0.2%, 0.3% and 0.5 wt% CNT-nanocomposites are presented in Fig. 3. The capacitive reactance is an indicator of the ability of a nanotube network to store electrical energy. It is known that when a uniform dispersion of CNTs is achieved in the matrix there is a consistency between resistive and capacitive phases, i.e. both phases decrease as the nanotube content increases. Indeed, values of both resistivity and capacitance decrease up to 0.1 wt% CNTs. Capacitance values of the 0.2 wt% mortar are higher by one order of magnitude, while the 0.3% and 0.5 wt% values stay approximately at the same levels. Figure 4 illustrates the capacitance and Young's modulus of mortars as a function of the CNT content.

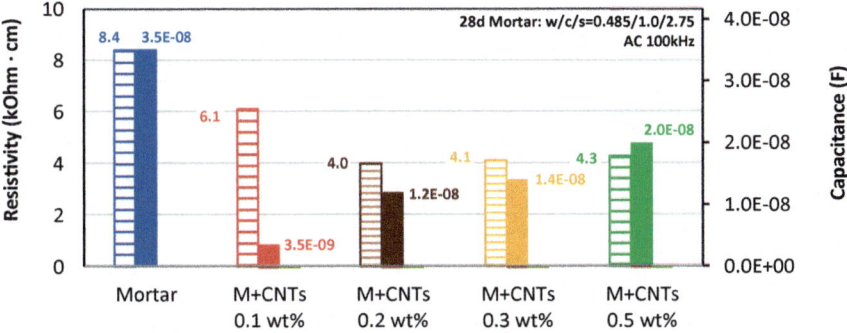

Fig. 3. Resistivity and capacitance of 28d mortars reinforced with CNTs at amounts of 0.1%, 0.2%, 0.3% and 0.5 wt%

It is observed that all CNT amounts increase the nanocomposites' Young's modulus from 50% to 89%. It is also observed that a relationship between the capacitive reactance and modulus exists that may provide valuable information on the actual CNT dispersion state in the matrix. The 0.1 wt% CNT reinforced mortar demonstrates a modulus of 27.4 GPa which corresponds to the highest improvement in stiffness (89%) and the lowest value of capacitance, 3.5E−09. This may be the result of an effective dispersion that leads to the formation of a continuous network of individual nanotubes. A few small size CNT agglomerates may exist; however, they have a small ability of storing the electrical charge, hence low values of both capacitance and resistivity are observed. As it was mentioned before, the 0.2 wt% mortar nanocomposites showed a sudden increase in capacitance of one order of magnitude. Interestingly, the modulus of the 0.2 wt% CNT reinforced mortars, compared to the 0.1 wt%, demonstrated a lower increase in modulus (70%). The 0.3% and 0.5 wt% mixes resulted in even lower increases of the modulus, 59% and 49%, respectively, and at the same time modest increases in capacitance values. Capacitance increases in percolative composites indicate an amplified electrical charge storage ability, probably due to the existence of either bigger in size, or more CNT agglomerates. These inclusions act as stress concentration areas that can definitely have an effect on the modulus development.

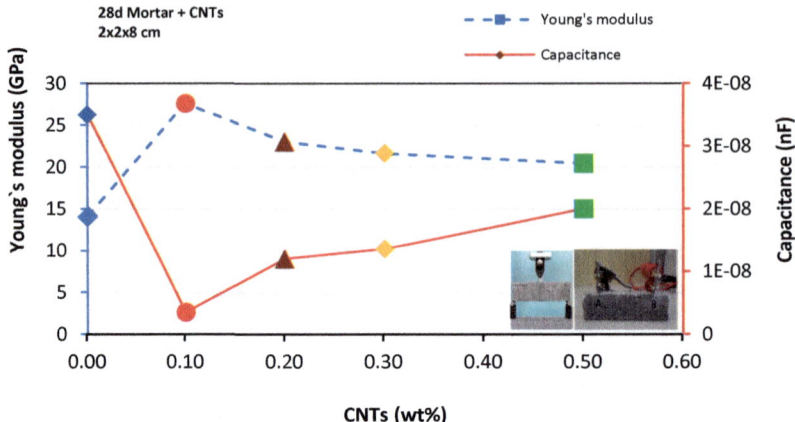

Fig. 4. Capacitance and Young's modulus of 28d mortars reinforced with CNTs at amounts of 0.1%, 0.2%, 0.3% and 0.5 wt%

4   Conclusions

In this study, a substantial enhancement of the Young's modulus associated with addition of different amounts of CNTs in mortars was demonstrated. The relationship of resistivity, capacitance and Young's modulus is crucial to determine the percolative behavior, the state of dispersion and the formation of a network of individual nanotubes in the mortar matrix. The resistivity results show that the critical amount 0.1 wt% of CNTs denotes the onset of percolation. The usefulness of the capacitance and modulus results in CNT reinforced cement based composites near percolation is discussed. In particular, it is shown that mortars reinforced with 0.1 wt% CNTs exhibit the highest modulus of elasticity and the lowest capacitive behavior. The addition of CNTs at amounts higher than the 0.1 wt% results in capacitance increases that indicate an amplified energy storage ability, probably due to the existence of either bigger in size or more CNT agglomerates. These inclusions act as stress concentration areas that can definitely have an effect on the modulus development.

Acknowledgements.   The authors would like to kindly acknowledge the company Glonatech S.A. for supplying CNTs.

References

1. Gdoutos, E.E., Konsta-Gdoutos, M.S., Danoglidis, P.A.: Portland cement mortar nanocomposites at low carbon nanotube and carbon nanofiber content: a fracture mechanics experimental study. Cement Concr. Compos. **70**, 110–118 (2016)
2. Danoglidis, P.A., Konsta-Gdoutos, M.S., Gdoutos, E.E., Shah, S.P.: Strength, energy absorption capability and self-sensing properties of multifunctional carbon nanotube reinforced mortars. Constr. Build. Mater. **120**, 265–274 (2016)

3. Konsta-Gdoutos, M.S., Danoglidis, P.A., Falara, M.G., Nitodas, S.F.: Fresh and mechanical properties, and strain sensing of nanomodified cement mortars: the effects of MWCNT aspect ratio, density and functionalization. Cement Concr. Compos. **82**, 137–151 (2017)
4. Li, J., Ma, P.C., Chow, W.S., To, C.K., Tang, B.Z., Kim, J.K.: Correlations between percolation threshold, dispersion state, and aspect ratio of carbon nanotubes. Adv. Func. Mater. **17**(16), 3207–3215 (2007)
5. Hsu, W.K., Kotzeva, V., Watts, P.C.P., Chen, G.Z.: Circuit elements in carbon nanotube-polymer composites. Carbon **42**, 1707–1712 (2004)
6. Shah, S.P., Konsta-Gdoutos, M.S., Metaxa, Z.S.: Highly concentrated carbon nanotube suspensions for cementitious Materials. United States Patent No. 9,499,439 (B2), 22 November 2016
7. Konsta-Gdoutos, M.S., Metaxa, Z.S., Shah, S.P.: Multi-scale mechanical and fracture characteristics and early-age strain capacity of high performance carbon nanotube/cement nanocomposites. Cement Concr. Compos. **32**(2), 110–115 (2010)
8. Azhari, F., Banthia, N.: Cement-based sensors with carbon fibers and carbon nanotubes for piezoresistive sensing. Cement Concr. Compos. **34**, 866–873 (2012)

Behavior of Partially and Fully FRP-Confined Circularized Square Columns (CSCs) Under Axial Compression

Yong-Chang Guo and Jun-Jie Zeng[✉]

School of Civil and Transportation Engineering,
Guangdong University of Technology, Guangzhou 510006, China
jjzeng@gdut.edu.cn

Abstract. Existing research has demonstrated that FRP confinement is effective for circular columns, whereas it is less effective for square columns. The lower FRP confinement effectiveness in a square column is predominantly attributed to the non-uniform FRP confinement in the column, while the concrete in an FRP-confined circular column is uniformly confined. An appropriate approach to enhancing the confinement effectiveness of FRP strengthening technique for square columns is to circularize a square column before FRP jacketing. This paper aims to study the compressive behavior of circularized square columns (CSCs). A total of 33 column specimens were prepared and tested under axial compression in this paper. The test results have indicated that the section circularization of square columns can significantly improve the effectiveness of FRP confinement, and strengthening square columns using section circularization in combination with partial FRP confinement is a promising and economical alternative to the fully FRP strengthening technique. Comparisons between the theoretical predictions and the test results were conducted, and the accuracy and reliability of existing partially FRP-confined concrete models were also examined.

Keywords: Partially FRP-confined concrete · Square column · Circularization
FRP · Confinement

1 Introduction

Square columns are more common than circular columns in practice because architects tend to prefer square columns because of their regular appearance and square columns are more easily constructed than circular columns due to the easier availability of square molds for concrete casting. For an efficient use of FRP in square columns, one of the most appropriate approaches is to implement an appropriate shape modification of square sections before FRP jacketing needs. For a square column shape-modified by circularization, the process is called section circularization (i.e., SC) and the resulting shape-modified column is called circularized square column (CSC) in the present study (Fig. 1). For ease of discussions, the concrete in the core square column is referred to as "core concrete" and the new concrete used for circularization as "peripheral concrete" hereafter (Fig. 1). In a CSC, the FRP jacket is loaded primarily in hoop tension from

© Springer International Publishing AG, part of Springer Nature 2019
E. E. Gdoutos (Ed.): ICTAEM 2018, SI 5, pp. 44–49, 2019.
https://doi.org/10.1007/978-3-319-91989-8_8

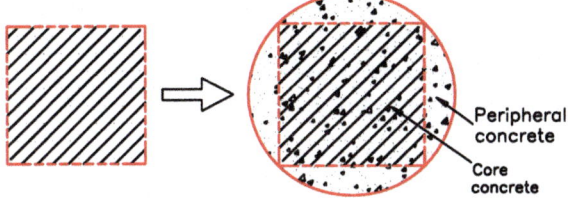

Fig. 1. Section circularization.

the beginning, resulting in higher confinement to the concrete compared with the square column with only rounded corners.

A number of studies were conducted on shape-modified columns from rectangular sections to elliptical/oval/curvilinear sections or from square sections to circular/curvilinear sections (e.g., Hadi et al. 2013; Pham et al. 2013; Yan and Pantelides 2011). However, in most of these studies, shape-modified columns were confined by longitudinally continuous FRP jackets (referred to as fully FRP-confined columns). Alternatively, the shape-modified columns can be confined using evenly spaced FRP strips along the longitudinal axis (referred to as partially FRP-confined columns (Fig. 2)). Although the vast majority of previous studies related to FRP-confined concrete columns focused on fully FRP-confined concrete, a number of studies were conducted on partially FRP-confined concrete (Barros and Ferreira 2008; Campione et al. 2015; Matthys et al. 2005; Park et al. 2008; Pham et al. 2015; Triantafyllou et al. 2015; Wang and Wu 2010; Wei et al. 2009; Wu et al. 2007), owing to its easier and faster application. Although both the SC before FRP jacketing and partially FRP confinement can enhance the strength and deformation capacities of concrete, the compressive behavior of partially FRP-confined CSCs has never been studied.

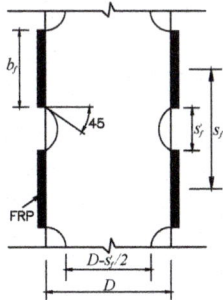

Fig. 2. Partially FRP-confined square column.

Against the above background, this paper aims to investigate axial compressive behavior of fully or partially-FRP confined CSCs. The test results in terms of the failure modes, stress-strain characteristics, and the ultimate axial stresses and ultimate strains were reported in detail. A theoretical analysis was conducted for predicting the compressive behavior of partially-FRP confined CSCs based on the FRP-confined concrete models (CNR-DT 200 R1 2013; fib 2001; Lam and Teng 2003). Through the comparisons between the theoretical predictions and the test results, the accuracy and reliability of these models were examined.

2 Experimental Program

In total, 33 concrete columns were prepared and tested to investigate the effects of vertical spacing ratio, FRP thickness, sectional shape and peripheral concrete strength. The vertical spacing ratio is defined as s'_f/s_f where s_f and s'_f are the center to center spacing of the adjacent FRP strips and the net spacing between the adjacent strips, respectively. These specimens were divided into three series (i.e., Series I, II and III), which totally consisted of 9 square columns (including one unconfined column tested as a reference column) and 24 CSCs. Two grades of concrete strength were applied for the peripheral concrete, and they were nominated as "C1" and "C2" respectively. All column specimens had a height of 595 mm and the corner radius of all square columns was 25 mm. Each column was given a name (Table 1) which started with "S" and "CS" representing "square column" and "circularized square column" respectively. For CSCs, the first part (i.e., CS) was followed by a number indicating the peripheral concrete strength ("1" for the peripheral concrete strength identical to the core concrete strength and "2" for the peripheral concrete strength higher than the core concrete strength). The second characteristic was a digital number representing the net spacing of the adjacent FRP strips. The last one was a number indicating FRP layers. The reference specimen was named "S-N" in which the characteristic "N" represents "no FRP confinement".

Table 1. Details of test specimens

Series	Specimen	Unconfined concrete strength (MPa)		Ultimate axial strain of unconfined concrete		s'_f (mm)	s'_f/s_f
		Core	Peripheral	Core	Peripheral		
I	S-N	24.3	/	0.0024		–	–
	S-0-1 (–2)	24.3	/	0.0024		0	0
	S-30-1 (–2)	24.3	/	0.0024		30	1/4
	S-45-1 (–2)	24.3	/	0.0024		45	1/3
	S-90-1 (–2)	24.3	/	0.0024		90	1/2
II	CS1-0-1 (–2, –3)	24.3	24.3	0.0024	0.0024	0	0
	CS1-30-1 (–2, –3)	24.3	24.3	0.0024	0.0024	30	1/4
	CS1-45-1 (–2, –3)	24.3	24.3	0.0024	0.0024	45	1/3
	CS1-90-1 (–2, –3)	24.3	24.3	0.0024	0.0024	90	1/2
III	CS2-0-1 (–2, –3)	24.3	60.3	0.0024	0.0022	0	0
	CS2-30-1 (–2, –3)	24.3	60.3	0.0024	0.0022	30	1/4
	CS2-45-1 (–2, –3)	24.3	60.3	0.0024	0.0022	45	1/3
	CS2-90-1 (–2, –3)	24.3	60.3	0.0024	0.0022	90	1/2

Normal strength concrete made of river sand and natural gravels was used to cast the test specimens in this experimental program. Standard cylinders were also cast with the same batch concrete to determine the mechanical properties of the unconfined concrete (f'_{co} and ε_{co}). After the specimens had been cured for 28 days in the laboratory environment, they were wrapped with carbon FRP (CFRP) sheets (i.e., strips) using a wet layup procedure, with fibers oriented along the hoop direction. A 150-mm long overlapping zone between the starting and the finishing end of each sheet was allowed to facilitate sufficient bonding of the FRP strip. An additional 50-mm width FRP strip was wrapped near each end of the column to avoid unexpected failure there.

The flat coupon tensile tests were conducted to determine the material properties of FRP sheets, following the recommends specified in the ASTM standard (ASTM D3039 2008). The average values of tensile strength, elastic modulus and rupture strain were determined as 4185.6 MPa, 259.7 GPa and 1.75%, respectively. Column compression test was conducted using a displacement controlled procedure with a rate of 0.4 mm/min.

3 Stress-Strain Responses

For ease of discussions, the specimens are divided into several groups with one or two of the column parameters of each group being identical. The stress-strain responses of the concrete in the test columns are shown in Fig. 3, with each figure consisting several curves as a group. In the figure, the experimental stress-strain curves were terminated at the point when the FRP rupture occurred, unless otherwise specified. In each sub-figure of Figs. 4 and 5, the cross-sectional shape, the concrete strength and the number of FRP layers are all identical, allowing for a direct comparison between the specimens with different net spacings of the adjacent FRP strips. It can be seen that all the axial stress-axial strain curves of the FRP-confined concrete have two stages consisting of an approximately linear elastic stage and a hardening/softening stage with gradually decreased stiffnesses. In general, the stress-strain curves show monotonically ascending (i.e., with a hardening stage); however, for some specimens confined with 90 mm net spacing strips (i.e., S-90-1, S-90-2, CS2-90-1, CS2-90-2, CS2-90-3), the stress-strain curves have a softening stage.

The net spacing of the adjacent FRP strips reflected the effectiveness of FRP confinement: an increase in the net spacing yielded a decrease in the ultimate axial stress, as indicated by Figs. 3, 4 and 5. That is because the effectiveness of FRP confinement was lower in a specimen with a larger net spacing. However, it is seen from Figs. 6–8 that the ultimate axial strain of the concrete and the FRP hoop rupture strain were approximately independent to the net spacing of the adjacent FRP strips. It is interesting to note from Fig. 6 that the ultimate axial strains of S-0-1 and S-0-2 (fully confined specimens) were smaller than those of S-30-1 and S-30-2 (partially confined specimens). This is probably because the axial deformation of the concrete between the adjacent FRP strips was much larger than that of the concrete wrapped with FRP strips. However, an increase in the net spacing of the FRP strips did not always lead to a corresponding increase in the ultimate axial strain of the concrete as the large net

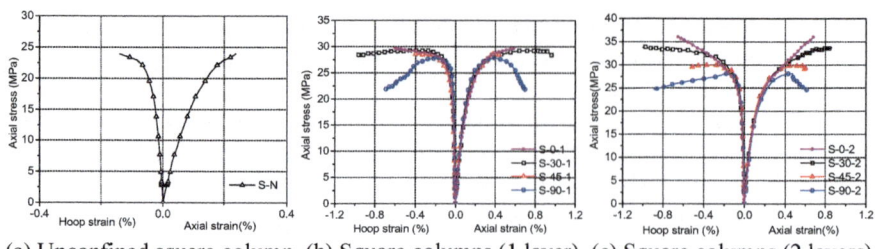

(a) Unconfined square column (b) Square columns (1 layer) (c) Square columns (2 layers)

Fig. 3. Stress-strain curves of concrete in the Series I columns

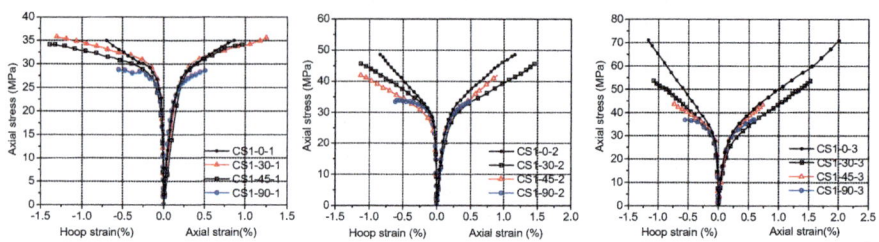

Fig. 4. Stress-strain curves of concrete in the Series II columns (1/2/3-layer)

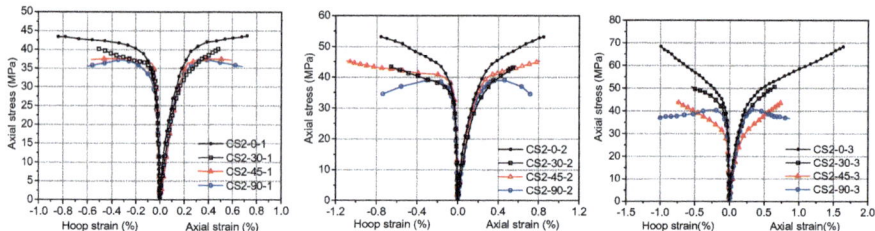

Fig. 5. Stress-strain curves of concrete in the Series III columns (1/2/3-layer)

spacing may induce an earlier failure of the column. More comparisons of the ultimate strengths and strains will be discussed in detail in the following section.

4   Conclusions

Based on the experimental and theoretical results presented in this paper, the following conclusions can be drawn:

(1) Significant strength and deformation increases are obtained for the FRP-confined CSCs compared to the fully FRP-confined square columns without circularization. The net spacing of the adjacent FRP strips reflects the effectiveness of FRP confinement: an increase in the net spacing leads to a decrease in the ultimate axial

stress. However, the ultimate axial strain of concrete and the hoop rupture strain of FRP are likely independent to the net spacing of the adjacent FRP strips.

(2) The combined use of the SC and the partially FRP strengthening technique saves the FRP material as much as 50% in the volumetric ratio, with strength and axial deformation capacities being comparable or even better than those of fully FRP-confined square columns.

(3) The combined use of the vertical confinement effectiveness coefficient provided by the design codes (CNR-DT 200 R1 2013; *fib* 2001) and Lam and Teng's (2003) model can provide accurate predictions for both the ultimate axial stress and the ultimate axial strain of the partially FRP-confined concrete.

References

ASTM D3039 Standard Test Method for Tensile Properties of Polymer Matrix Composite Materials (D3039M), West Conshohocken, USA (2008)

Barros, J.A.O., Ferreira, D.R.S.M.: Assessing the efficiency of CFRP discrete confinement systems for concrete cylinders. J. Compos. Constr. ASCE **12**(2), 134–148 (2008)

CNR-DT 200 R1 Guide for the Design and Construction of Externally Bonded FRP Systems for Strengthening Existing Structures, Advisory Committee on Technical Recommendations For Construction, National Research Council, Rome, Italy (2013)

Campione, G., La Mendola, L., Monaco, A., Valenza, A., Fiore, V.: Behavior in compression of concrete cylinders externally wrapped with basalt fibers. Compos. Part B Eng. **69**, 576–586 (2015)

fib. Externally Bonded FRP Reinforcement for RC Structures, The International Federation for Structural Concrete, Lausanne, Switzerland (2001)

Hadi, M.N.S., Pham, T.M., Xu, L.: New method of strengthening reinforced concrete square columns by circularizing and wrapping with fiber-reinforced polymer or steel straps. J. Compos. Constr. ASCE **17**(2), 229–238 (2013)

Lam, L., Teng, J.G.: Design-oriented stress-strain model for FRP-confined concrete. Constr. Build. Mater. **17**(6–7), 471–489 (2003)

Matthys, S., Toutanji, H., Audenaert, K., Taerwe, L.: Axial behavior of large-scale columns confined with fiber-reinforced polymer composites. ACI Struct. J. **102**(2), 258–267 (2005)

Park, T.W., Na, U.J., Chung, L., Feng, M.Q.: Compressive behavior of concrete cylinders confined by narrow strips of CFRP with spacing. Compos. Part B Eng. **39**(7–8), 1093–1103 (2008)

Pham, T.M., Doan, L.V., Hadi, M.N.: Strengthening square reinforced concrete columns by circularisation and FRP confinement. Constr. Build. Mater. **49**, 490–499 (2013)

Saljoughian, A., Mostofinejad, D.: Axial-flexural interaction in square RC columns confined by intermittent CFRP wraps. Compos. Part B Eng. **89**, 85–95 (2016)

Triantafyllou, G.G., Rousakis, T.C., Karabinis, A.I.: Axially loaded reinforced concrete columns with a square section partially confined by light GFRP straps. J. Compos. Constr. **19**(1), 1–15 (2015). 04014035

Wei, H., Wu, Z., Guo, X., Yi, F.: Experimental study on partially deteriorated strength concrete columns confined with CFRP. Eng. Struct. **31**(10), 2495–2505 (2009)

Yan, Z., Pantelides, C.P.: Concrete column shape modification with FRP shells and expansive cement concrete. Constr. Build. Mater. **25**(1), 396–405 (2011)

Stress-Strain Behavior of Circular Concrete Columns Partially Wrapped with FRP Strips

Jun-Jie Zeng, Li-Juan Li, and Yong-Chang Guo[✉]

School of Civil and Transportation Engineering, Guangdong University of
Technology, Guangzhou 510006, China
guoyc@gdut.edu.cn

Abstract. Fiber-reinforced polymer (FRP) jacketing or wrapping has become
an attractive strengthening technique for concrete columns. Within this
strengthening technique, FRP composites are wrapped around the concrete
column with the fibers in the jacket being oriented in the hoop direction. In
practice, the FRP jackets can be either continuous or discontinuous along the
column height and thus the resulting column is referred to as fully or partially
wrapped FRP-confined concrete columns. Existing research has demonstrated
that the partially strengthening technique by discrete FRP strips is a promising
and economic alternative to the fully FRP strengthening technique. Although a
number of experimental investigations have been conducted on partially
wrapped FRP-confined concrete columns, the stress-strain behavior of
FRP-confined concrete in partially wrapped concrete columns is not yet
understood. This paper presents an experimental program to investigate the
axially compressive behavior of circular concrete columns partially wrapped
with FRP strips. The test results are presented and compared with the predictions
from a typical analysis-oriented stress-strain model to examine its reliability and
accuracy. It has been demonstrated that the model provides reasonably accurate
predictions of the ultimate axial stress of partially FRP-confined concrete while
it usually underestimates the ultimate axial strain.

Keywords: Partially FRP-confined concrete · FRP
Analysis-oriented stress-strain model · Confinement

1 Introduction

Existing research has demonstrated that the fully FRP confinement can significantly
enhance the compressive strength and deformation capacities of circular concrete
columns (Lam and Teng 2003; Matthys et al. 2005; Pham et al. 2013; Lin and Teng
2017). Alternatively, the concrete column can be wrapped with longitudinally discrete
(i.e., spaced) FRP strips/hoops (Fig. 1), which is referred to as a partially FRP-confined
concrete column. Although most of the existing studies are related to fully
FRP-confined concrete columns, partially FRP-confined concrete columns have also
been demonstrated to process an adequate increase in strength and a remarkable
increase in axial deformation capacity compared with their counterparts (i.e.,
un-confined concrete columns) (e.g., Barros and Ferreira 2008; Park et al. 2008;

© Springer International Publishing AG, part of Springer Nature 2019
E. E. Gdoutos (Ed.): ICTAEM 2018, SI 5, pp. 50–54, 2019.
https://doi.org/10.1007/978-3-319-91989-8_9

Wei et al. 2009; Campione et al. 2015; Pham et al. 2015; Zeng et al. 2017). In addition, strengthening columns with discrete FRP strips is expected to be able to avoid FRP buckling failure which will easily occurred in concrete-filled FRP tubes as the axial stiffness of the FRP tube cannot be neglected. Also, less FRP materials are needed for partially FRP-confined concrete columns and thus partial FRP strengthening can be applied easier and faster than full FRP strengthening (Pham et al. 2015; Zeng et al. 2017).

The interest to understand the behavior of partially FRP-confined concrete has led to a few experimental studies on the behavior of concrete in partially FRP-confined columns (Barros and Ferreira 2008; Park et al. 2008; Campione et al. 2015; Pham et al. 2015; Triantafyllou et al. 2015; Saljoughian and Mostofinejad 2016; Zeng et al. 2017). The confinement in the axially loaded circular columns fully wrapped by FRP jackets is uniform. However, as for the confined concrete in partially FRP-confined circular columns, the confinement is non-uniform (Fig. 1) for the section at a given horizontal level within the bare concrete between two FRP strips (Mander et al. 1988). The use of discrete FRP strips results in the less efficient confinement to the concrete between the two adjacent FRP strips. The confinement mechanism is similar to the concrete confined by steel hoops or spirals, in which the reduced confinement effect between two adjacent strips can be considered through the "arching effect". Based on the arching action assumption, a parabola with an initial slope of 45° covering a clear vertical spacing (s_f') between two FRP strips is defined to separate the effectively confined part from the ineffectively confined part of the concrete core (Fig. 1). As illustrated in Fig. 2, the ineffective confinement area is located between two adjacent FRP strips. The difference of confinement mechanisms between fully and partially FRP-confined concrete implies different behavior between them, which may lead to the difference in stress-strain curves between the confined concrete in partially FRP-confined concrete columns and that in fully FRP-confined concrete columns. As a result, the reliability and accuracy of existing theoretical models for fully FRP-confined concrete need to be carefully examined for their applicability for partially FRP-confined concrete.

To this end, an experimental program was conducted to study the axially compressive behavior of circular columns wrapped with CFRP strips. In total 60 columns were prepared and tested. The main test variable examined in this experimental program included the clear spacing and width of FRP strips. The test results, in term of the effects of the clear spacing and width of FRP strips on the axial compressive behavior of partially FRP-confined concrete, were presented and discussed. A widely accepted analysis-oriented stress-strain model proposed by Teng et al. (2007) was used to predict the stress-strain behavior of partially FRP-confined concrete and its reliability and accuracy were examined by the comparisons between the test results and the model predictions.

2 Experimental Program

Totally sixty FRP-confined cylindrical column specimens were tested to investigate the effects of FRP strip width, clear spacing and FRP thickness. All specimens had a diameter of 150 mm and a height of 300 mm. Note that the clear spacing and width of

FRP strips are the main parameters studied in the current study and they represent the coefficient of FRP vertical efficiency. Note that the vertical confinement effectiveness coefficient, which is a coefficient representing the coefficient of FRP vertical efficiency, is given by $\left(1 - s'_f/2D\right)^2$ where s'_f is the clear spacing of two adjacent FRP strips and D is the diameter of the column (*fib* 2001). The design of strip number and strip width led to several different clear spacing ratios in present study. It is noted that the ratio between the clear spacing of two adjacent FRP strips and the column diameter is referred to as clear spacing ratio (i.e., s'_f/D), as shown in Fig. 1. Each column has two nominally identical specimens.

The flat coupon tensile tests were conducted to determine the material properties of FRP sheets, following the recommends specified in the ASTM standard (ASTM D3039 2008). The average values of tensile strength, elastic modulus and rupture strain were determined as 4185.6 MPa, 259.7 GPa and 1.75%, respectively. Column compression test was conducted using a displacement controlled procedure with a rate of 0.4 mm/min.

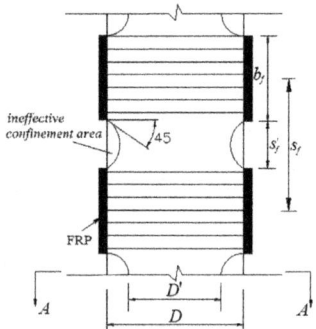

Fig. 1. Partially FRP-confined concrete column.

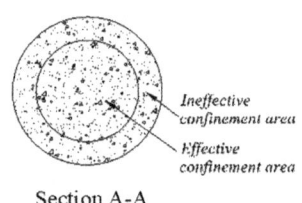

Section A-A

Fig. 2. Effective confinement area.

3   Conclusions

The test results of 60 circular columns fully or partially wrapped with FRP strips are presented in this paper. A widely accepted analysis-oriented model proposed by Teng et al. (2007) is used to predict the stress-strain behavior of partially FRP-confined concrete, and the reliability and accuracy of the model are then examined through the comparisons between the test results and the model predictions. Based on the study presented in this paper, the following conclusions can be drawn:

(1) Significant increases in the strength and axial deformation capacity are obtained for partially FRP-confined concrete. Similar to fully wrapped columns, partially wrapped columns are mostly failed due to the rupture of FRP strips. However, an increase in the clear FRP strip spacing may lead to the crushing failure of the bare concrete between two adjacent strips.

(2) A three-segment stress-strain curve can predict the behavior of partially FRP-confined concrete better compared with a two-segment stress-strain curve which normally captures the behavior of fully FRP-confined concrete well. This three-segment stress-strain curve is consisted of a first linear ascending segment, a transition segment and a second ascending/descending segment.

(3) The stress-strain curves of FRP-confined concrete in fully wrapped columns show monotonically ascending behavior owing to the sufficient FRP confinement. For partially FRP-confined concrete, the slope of the third segment is increased with the width of FRP strips while it is nearly independent to the FRP strip thickness. The normalized axial strength increases with the width of FRP strips for specimens confined with four or five FRP strips while it seems to be independent to the width of FPR strips for specimens confined with only three FRP strips. The ultimate axial strain of concrete increases with the width of FRP strips. The normalized axial strength and ultimate strain increase with the FRP thickness. An increase in the clear spacing of adjacent FRP strips leads to a decrease in the axial strength while the ultimate axial strain of concrete and the hoop rupture strain of FRP are approximately independent to the clear spacing of adjacent FRP strips.

(4) The equation for calculating the minimum amount of FRP is conservative. There may exist an optimum value of s'_f/d around 0.5, leading to the monotonic ascending stress-strain behavior not easily being obtained (i.e., $f'_{cu}/f'_{co} > 1$).

(5) Teng et al.'s (2007) model is reasonably accurate for predicting the axial strength of partially FRP-confined concrete, providing that the confinement effectiveness coefficients (*fib* 2001; CNR-DT 200 R1 2013) have properly been considered. However, the predicted ultimate strain is significantly underestimated and the stress at the turning point is also underestimated by Teng et al.'s (2007) model.

References

ASTM D3039: Standard Test Method for Tensile Properties of Polymer Matrix Composite Materials (D3039M), West Conshohocken, USA (2008)

Barros, J.A.O., Ferreira, D.R.S.M.: Assessing the efficiency of CFRP discrete confinement systems for concrete cylinders. J. Compos. Constr. **12**(2), 134–148 (2008)

CNR-DT 200 R1: Guide for the Design and Construction of Externally Bonded FRP Systems for Strengthening Existing Structures, Advisory Committee on Technical Recommendations For Construction, National Research Council, Rome, Italy (2013)

Campione, G., La Mendola, L., Monaco, A., Valenza, A., Fiore, V.: Behavior in compression of concrete cylinders externally wrapped with basalt fibers. Compos. Part B Eng. **69**, 576–586 (2015)

fib: Externally Bonded FRP Reinforcement for RC Structures. The International Federation for Structural Concrete, Lausanne, Switzerland (2001)

Hadi, M.N.S., Pham, T.M., Xu, L.: New method of strengthening reinforced concrete square columns by circularizing and wrapping with fiber-reinforced polymer or steel straps. J. Compos. Constr. **17**(2), 229–238 (2013)

Lam, L., Teng, J.G.: Design-oriented stress-strain model for FRP-confined concrete. Constr. Build. Mater. **17**(6–7), 471–489 (2003)

Lin, G., Teng, J.G.: Three-dimensional finite-element analysis of FRP-confined circular concrete columns under eccentric loading. J. Compos. Constr. **21**(4), 04017003 (2017)

Mander, J.B., Priestley, M.J., Park, R.: Theoretical stress-strain model for confined concrete. J. Struct. Eng. **114**(8), 1804–1826 (1988)

Matthys, S., Toutanji, H., Audenaert, K., Taerwe, L.: Axial behavior of large-scale columns confined with fiber-reinforced polymer composites. ACI Struct. J. **102**(2), 258–267 (2005)

Park, T.W., Na, U.J., Chung, L., Feng, M.Q.: Compressive behavior of concrete cylinders confined by narrow strips of CFRP with spacing. Compos. Part B Eng. **39**(7–8), 1093–1103 (2008)

Pham, T.M., Doan, L.V., Hadi, M.N.: Strengthening square reinforced concrete columns by circularisation and FRP confinement. Constr. Build. Mater. **49**, 490–499 (2013)

Pham, T.M., Hadi, M.N.S., Youssef, J.: Optimized FRP wrapping schemes for circular concrete columns under axial compression. J. Compos. Constr. **19**(6), 04015015 (2015)

Saljoughian, A., Mostofinejad, D.: Axial-flexural interaction in square RC columns confined by intermittent CFRP wraps. Compos. Part B Eng. **89**, 85–95 (2016)

Teng, J.G., Huang, Y.L., Lam, L., Ye, L.P.: Theoretical model for fiber-reinforced polymer-confined concrete. J. Compos. Constr. **11**(2), 201–210 (2007)

Triantafyllou, G.G., Rousakis, T.C., Karabinis, A.I.: Axially loaded reinforced concrete columns with a square section partially confined by light GFRP straps. J. Compos. Constr. **19**(1), 1–15 (2015). 04014035

Wei, H., Wu, Z., Guo, X., Yi, F.: Experimental study on partially deteriorated strength concrete columns confined with CFRP. Eng. Struct. **31**(10), 2495–2505 (2009)

Zeng, J.J., Guo, Y.C., Gao, W.Y., Li, J.Z., Xie, J.H.: Behavior of partially and fully FRP-confined circularized square columns under axial compression. Constr. Build. Mater. **152**, 319–332 (2017)

Improving the Tribological Properties of Ti6Al4V Alloy with Multi-walled Carbon Nanotube Additions

Adewale Adegbenjo[1](✉), Peter Olubambi[1], and Johannes Potgieter[2]

[1] Centre for Nanoengineering and Tribocorrosion, School of Mining, Metallurgy and Chemical Engineering, University of Johannesburg, Johannesburg, South Africa
waleeleect@gmail.com
[2] School of Chemical and Metallurgical Engineering, University of the Witwatersrand, Johannesburg, South Africa

Abstract. The effects of varied weight fractions of multi-walled carbon nanotubes (MWCNTs) on the dry sliding wear characteristics of Ti6Al4 V were investigated in this study. Dry sliding wear tests were conducted at three applied load levels of 5, 15 and 25 N on spark plasma sintered unreinforced Ti6Al4V alloy and MWCNTs reinforced Ti6Al4V composites containing 1, 2 and 3 wt% MWCNTs respectively sintered at 1000 °C. The ball-on-flat test configuration with tungsten carbide (WC) as the counterface material was used during the tests. Wear scars and debris were characterized by scanning electron microscopy (SEM) and energy dispersive X-ray spectrometry (EDX) techniques. It was observed that the wear resistances and coefficients of friction for the composites were significantly improved over that of the unreinforced Ti6Al4V alloy. Worn surface analysis showed the prevalent wear mechanisms at low and high applied loads were adhesive and abrasive wears respectively. Wear debris analysis by EDX showed the presence of tungsten (W), which suggests a transfer layer of the counterface material. Wear resistance enhancement in the composites was directly related to the extent of MWCNTs dispersion within Ti6Al4V matrix, the interfacial bond strength between the matrix and the reinforcement, as well as the presence of hard TiC interfacial product.

Keywords: Ti6Al4V alloy · Carbon nanotubes · Dry sliding wear

1 Introduction

The Ti6Al4V alloy has been a toast of the aerospace, automobile and chemical industries owing to its low specific gravity, high specific strength, excellent corrosion resistance and comprehensive mechanical properties [1]. Nevertheless, the application of the alloy has been largely limited in these industries due to its inferior hardness and poor tribological performance in structural components.

To this end, researchers have explored various methods to proffer a lasting solution to the characteristic tribological setbacks in the applications of Ti6Al4V. Among the methods used in the past are coatings, surface treatments, ion implantation, oxygen

© Springer International Publishing AG, part of Springer Nature 2019
E. E. Gdoutos (Ed.): ICTAEM 2018, SI 5, pp. 55–61, 2019.
https://doi.org/10.1007/978-3-319-91989-8_10

diffusion and solid film lubrication [2, 3]. Of recent, some researchers have recommended the use of carbon nanotubes (CNTs) as ideal, high performance, ultra-high strength reinforcements for metal matrices. CNTs reinforced metal matrices with enhanced mechanical/thermal characteristics, low wear rates and significantly improved coefficient of friction have also been reported [4].

However, only few studies are available in literature till date on the tribological behavior of metal matrices reinforced with MWCNTs. Also, little attention has been given to the understanding of the tribological characteristics of Ti matrices reinforced with MWCNTs. It is intended in this study that the development of MWCNTs/Ti6Al4V composites will open new possibilities for the application of Ti6Al4V as tribological components in the nearest future. Hence, this present study investigated the dry sliding wear characteristics of spark plasma sintered MWCNTs/Ti6Al4V composites under varied applied loads, with a focus on understanding the effect(s) of MWCNTs weight fractions on the tribological properties of Ti6Al4V matrices.

2   Experimental

The Dry sliding wear tests were conducted on Ti6Al4V and MWCNTs/Ti6Al4V composites sintered at 1000 °C under applied loads of 5, 15 and 25 N respectively, at room temperature and humidity employing a UMT-2-CETR tribometer (Bruker Nano Inc., Campbell, CA). The dynamic normal load (FZ), depth of wear track (Z), friction force (FX), and the coefficient of friction (μ) were monitored from the tribometer set-up. The ball-on-flat test mode (Fig. 1) was used during the tests in order to simulate the reciprocating sliding wear conditions to which Ti6Al4V materials are subjected in real life situation.

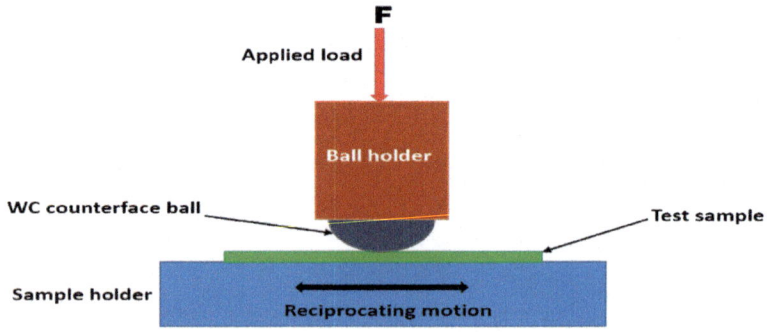

Fig. 1. Schematic diagram of a typical ball-on-flat tribometer set up.

Sectioned mirror polished representative samples were made from the as-received sintered samples. These were etched with Kroll's reagent and were later examined under FESEM. Wear test samples were cut into (20 × 20 × 3) mm dimensions and prepared to mirror polished surfaces following the conventional metallographic procedures. Before each test, the ball and sample surfaces were cleaned with acetone and allowed to dry in air to remove dust and any possible solid contaminants. Wear depths and coefficient of friction (COF) were monitored continuously for 1000 s as the WC counterface ball (Ø10 mm) slides against the test sample in a reciprocating motion at a frequency of 5 Hz and reciprocating speed of 10 mm/s. A stroke length of 2 mm was used during the tests. At least three triplicate tests were run at each test condition to ensure good repeatability in both the wear and friction results. The wear volume on the scar surface was calculated using the single-trace analysis method as stated in Eq. 1. V_w is the wear volume, R_s is radius of scar, W is width of scar while L_s and Z_w are the stroke length and wear depth respectively.

$$V_w = L_s \left[R_s^2 Sin^{-1} \left(\frac{W}{2R_s} \right) - \left(\frac{W}{2} \right) (R_s - Z_w) \right] + \frac{\pi}{3} Z_W^2 (3R_S - Z_W) \qquad (1)$$

This method has been reported to have a good agreement with the 3D analysis technique, giving more reliable results than the mass loss and 2D analysis methods. The method is simple, as only the scar size calculation and a once-off profiling trace on the scar surface is needed. Wear scar and debris analyses were carried out using a high resolution FESEM.

3 Results and Discussions

Table 1 shows the properties of the as-received, sintered Ti6Al4V and MWCNTs/Ti6Al4V samples used for the dry sliding wear tests. All the samples had densities almost equal to their theoretical densities. Hence, the samples are presumed to be fully dense, none had relative density less than 95%. However, it could be observed that increased MWCNTs content led to a decline in the relative densities of the composites, whereas their measured hardness values were improved. These were attributed to the difficulty in achieving homogeneous dispersion of MWCNTs and the presence of TiC in the composites respectively [5, 6].

Table 1. Density and microhardness properties of Ti6Al4V and Ti6Al4V/MWCNTs samples sintered at 1000 °C used for dry sliding wear test.

MWCNTs (wt%)	Density[a] (g/cm³)	Density[b] (g/cm³)	Rel. Density (%)	Microhardness HV$_{0.5}$
0	4.43	4.41 ± 0.01	99.60 ± 0.03	362.19 ± 13.45
1	4.41	4.37 ± 0.01	99.10 ± 0.02	378.00 ± 8.84
2	4.38	4.31 ± 0.02	98.40 ± 0.04	395.73 ± 6.94
3	4.36	4.27 ± 0.01	97.90 ± 0.02	397.58 ± 4.88

[a] Rule of mixture (ρ Ti6Al4V = 4.43 g/cm³ and ρ MWCNT = 2.1 g/cm³),
[b] Archimedes principle

The SEM images of the sintered samples are presented in Fig. 2. The unreinforced Ti6Al4V (Fig. 2a) showed a fully lamellar microstructure with the characteristic basket weave (Widmanstätten) α + β network structure of Ti6Al4V.

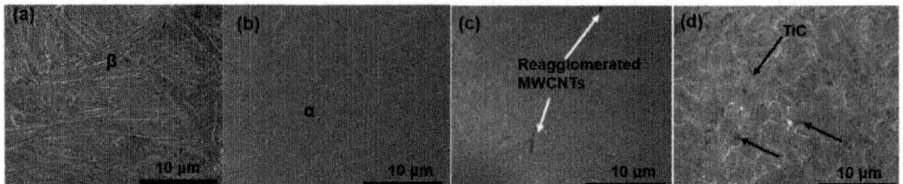

Fig. 2. SEM images of (a) Ti6Al4V and MWCNTs reinforced Ti6Al4V composites with varied weight fractions of the reinforcement (b) 1 wt% (c) 2 wt% and (d) 3 wt%; sintered at 1000 °C.

This microstructure was transformed to equi-axed structures (Fig. 2b–d) with MWCNTs addition into the metal matrices. No pores were seen in the micrographs. The 2 wt% MWCNTs composite (Fig. 2c) had re-agglomerated MWCNTs, while that with 3 wt% MWCNTs (Fig. 2d) had clustered titanium carbide (TiC) interfacial products in their microstructures. This is why this sample had the highest hardness as seen in Table 1.

3.1   Wear Volume Loss

The variation of the observed wear volume losses in the consolidated Ti6Al4V and MWCNTs/Ti6Al4V composites is shown in Fig. 3. It is evident that the developed composite materials had improved wear resistance over that of the unreinforced Ti6Al4V alloy. This was attributed to the higher hardness exhibited by the composites as seen in Table 1. Nevertheless, wear losses were found to increase with increased applied load in both Ti6Al4V and the composites.

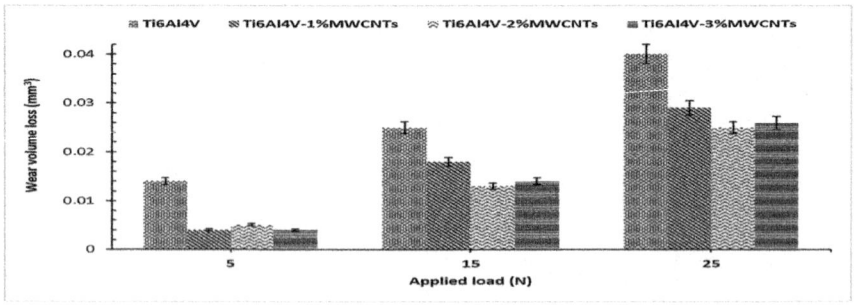

Fig. 3. Wear volume loss sintered Ti6Al4V and MWCNTs/Ti6Al4V composites.

At 5 N applied load, the 3 wt% and the 1 wt% MWCNTs composites had the same wear volume loss, while the 2 wt% MWCNTs had higher wear loss that both of them. However, the wear resistance of the 3 wt% MWCNTs composite dropped below that of the 2 wt% MWCNTs composite at higher applied loads. This suggests that the interfacial hard TiC phases were not able to withstand the wear effect at higher loads as they soon break away due to suspected weak interfacial bonds between the matrix and the reinforcement. Hence, the matrix Ti6Al4V could not successfully transfer the wear load to the MWCNTs reinforcement at these higher applied loads. The inadequate dispersion of MWCNTs in the 3 wt% MWCNTs composite which was originally the reason for the occurrence of interfacial reactions was thought to be responsible for this.

The worn surfaces of the samples under 5 and 25 N applied loads as observed by SEM is presented in Figs. 4 and 5 respectively. The worn surface images are consistent with the wear volume loss variation in the samples as seen earlier in Fig. 3. The surfaces under 5 N load exhibited adhesive wear features while at 25 N, the predominant features were that of abrasive and oxidative wears. The worn surfaces also supported that severe wear activities took place at the higher load and that the composite materials had improved wear resistances over that of Ti6Al4V alloy. Analysis of the wear debris at 5 and 25 N loads (images not shown) also supported that adhesive and abrasive wear mechanisms were predominant at low and higher applied loads respectively.

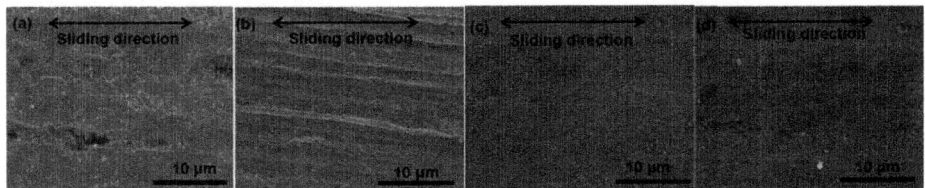

Fig. 4. Worn surfaces sintered of (a) Ti6Al4V and MWCNTs/Ti6Al4V composites containing varied weight fractions of MWCNTs (b) 1 wt%, (c) 2 wt% and (d) 3 wt% tested under 5 N applied load.

The wear debris of the composite samples under both applied loads were mostly powdery and composed of smaller sized flake-like chips, which suggests they experienced less severe wear compared to the unreinforced Ti6Al4V alloy.

Figure 6a is the SEM image an area spotted within the debris of the 3 wt% MWCNTs composite, which was tested under 5 N. The cylindrical structure of MWCNTs were maintained. EDS analysis of points 2 and 4 shown respectively in Fig. 6b and c showed these points are rich in Ti6Al4V constituent elements and carbon from the MWCNTs reinforcement or TiC respectively. Traces of tungsten (W) is evident at point 2, suggesting that the wear debris contained a transferred element from the WC counterface material.

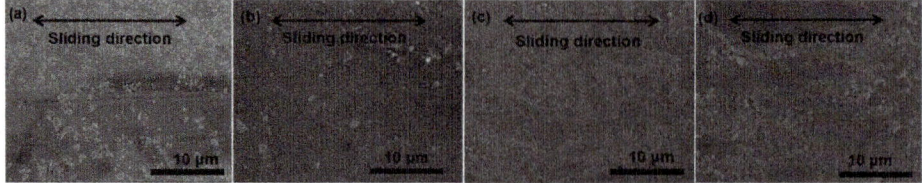

Fig. 5. Worn surfaces sintered of (a) Ti6Al4V and MWCNTs/Ti6Al4V composites containing varied weight fractions of MWCNTs (b) 1 wt%, (c) 2 wt% and (d) 3 wt% tested under 25 N applied load.

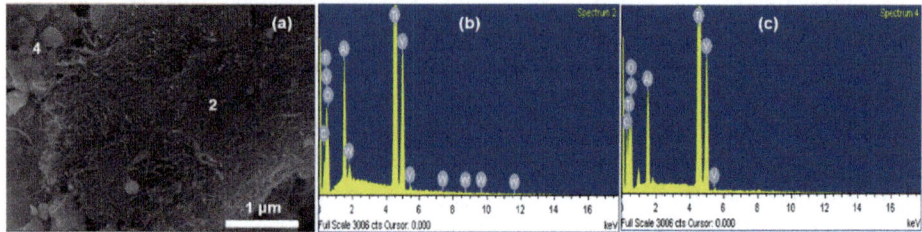

Fig. 6. SEM/EDS analysis of an area within the debris of the 3 wt% MWCNTs composite tested under 5 N applied load.

3.2   Coefficient of Friction (COF)

The variation of COF in sintered Ti6Al4V and MWCNTs/Ti6Al4V composites shown in Fig. 7 confirmed earlier positions that the composites had better wear characteristics than the unreinforced Ti6Al4V. All the composites exhibited enhanced COF over Ti6Al4V. However, the 1 wt% MWCNTs composite had the comparatively best COF performance as its COF continued to reduce with increase in the applied load, exhibiting the least COF under the highest applied load. The COF for the 3 wt% MWCNTs composite on the other hand deteriorated with increase in applied load.

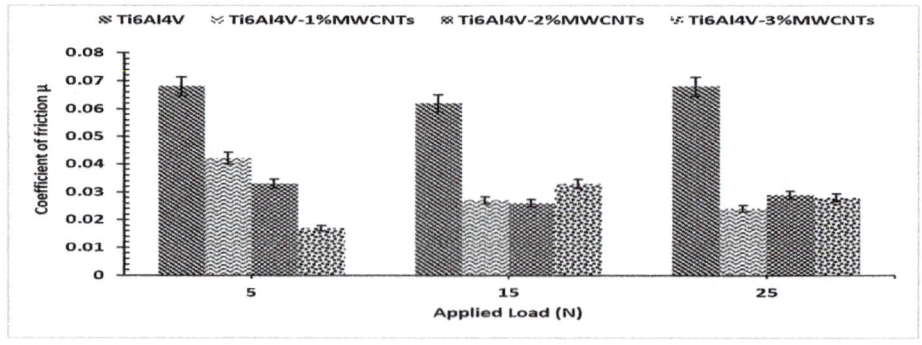

Fig. 7. COF variation in sintered Ti6Al4V and MWCNTs/Ti6Al4V composites.

4 Conclusions

Addition of MWCNTs to Ti6Al4V matrices improved their wear resistance and COF performances. However, the improved tribological performance was found to be directly dependent on the extent of MWCNTs dispersion with the metal matrix. The retention of MWCNTs cylindrical structure within the sintered composites acted as solid lubricants which enhanced the COF performance of the composites.

References

1. Williams, J.C., Starke, E.A.: Progress in structural materials for aerospace systems. Acta Mater. **51**(19), 5775–5799 (2003)
2. Dong, A.H., Bell, T.: Enhanced wear resistance of titanium surfaces by a new thermal oxidation treatment. Wear **238**(2), 131–137 (2000)
3. Met, A.C., Vandenbulcke, L., Catherine, M.S.: Friction and wear characteristics of various prosthetic materials sliding against smooth diamond-coated titanium alloy. Wear **255**(7–12), 1022–1029 (2003)
4. Bakshi, A.S., Lahiri, D., Agarwal, A.: Carbon nanotube reinforced metal matrix composites-a review. Int. Mater. Rev. **55** (2010)
5. Wang, F.C., Zhang, Z.H., Sun, Y.J., Liu, Y., Hu, Z.Y., Wang, H., Korznikov, A.V., Korznikova, E., Liu, Z.F., Osamu, S.: Rapid and low temperature spark plasma sintering synthesis of novel carbon nanotube reinforced titanium matrix composites. Carbon **95**, 396–407 (2015)
6. Feng, X., Sui, J., Cai, W., Liu, A.: Improving wear resistance of TiNi matrix composites reinforced by carbon nanotubes and in situ TiC. Scripta Mater. **64**(9), 824–827 (2011)

Numerical Analysis of Delamination in Composite Structures Using Strain Measurements from Fiber Bragg Gratings Sensors

Valeriy Matveenko[1,2], Grigoriy Serovaev[1,2],
and Mikhail Tashkinov[1(✉)]

[1] Perm National Research Polytechnic University,
Komsomolsky Avenue 29, 614990 Perm, Russia
m.tashkinov@pstu.ru
[2] Institute of Continuous Media Mechanics,
Academician Korolev Street, 1, 614013 Perm, Russia

Abstract. The possibilities of strain measurements in composite structures using optical fiber strain sensors based on Bragg gratings are demonstrated. Specifics of interaction of the sensors with the environment were considered. Indications of optical fiber sensors can be used to predict mechanical behavior of composites structures as well as internal defects development based on numerical models. An approach to numerical modeling for identification of the most critical zones in terms of the appearance and development of delamination defects in polymer composite structures is proposed. Computational results for the case studies are presented.

Keywords: Laminate composites · Optical fiber strain sensors
Bragg gratings · Delamination

1 Introduction

In the recent years, research in smart materials and their fields of applications had been developing very actively. The idea of creating materials that along with prescribed basic functional provide real-time information about their internal state is highly relevant for polymer composite materials and structures. The concept of smart composite materials with self-diagnostic functions had found its implementation with involvement of optical fiber sensors with Bragg gratings (FBG sensors), which, in particular, are able to measure various physical and mechanical quantities [1–4]. Embedded into composites, optical fiber sensors under severe loading conditions have advantages, including sensitivity, in comparison with other types of sensors. Besides, they are able to withstand strains equal to strains of composite laminate, are affordable and easy to manufacture, are immune to electrical interference.

The proposed technique for predicting strength and mechanical behavior of composite structures is to compare the set of measurements from FBG sensors with the results of numerical modelling with consideration of the features of the microstructure, including occurrence and development of defects. The computational component of the

© Springer International Publishing AG, part of Springer Nature 2019
E. E. Gdoutos (Ed.): ICTAEM 2018, SI 5, pp. 62–67, 2019.
https://doi.org/10.1007/978-3-319-91989-8_11

technique provides finite element simulations of mechanical behavior and failure of composite structure. The experimental part is based on strains measurements obtained by optical fiber strain sensors on Bragg gratings fixed on the surface or embedded between the layers of composite.

2 Strain Measurements Using Fiber Bragg Gratings

2.1 Operating Principles

The principle of work of strain sensors based on Bragg gratings is as follows. The interrogator generates and transmits into the optical fiber a broadband light signal in a given wavelength interval. The Bragg grating reflects some specific wavelength of this signal. The main part of the reflected optical signal has a wavelength of λ, the resonant wavelength of the reflected spectrum. The magnitude of this wave is directly proportional to the effective refractive index n and the geometric length of the Bragg grating period L. This dependence has determined the possibility of recording the change in the relative length of the Bragg grating period from the analysis of the change in the relative magnitude of the resonance wavelength of the reflected spectrum. The dependence of the change in the wavelength of the reflected spectrum on the longitudinal deformation of the optical fiber ε and on the temperature change ΔT in the linear approximation has the following form:

$$\Delta \lambda_{Br} = 2nL\left(\left\{1 - \left(\frac{n^2}{2}\right)[p_{12} - v(p_{11} + p_{12})]\right\}\varepsilon + \left[\alpha + \frac{1}{n}\frac{dn}{dT}\right]\Delta T\right) \tag{1}$$

where n is refractive index of the optical fiber, p_{11} and p_{12} are the Pokkels coefficients of strain-optical tensor, α is the temperature expansion coefficient of quartz core of the optical fiber, v is Poisson coefficient.

The results obtained earlier showed that reliable information on deformations recorded by built-in strain sensors on Bragg gratings in a complex stress state is possible only if the strain tensor component along the fiber is significantly exceeds the tensor components in the plane perpendicular to the optical fiber [5, 6]. Then, at a constant temperature, the relationship between the change in the wavelength of the reflected spectrum can be determined in terms of the deformation coefficient K:

$$\Delta \lambda_{Br} = K\varepsilon \tag{2}$$

It was established that for a quartz optical fiber that does not interact at the location of the Bragg grating with other solid bodies, the deformation coefficient has the value of K = 0.78.

2.2 Sensors Interaction with the Environment

If an optical fiber with Bragg gratings interacts with a surface of a material (for example, through an adhesive bond), or is embedded in a material (for instance, between composite plies), the value of K should be refined. With respect to the values of the strains obtained by this system at the locations of the sensors, a calibration of the strain values is being performed to find a calibration coefficient, each for specific application. Consideration of the calibration factor makes it possible to increase the accuracy of the registered strains by 5%.

The interface between the sensor and the surface under investigation on the scale of the characteristic dimensions of the sensor represents a structure consisting of a set of elements made of materials with various physical, mechanical, and thermophysical properties. To obtain a thermal compensation dependence of such complex spatial structures, mathematical modeling of the deformational response of these structures to temperature effects, including nonstationary ones, can be implemented. This circumstance is especially important for estimating the characteristic times at which the uniformity of the temperature field in the vicinity of the sensor location is achieved.

The mathematical formulation of the problem consists of the temperature boundary value problem and the problem of the deformational response to the temperature effect. Both of these problems are formulated for a system consisting of an optical fiber strain sensor matched with a fragment of the material of the measured object. The problems are solved numerically by the finite element method. As a result of the solution, the temperature response of the compensation sensor is determined as a graph of the deformation of the compensation sensor against temperature changes.

3   Interrelation of Indications of Optical Fiber Sensors with Mathematical Models of Delamination in Composite Structures

Strain monitoring data in composite structures can be used to predict mechanical behavior and defect development based on numerical models.

As a case study, a construction in the form of a plate bent at right angle with a step-length-varying width was chosen (Fig. 1). The object of the study is modeled as a homogeneous anisotropic elastic body with the mechanical characteristics given in Table 1.

Table 1. Mechanical characteristics for numerical model

$E_x = 22$ ГПа	$\upsilon_{xy} = 0.14$	$G_{xy} = 4$ ГПа
$E_y = 22$ ГПа	$\upsilon_{yz} = 0.18$	$G_{yz} = 3$ ГПа
$E_z = 6$ ГПа	$\upsilon_{xz} = 0.13$	$G_{xz} = 3$ ГПа

The transfer of load and fixing of the sample is carried out using a special testing rig shown in Fig. 1. The lower tooling element remains stationary, 5 mm vertical movements are assigned to the upper element, as a result of which the sample and the tool elements interact at the surfaces, marked with color in the Fig. 1. The coefficient of friction on contact is 0.1. As a result of application of the above-described boundary

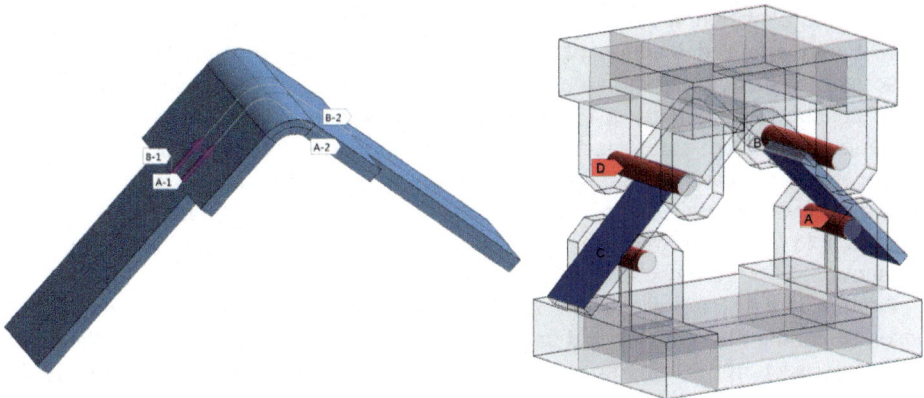

Fig. 1. A structural element with the supposed location of optical fibers, as well as loading setup for a four-point bending.

conditions, the sample bends - the upper surface is in the compression state, the lower one is the stretching.

Based on the results of a numerical calculation, it is possible to determine the region where debonding is most probable for a given type of loading. The critical zone was assessed according to the highest normal stress in the plane perpendicular to the sample plane. The maximum value of the stress is achieved in the central part of the sample. This zone is most susceptible to the appearance of defects (Fig. 2), where it is necessary to control strain levels with the help of FBG sensors.

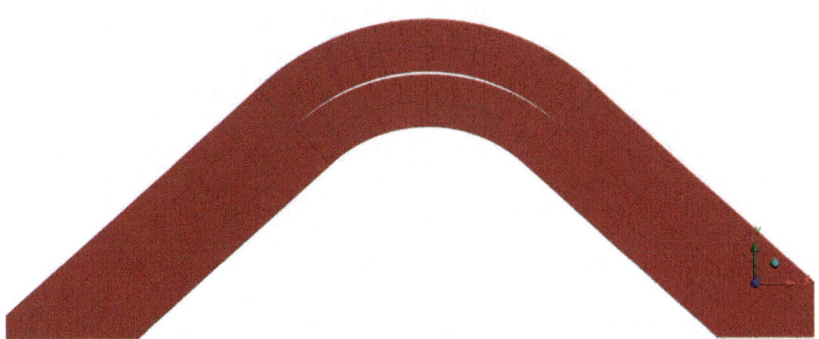

Fig. 2. Finite-element model of a sample with debonding

In order to estimate the response of strains measured by FBG sensor in the presence of a defect in a composite structure, models with a delamination of fixed size were created. Delaminations with sizes of 10, 20, 30 and 45 mm along the width of the sample were considered. For all models, analysis of distribution of strains along the lines of the supposed location of optical fibers is studied: on the upper and lower surfaces of the sample (as shown on Fig. 1). The longitudinal deformation component, which can be measured by a FBG sensor, is considered. The results of numerical simulation showed a significant change in the longitudinal component of the deformations upon appearance and increase of debonding. With this specific geometric configuration of the test sample, it is preferable to arrange the optical fibers at the center of the sample length, on the upper surface of the sample and on the bottom. This zone was shown to be the most sensitive to presence of delamination.

Figures 3 and 4 represents the strain curves that can be measured by the FBG sensors when the fibers are located on the respective surfaces in the middle of the width of the cross section of the sample in the presence of defects of different lengths in the zone of maximum stress. These results can be a ground for recommendations on the installation of sensors. One of them should be located in the delamination zone, and the other in the zone where the effect of the defect on the strain level is negligible. With the same type of loading, the ratio of the measurements of the sensors in the absence of a delamination will be a certain constant. In the case of debonding occurrence, the value of this constant changes. The increase in the magnitude of this deviation will characterize the growth of delamination.

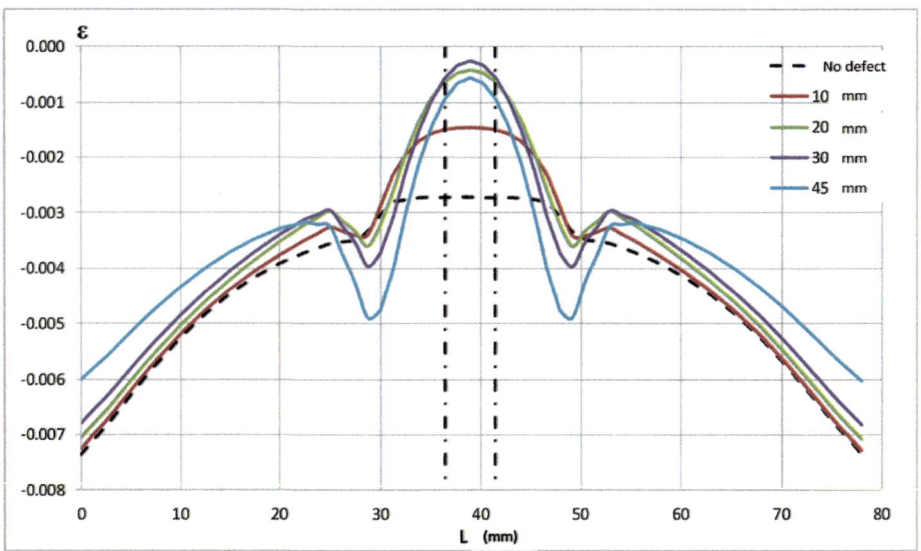

Fig. 3. Distribution of the strain on the upper surface of the specimen along the line over the width of the section, depending on the size of the delamination

The growth kinetics of the debonding can be correlated with the dependence of the strain measurements on the size of the defect. This ideology of defect registration will be extremely useful in the cyclical loading of structures and delamination propagation due to fatigue failure.

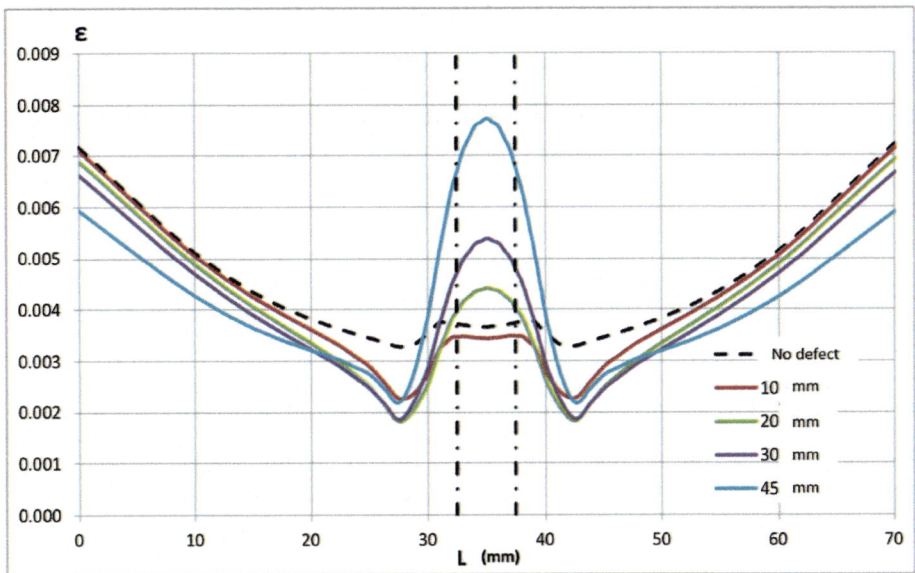

Fig. 4. The distribution of strain on the lower surface of the specimen along the line over the width of the section

Acknowledgments. The research was performed at the Perm National Research Polytechnic University, with the support of the Russian Science Foundation (project № 15-19-00243).

References

1. Frieden, J., Cugnoni, J., Botsis, J., Gmür, T., Ćorić, D.: High-speed internal strain measurements in composite structures under dynamic load using embedded FBG sensors. Compos. Struct. **92**, 1905–1912 (2010). https://doi.org/10.1016/j.compstruct.2010.01.007
2. Yashiro, S., Okabe, T., Toyama, N., Takeda, N.: Monitoring damage in holed CFRP laminates using embedded chirped FBG sensors. Int. J. Solids Struct. **44**, 603–613 (2007). https://doi.org/10.1016/j.ijsolstr.2006.05.004
3. Zhao, X., Gou, J., Song, G., Ou, J.: Strain monitoring in glass fiber reinforced composites embedded with carbon nanopaper sheet using Fiber Bragg Grating (FBG) sensors (2009)
4. Beukema, R.P.: Embedding technologies of FBG sensors in composites : technologies, applications and practical use. In: 6th European Workshop on Structural Health Monitoring, pp. 1–8 (2012)
5. Matveenko, V.P., Serovaev, G.S.: Measurement of strains by optical fiber Bragg grating sensors embedded into polymer composite material. Struct. Control Health Monit. **25**(4), 1–11 (2017). https://doi.org/10.1002/stc.2118
6. Anoshkin, A.N., Voronkov, A.A., Kosheleva, N.A., Matveenko, V.P., Serovaev, G.S., Spaskova, E.M., Shardakov, I.N., Shipunov, G.S.: Measurement of inhomogeneous strain fields by fiber optic sensors embedded in a polymer composite material. Mech. Solids. **51**, 542–549 (2016). https://doi.org/10.3103/s0025654416050058

Combined Anisotropic Viscodamage-Viscoplasticity Model for Rock Under Dynamic Loading

Timo Saksala$^{(\boxtimes)}$

Tampere University of Technology, PO Box 600, 33101 Tampere, Finland
timo.saksala@tut.fi

Abstract. This paper deals with numerical modelling of rock fracture under dynamic loading. For this end, an anisotropic viscodamage-viscoplasticity model for rock is developed. In the viscodamage part of the model, the Rankine criterion indicates the tensile stress states leading to rate-dependent anisotropic damaging. The anisotropy of damage is based on the compliance damage tensor given by the dyadic product of the gradient of the Rankine criterion with itself. In compression, the inelastic deformation and compressive strength degradation is governed by a Mohr-Coulomb viscoplasticity model. The model performance is first demonstrated at the material point level using a two-element model. Then, the dynamic Brazilian disc test, involving both shear and tensile fracture types along with strain rate hardening, on rock is simulated as a laboratory sample level problem.

Keywords: Viscoplasticity · Anisotropic damage · Rock fracture
Finite element method

1 Introduction

Brittle materials such rock and concrete exhibit both stiffness and strength degradation as well as irreversible strains [1]. Irreversible deformation takes place especially in compression while in tension it can be neglected. Therefore, a constitutive model with damage and plasticity component is needed to correctly predict the behavior of these materials. Moreover, microcracking of brittle materials have usually loading induced orientation which results in damage induced anisotropy. Thereby, an anisotropic damage model is often preferable over an isotropic one, see [2] for concrete and [3] for rock as examples. Brittle materials are also strain rate-sensitive which is realized as higher strengths upon increased loading rates both in compression and tension [4].

In this paper, a constitutive model for rock capable of accounting for rate-dependent anisotropic damage in tension and rate-dependent strength degradation accompanied with inelastic strain in compression is presented. The stress states leading to damaging in tension are indicated with the Rankine criterion while a Mohr-Coulomb viscoplasticity model governs the failure in compression. Numerical examples are solved in order to demonstrate the performance of the model.

© Springer International Publishing AG, part of Springer Nature 2019
E. E. Gdoutos (Ed.): ICTAEM 2018, SI 5, pp. 68–73, 2019.
https://doi.org/10.1007/978-3-319-91989-8_12

2   Theory of the Model

A brief description of the viscoplasticity and viscodamage concepts are first given in general form. Then, the specific morel is specified. Under the infinitesimal kinematics assumption, justified by the brittle nature of rock materials, the strain can be decomposed into

$$\varepsilon = \varepsilon_e + \varepsilon_{vp} + \varepsilon_{vd} \tag{1}$$

where ε_e, ε_{vp}, ε_{vd} are the elastic, viscoplastic and viscodamage parts of the total strain, respectively.

2.1   Viscodamage and Viscoplasticity Consistency Approach

Unlike the classical viscoplasticity formulations by Perzyna and Duvaut-Lions, the more recent viscoplastic consistency model [5] utilizes the consistency condition. The main components of such a model are [3]

$$f_{vp} = f(\sigma, \kappa_{vp}, \dot{\kappa}_{vp}) = \hat{f}_{vp}(\sigma) - (\sigma_y + h_{vp}\kappa_{vp} + s_{vp}\dot{\kappa}_{vp})$$

$$\dot{\varepsilon}_{vp} = \dot{\lambda}_{vp}\frac{\partial g_{vp}}{\partial \sigma}, \quad \dot{\kappa}_{vp} = \dot{\lambda}_{vp}k_{vp}(\sigma, \kappa_{vp}) \tag{2}$$

$$f_{vp} \leq 0, \quad \dot{\lambda}_{vp} \geq 0, \quad \dot{\lambda}_{vp}f_{vp} = 0$$

where $\hat{f}_{vp}(\sigma)$ is a yield function, given function of stress, σ_y is the yield stress, h_{vp} and s_{vp} are the plastic and viscosity moduli, g_{vp} is the viscoplastic potential, κ_{vp}, $\dot{\kappa}_{vp}$ are the internal variable and its rate, respectively, $\dot{\lambda}_{vp}$ is the viscoplastic increment, and $k_{vp}(\sigma, \kappa_{vp})$ is function that relates κ_{vp} to $\dot{\lambda}_{vp}$.

The compliance damage concept can also be cast in the consistency format. The basic ingredients of this model are [3]

$$f_{vd} = f(\sigma, \kappa_{vd}, \dot{\kappa}_{vd}) = \hat{f}_{vd}(\sigma) - (\sigma_f + h_{vd}\kappa_{vd} + s_{vd}\dot{\kappa}_{vd})$$

$$\dot{\mathbf{D}} = \frac{\dot{\lambda}_{vd}}{\hat{f}_{vd}(\sigma)}\frac{\partial f_{vd}}{\partial \sigma} \otimes \frac{\partial f_{vd}}{\partial \sigma}, \quad \dot{\kappa}_{vd} = \dot{\lambda}_{vd}k_{vd}(\sigma, \kappa_{vd}) \tag{3}$$

$$f_{vd} \leq 0, \quad \dot{\lambda}_{vd} \geq 0, \quad \dot{\lambda}_{vd}f_{vd} = 0$$

where f_{vd} is the viscodamage loading function, σ_f is the fracture stress indicating the damage initiation stress, $\mathbf{D}$ is the (fourth order) compliance tensor, whereas the meanings of the rest of symbols are equivalent to the corresponding ones related to the viscoplastic model. It should be noted that the damage loading function $\hat{f}_{vd}(\sigma)$ is a homogeneous function of degree one, i.e. so that $\partial\hat{f}_{vd}(\sigma)/\partial\sigma : \sigma = \hat{f}_{vd}(\sigma)$.

In the present compliance damage format, the current value of compliance tensor, which evolves by damage from the initial value $\mathbf{D}_{(t\,=\,0)} = \mathbf{0}$, is added to the elastic compliance to obtain: $\mathbf{E}_d = (\mathbf{C}_e^{-1} + \mathbf{D})^{-1}$ with $\mathbf{E}_d$ and $\mathbf{C}_e$ being the damaged stiffness and the standard elasticity tensor, respectively. Finally, the relation between the damage compliance and the damage deformation reads $\boldsymbol{\varepsilon}_{vd} = \mathbf{D} : \boldsymbol{\sigma}$.

2.2   The Present Model Based on Rankine and Mohr-Coulomb Criteria

In the present model, the classical Rankine criterion serves as the viscodamage tensile loading function and the Mohr-Coulomb criterion is the viscoplastic yield function in compression. Written in the xy-stress space, these criteria are

$$f_R(\boldsymbol{\sigma}, \lambda_R, \dot{\lambda}_R) = \underbrace{\frac{1}{2}(\sigma_x + \sigma_y) + \sqrt{\left(\frac{\sigma_x - \sigma_y}{2}\right)^2 + \sigma_{xy}^2}}_{\hat{f}_{vd}(\boldsymbol{\sigma})} - \sigma_t(\lambda_R, \dot{\lambda}_R)$$

$$f_{MC}(\boldsymbol{\sigma}, \lambda_{MC}, \dot{\lambda}_{MC}) = \underbrace{\frac{k-1}{2}(\sigma_x + \sigma_y) + (k+1)\sqrt{\left(\frac{\sigma_x - \sigma_y}{2}\right)^2 + \sigma_{xy}^2}}_{\hat{f}_{vp}(\boldsymbol{\sigma})} - \sigma_c(\lambda_{MC}, \dot{\lambda}_{MC})$$

$$\sigma_t(\lambda_R, \dot{\lambda}_R) = \sigma_{t0} + h_R \lambda_R + s_R \dot{\lambda}_R, \quad h_R = -g\sigma_{t0}\exp(-g\kappa_R)$$

$$\sigma_c(\lambda_{MC}, \dot{\lambda}_{MC}) = \sigma_{c0} + h_{MC}\lambda_{MC} + s_{MC}\dot{\lambda}_{MC}, \quad k = (1+\sin\phi)/(1-\sin\phi)$$

$$\text{(4)}$$

where $\boldsymbol{\sigma}$ is the stress tensor, σ_{t0} and σ_{c0} are the tensile and compressive strengths, φ is the internal friction angle, h_R and h_{MC} are the softening moduli in tension and compression, respectively, and s_R and s_{MC} are the constant viscosity moduli. The softening moduli are defined with the specific mode I and II fracture energies as $g = \sigma_{t0}l_e/G_{Ic}$ and $h_{MC} = -\sigma_{c0}^2 l_e/2G_{IIc}$ with l_e being a characteristic length of a finite element. Moreover, non-associated flow rule with a potential, g_{MC}, similar in form to f_{MC} but using the dilatation angle in the definition of k instead of the internal friction angle is employed. Next, the stress integration with respect to the combined model is considered.

2.3   Stress Return Mapping with the Coupled Model

The stress return mapping with the combined viscodamage-viscoplastic consistency model can be performed with the standard methods of multisurface plasticity (i.e. the elastic predictor-plastic corrector split) due to the fact the consistency is enforced in both of the components, see [3, 5] for details. The trial stress is calculated by $\boldsymbol{\sigma}_{trial}^{t+\Delta t} = \mathbf{C}_e : (\boldsymbol{\varepsilon}^{t+\Delta t} - \boldsymbol{\varepsilon}_{vp}^t - \boldsymbol{\varepsilon}_{vd, n=0}^{t+\Delta t}) = (\mathbf{C}_e^{-1} + \mathbf{D}_n)^{-1} : (\boldsymbol{\varepsilon}^{t+\Delta t} - \boldsymbol{\varepsilon}_{vp}^t)$ where the relation between the initial viscodamage strain and the trial stress, $\boldsymbol{\varepsilon}_{vd, n=0}^{t+\Delta t} = \mathbf{D}_n : \boldsymbol{\sigma}_{trial}^{t+\Delta t}$, has been used. Assuming that both the Rankine viscodamage and Mohr-Coulomb viscoplastic criteria are violated by the trial stress, i.e. $f_R > 0 \,\& f_{MC} > 0$, the update formulae read

$$[\delta\lambda_R \; \delta\lambda_{MC}]^T = \mathbf{G}_n^{-1}[f_R \; f_{MC}]_n^T$$

$$\varepsilon_{n+1}^{vp} = \varepsilon_n^{vp} + \delta\lambda_{MC}\frac{\partial g_{MC}}{\partial \boldsymbol{\sigma}_n}, \quad \mathbf{D}_{n+1} = \mathbf{D}_n + \frac{\delta\lambda_R}{\hat{f}_{vd}(\boldsymbol{\sigma})}\frac{\partial f_R}{\partial \boldsymbol{\sigma}_n} \otimes \frac{\partial f_R}{\partial \boldsymbol{\sigma}_n}$$

$$\boldsymbol{\sigma}_{n+1} = \boldsymbol{\sigma}_n - \left(\mathbf{C}_e^{-1} + \mathbf{D}_n\right)^{-1} : \left(\delta\lambda_R\frac{\partial f_R}{\partial \boldsymbol{\sigma}_n} + \delta\lambda_{MC}\frac{\partial g_{MC}}{\partial \boldsymbol{\sigma}_n}\right)$$

$$\Delta\lambda_{R,n+1} = \Delta\lambda_{R,n} + \delta\lambda_R, \quad \Delta\lambda_{MC,n+1} = \Delta\lambda_{MC,n} + \delta\lambda_{MC}$$

$$\lambda_{R,n+1} = \lambda_{R,n} + h_{R,n}\delta\lambda_R + s_R\frac{\Delta\lambda_{R,n+1}}{\Delta t}, \quad \lambda_{MC,n+1} = \lambda_{MC,n} + h_{MC,n}\delta\lambda_{MC} + s_{MC}\frac{\Delta\lambda_{MC,n+1}}{\Delta t}$$

$$(5)$$

with

$$\mathbf{G}_n = \begin{bmatrix} \frac{\partial f_R}{\partial \sigma} : \mathbf{E}_d : \frac{\partial f_R}{\partial \sigma} + h_{R,n} + \frac{s_R}{\Delta t} & \frac{\partial f_R}{\partial \sigma} : \mathbf{E}_d : \frac{\partial g_{MC}}{\partial \sigma} \\ \frac{\partial f_{MC}}{\partial \sigma} : \mathbf{E}_d : \frac{\partial f_R}{\partial \sigma} & \frac{\partial f_{MC}}{\partial \sigma} : \mathbf{E}_d : \frac{\partial g_{MC}}{\partial \sigma} + h_{MC,n} + \frac{s_{MC}}{\Delta t} \end{bmatrix} \quad (6)$$

If one of the increments $\Delta\lambda_R$, $\Delta\lambda_{MC}$ becomes negative during this iteration, the case is not a genuine corner plasticity case. In such a situation, the iteration is restarted keeping only the surface with positive increment active.

3 Numerical Examples

The numerical simulations demonstrating the performance of the present model are carried out here. The equations of motion are solved with an explicit time marching scheme. The model for dynamic Brazilian disc test is depicted in Fig. 1.

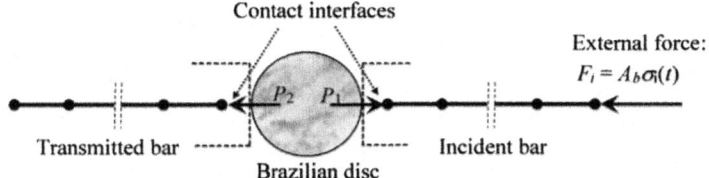

Fig. 1. Computational model for dynamic Brazilian disc test simulation.

The compressive stress wave induced by the impact of the striker bar is simulated as an external stress pulse, $\sigma_i(t) = A\sin(\omega t)$ where A is the amplitude and $\omega = 2\pi/T$ with $T = 160$ μs. The incident and transmitted bars are modeled with two-node standard bar elements and the Brazilian disc is meshed with the six-node triangular elements. Finally, the contacts between the bars and the disc are modeled by imposing kinematic (impenetrability) constraints.

The material properties representative for a hard rock, such as Kuru gray granite, and the model parameters used in the simulations are: Young's modulus $E = 67$ GPa, Poisson's ratio $\nu = 0.26$, material density $\rho = 2600$ kg/m^3, tensile strength $\sigma_{t0} = 11.4$ MPa, compressive strength $\sigma_{c0} = 235$ MPa, mode I fracture energy

G_{Ic} = 50 N/m, mode II fracture energy G_{IIc} = 5000 N/m, internal friction angle φ = 50°, dilatation angle ψ = 5°, and viscosity moduli (if not otherwise explicitly said) s_R = 0.05 MPa · s, s_{MC} = 0.05 MPa · s.

3.1   Material Point Level Simulations

The model prediction is tested in uniaxial tension and compression with different values of viscosity as well as in a load reversal program using a 2-element model. The results are shown in Fig. 2.

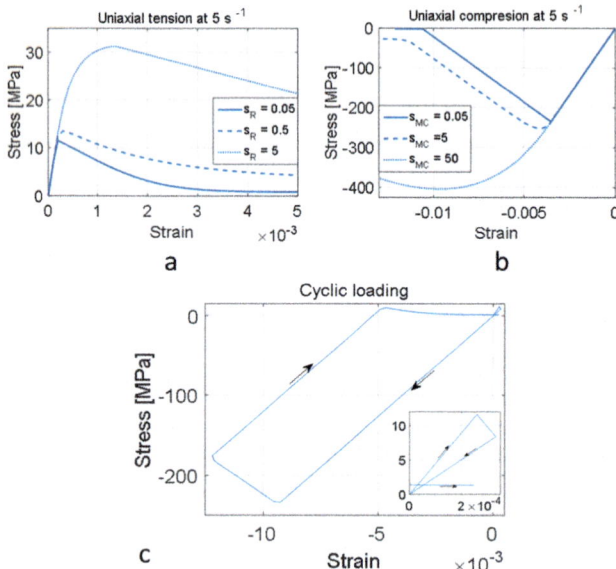

Fig. 2. Model predictions at the material point level: effect of viscosity uniaxial tension (a), compression (b), and the response in load reversals (c).

The results show that the model captures the strain-rate hardening effect at a given strain rate by adjusting the viscosity moduli. Under cyclic loading, stiffness degradation takes place under tension while strain softening governs the post-peak response in compression. It should be noted that the unilateral nature, or stiffness recovery, of crack closure is not taken into account in the present model.

3.2   Dynamic Brazilian Disc Simulations

The dynamic Brazilian disc test on rock is simulated as a laboratory sample level dynamic problem. The model in Fig. 1 is used with the amplitude of A = 50 MPa for the external loading. The radius of the steel bars is 11 mm while the BD disc diameter D = 41 mm and the thickness h = 16 mm. The simulation results are shown in Fig. 3. The tensile strength is calculated with the average of the maximum contact forces, $P = (P_1 + P_2)/2$, by $\sigma_T = 2P/\pi hD$ giving about 13 MPa.

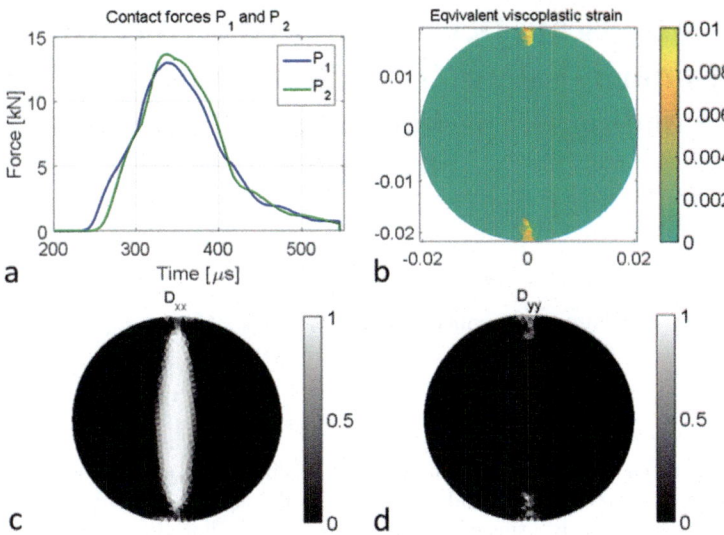

Fig. 3. Dynamic Brazilian test simulation results: Contact forces (a), equivalen viscoplastic strain (b), damage D_{xx} (c), and damage D_{yy} (c) distributions.

According to the results in Fig. 3a, the "dynamic equilibrium" is well achieved since there are no significant deviations between the contact forces. The disc failure mode is the typical axial splitting as attested in the distribution of damage component D_{xx} (Fig. 3c) accompanied with some shear failures at the contact areas, as can be observed from the equivalent viscoplastic strain distribution in Fig. 3b. It can be concluded that the model can predict, in the continuum sense, some of the rock failure characteristics under dynamic loading.

References

1. Brady, B.H.G., Brown, E.T.: Rock Mechanics for Underground Mining. Springer, Berlin (1993). https://doi.org/10.1007/978-1-4020-2116-9
2. Hansen, E., Willam, K., Carol, I.: A two-surface anisotropic damage/plasticity model for plain concrete. In: de Borst, R., et al. (eds.) Proceedings of Framcos-4 Conference Paris (Fracture Mechanics of Concrete Structures), Balkema (2001)
3. Saksala, T., Ibrahimbegovic, A.: Anisotropic viscodamage-viscoplastic constitutive model with a parabolic cap for rocks with brittle and ductile behavior. Int. J. Rock Mech. Min. Sci. **70**, 460–473 (2014)
4. Zhang, Q.B., Zhao, J.: A review of dynamic experimental techniques and mechanical behaviour of rock materials. Rock Mech. Rock Eng. **47**, 1411–1478 (2014)
5. Wang, W.M., Sluys, L.J., De Borst, R.: Viscoplasticity for instabilities due to strain softening and strain-rate softening. Int. J. Numer. Meth. Eng. **40**, 3839–3864 (1997)

Multiscale Defects Induced Criticality in Damage-Failure Transition Under Intensive Loading (Experimental and Theoretical Results)

Oleg Naimark$^{(\boxtimes)}$

Institute of Continuous Media Mechanics UB RAS,
Acad.Korolev str. 1, 614013 Perm, Russia
naimark@icmm.ru

Abstract. Defects kinetics is analyzed as specific type of the criticality in out-of-equilibrium system "solid with defects" – the structural-scaling transition. Defect induced mechanisms of structural relaxation are linked to the generation of different types of the collective modes of defects, that have the nature of self-similar solutions: auto-solitary waves related to multiscale plastic strain localization and blow-up dissipative structures providing the damage localization kinetics. The subjection of solid responses to mentioned self-similar solutions is analyzed both theoretically and experimentally with the goals to link strain and damage localization effects under adiabatic shear failure, splitting of shock wave fronts, the power law universality of fragmentation statistics and failure wave initiation and propagation with specific types of multiscale collective modes of defects. The data of high resolution experiments is analyzed according to developed theoretical approach.

Keywords: Damage-failure transition · Defects induced criticality
Dynamic and shock wave loading

1 Introduction

Much interest has been shown in fundamental problems of failure due to mesoscopic nature of this phenomenon related to the multiscale damage accumulation (microcracks, microshears). Solid with these defects reveals the properties of out-of-equilibrium systems and field description of deformed solids should be consistent with theory of defects and physics of damage-failure transition in terms of continuous variables characterizing the behavior of defect ensembles. Statistical approach for the description of out-of-equilibrium states of solid with defects was developed in [1] and allowed the formulation of the field description. Mechanisms of multiscale structural relaxation induced by mesoscopic defects is the consequence of characteristic form of out-of-equilibrium free energy, corresponding free energy release in terms of internal variables and realizes as specific type of critical phenomena in solids with defects – structural-scaling transition. Internal variables represent the defect induced strain (defect density tensor) and the susceptibility of material to the defects

© Springer International Publishing AG, part of Springer Nature 2019
E. E. Gdoutos (Ed.): ICTAEM 2018, SI 5, pp. 74–78, 2019.
https://doi.org/10.1007/978-3-319-91989-8_13

growth in terms of the ratio of two characteristic scales: mean size of defects and spacing between defects as parameters of defects interaction. Nonlinearity of free energy release provides the qualitative changes in the kinetics for defect density tensor (defect induced strain) leading to the nucleation of damage localization areas (with characteristic fractography images on the fracture surface). The sizes of damage localization areas and temporal dynamics of defects accumulation are given by the self-similar "blow-up" solution for damage evolution equation. These self-similar solutions are localized on characteristic lengths and represent the set of collective modes of defects. The subjection of solid behavior to the set of these collective modes leads to qualitative changes of crack dynamics (transition from steady to branching regime of crack propagation), scale universality of fragmentation statistics and self-similar scenario of failure under dynamic and shock wave loading [2].

2   Collective Modes of Defects

One of the key results of the statistical approach and statistically based phenomenology are the establishment of two "order parameters" responsible for the structure evolution – the defect density tensor p_{ik} and the structural scaling parameter $\delta = (R/r_0)^3$, which represents the ratio of the spacing between defects and characteristic size of defects [6, 7]. The value of structural-scaling parameter characterizes the current susceptibility of material to the nucleation and growth of defects. Dependence of solid responses on structural-scaling parameter reflects important feature of damage kinetics, statistical self-similarity [1], that was established for microshear (microcrack) distribution function for different stages of damage accumulation represented in dimensionless (self-similar) coordinates. Statistically predicted non-equilibrium free energy F represents generalization of the Ginzburg-Landau expansion in terms of mentioned order parameters – the defect density tensor (defect induced deformation $p(x) = p_{zz}$ in uni-axial loading in z-direction) and structural scaling parameter δ:

$$F = 1/2A\,(\delta, \delta_*)p^2 - 1/4Bp^4 - 1/6C\,(\delta, \delta_c)p^6 - D\sigma p + \chi\,(\partial p/\partial x)^2, \qquad (1)$$

The damage evolution is determined by the kinetic equations for the defect density p and structural-scaling parameter δ

$$\dot{p} = -\Gamma_p \frac{\Delta F}{\Delta p}, \quad \dot{\delta} = -\Gamma_\delta \frac{\partial F}{\partial \delta}, \qquad (2)$$

where Γ_p, Γ_δ are the kinetic coefficients, $\Delta(\ldots)/\Delta t$ is the variation derivative. Kinetic equations (Eq. 2) and the equation for the total deformation $\varepsilon = \hat{C}\sigma + p$ ($\hat{C}$ is the component of the elastic compliance tensor) represent the constitutive equations of materials with mesodefects. Material responses include the generation of characteristic collective modes – the autosolitary waves in the range of $\delta_c < \delta < \delta_*$ and the "blow-up" dissipative structure in the range $\delta < \delta_c \approx 1$. The generation of these collective modes under the loading provides the change of the system symmetry according to the group properties of equations in corresponding ranges of structural-scaling parameter and initiates specific mechanisms of damage-failure transition on the scales of damage localization with the blow-up defect kinetics.

The damage localization kinetics follows to the "blow-up" self-similar solution [1]

$$p = g(t)f(\xi), \xi = x/L_H, g(t) = G(1 - t/\tau_c)^{-m}, \tag{3}$$

and can be considered as the precursor of crack nucleation.

The parameters in Eq. 3 are: τ_c is the so-called "peak time" ($p \to \infty$ at $t \to \tau_c$ for the self-similar profile $f(\xi)$ localized on the scale L_H, $G > 0$, $m > 0$ are the parameters of non-linearity, which characterize the free energy release rate for $\delta < \delta_c$. The self-similar solution Eq. 3 describes the blow-up damage kinetics for $t \to \tau_c$ (on the set of spatial scales $L_H = kL_c$, $k = 1, 2, \ldots K$, where L_c and L_H corresponds to the so-called "simple" and "complex" blow-up dissipative structures. Generation of the complex blow-up dissipative structures appears when the distance L_S between simple structures approaches to the scale L_c. Similar scenario of the "scaling transition" proceeds for the blow-up structures of different complexity to involve in damage localization the larger scales of material.

The description of damage kinetics as the structural-scaling transition allowed the consideration of solid with defects as a dynamic system with spatial-temporal degrees of freedom (corresponding to the set of blow-up dissipative structures of different complexity). Stochastic behavior in this case can be linked with the dynamics of out-of-equilibrium system with the features of flicker noise, or $1/f$- statistics [3]. The self-similar nature of mentioned collective modes associated with damage localization zones has the importance in the case of dynamic loading, when the "excitation" of these modes can lead to the subjection of failure to the dynamics of these modes.

3 Fragmentation Statistics. Resonance Excitation of Failure (Failure Waves)

Fragmentation statistics was studied during in situ experiments for impact loaded fused quartz rods and fracture luminescence recording to analyze the temporal sequences of failure hotspots initiation and the following study of fragmentation statistics for recovered samples after the fragment weighing [3, 4]. Temporal fracture luminescence events and fragment size distribution demonstrated the power law statistics (the flicker or $1/f$- noise) that is characteristic for the out-of-equilibrium critical systems revealing the so-called self-organized criticality (SOC).

It is interesting the limit case of failure revealing the temporal-spatial independence of failure evolution on stress. Namely this situation is observed in experiment for failure wave initiation [4].

Experimental study of failure wave generation and propagation was realized for the symmetric Taylor test on fused-quartz rods. Figure 1 shows the processing of a high-speed photography (upper picture) for the flyer rod travelling at 534 m/s at impact.

Three dark zones correspond to the image of impact surface (green triangle), failure wave (red square) and (blue diamond) the shock wave. The initial slope for the failure

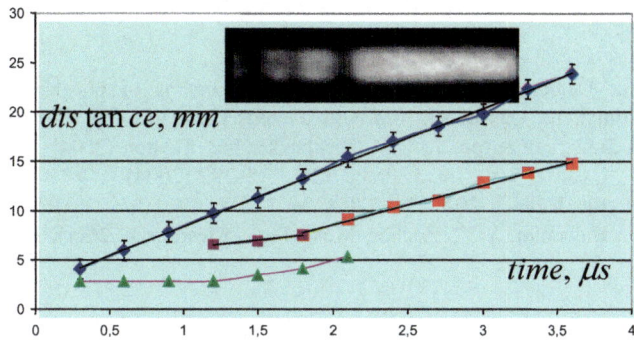

Fig. 1. The Taylor test data

wave gives the front velocity $V_{fw} \approx 1.57$ km/s that is close to traditionally measured in the plate impact test [3, 4].

However, the experiment revealed the increase of failure front velocity up to the value $V_{fw} \approx 4$ km/s. Approaching of failure wave front velocity to the shock front velocity supports theoretically based result concerning the failure wave nature as "delayed failure" with the limit of "delay time" corresponding to the "peak time" t_c in the self-similar solution, Eq. 3. Theoretical analysis of this situation allowed the interpretation of damage kinetics as the "resonance excitation" of blow-up damage localization modes.

4 Summary

It is shown that the process of damage-failure transition can be considered as specific type of criticality in out-of-equilibrium system "solid with defects" and wide range constitutive model was proposed as the generalization of the Ginzburg-Landau approach in terms of independent field variables describing typical mesoscopic defects (microshears, microcracks). Specific types of the collective modes of defects were established as self-similar solution of the evolution equation for mentioned damage parameter. The set of blow-up self-similar collective modes of defects can be considered as the independent variables provided universality of nonlinear dynamics of damage-failure transition: transition from steady-state crack propagation to the branching regime with pronounced intermittency in crack propagation velocity, "resonance" excitation of damage localization in shocked materials ("dynamic branch" under spall failure, failure waves), spatial-temporal power law universality in dynamic fragmentation. Original experimental data supported the assumption concerning the role of multiscale blow-up collective modes of defects on self-similar responses of materials in wide range of load intensity.

Acknowledgments. Research was supported by the project n. 17-01-00867 a of the Russian Foundation for Basic Research.

References

1. Naimark, O.B.: Defect induced transitions as mechanisms of plasticity and failure in multifield continua. In: Capriz, G., Mariano, P. (eds.) Advances in Multifield Theories of Continua with Substructure, pp. 75–114. Birkhauser Inc., Boston (2004)
2. Naimark, O.B., Uvarov, S.V.: Int. J. Fract. **128**, 285–292 (2004)
3. Naimark, O.B.: Int. J. Fract. **202**, 271–279 (2016)
4. Naimark, O.B., Bayandin, Y.V., Zocher, M.A.: Phys. Mesomech. **20**(1), 10–30 (2017)

An Approximate Solution for Plane Strain Rolling of Viscoplastic Sheets

Elena Lyamina[✉][iD]

Ishlinsky Institute for Problems in Mechanics RAS, pr. Vernadskogo 101-1,
Moscow 119526, Russia
lyamina@inbox.ru

Abstract. This paper presents a generalization of the widely used approach of Orowan for calculating the roll pressure in flat rolling of homogeneous rigid perfectly strips on viscoplastic sheets. In contrast to the original approach, the approach for viscoplastic sheets requires the analysis of the velocity field since the model is rate dependent. The present paper employs the solution for plane strain compression of a viscoplastic block for finding the through thickness distribution of stress and velocity. Strain rate hardening is taken into account by means of the Herschel-Bulkley model.

Keywords: Orowan approach · Flat rolling · Viscoplastic material

1 Introduction

Several analytical models for analysis of the process of sheet rolling are available in the literature. Reviews of these models can be found in [1, 2]. The model proposed by Orowan [3] results in solutions that show good agreement with experiment [4–6]. This model is often used to verify other theories of rolling [7, 8]. In particular, the solution [8] based on a series expansion shows a better agreement with the solution based on the model [3] as the number of terms in the series expansion increases. Moreover, there are very fast computational methods [9–11] for solving the equation that results from the model [3], which is of special importance when calculations are to be performed in real time. The model [3] is based on an analytic stress solution for a wedge of rigid perfectly plastic material derived in [12]. Therefore, the original method [3] is applicable only if the problem is statically determinate. In many cases the latter condition is not satisfied and a velocity solution is required. For example, an analysis of rolling of a three layer sheet based on Orowan's approach has been carried out in [13]. The stress and velocity equations are also coupled in viscoplasticity. The objective of the present paper is to extend the model [3] to analysis of rolling of viscoplastic sheets. It is worthy of note that other popular approximate methods for analysis of metal forming processes are not applicable in this case. In particular, the slab method cannot be used because the yield stress depends on the equivalent strain rate and the method does not consider the velocity field. The upper bound method cannot be used because the friction stress depends on the velocity field.

2 Compression of a Viscoplastic Strip

A semi-analytical solution to the problem of the plane strain compression of a vis-coplastic strip between parallel platens has been provided in [14]. In this section, this solution is summarized for the subsequent use in analysis of plane strain rolling of sheets. The Cartesian coordinate system (x, y) is shown in Fig. 1 for plates of width $2L$ and instantaneous gap $2H$. The velocity of each plate is U. The process has two axes of symmetry, $x = 0$ and $y = 0$. It is therefore sufficient to consider the domain $x \geq 0$ and $y \geq 0$. The plane strain yield criterion is

$$\left(\sigma_{xx} - \sigma_{yy}\right)^2 + 4\sigma_{xy}^2 = 4k^2 \tag{1}$$

where σ_{xx}, σ_{yy} and σ_{xy} are the Cartesian stress components and k is the shear yield stress. It is assumed that k depends on the equivalent strain rate, ξ_{eq}, defined as

$$\xi_{eq} = \sqrt{\frac{3}{2}\left(\xi_{xx}^2 + \xi_{yy}^2 + 2\xi_{xy}^2\right)} \tag{2}$$

where ξ_{xx}, ξ_{yy} and ξ_{xy} are the Cartesian components of the strain rate tensor. The associated flow rule reads

$$\xi_{xx} = \lambda\left(\sigma_{xx} - \sigma_{yy}\right), \quad \xi_{yy} = -\lambda\left(\sigma_{xx} - \sigma_{yy}\right), \quad \xi_{xy} = 2\lambda\sigma_{xy}, \quad \lambda > 0. \tag{3}$$

Let u_x and u_y be the Cartesian velocities. The boundary conditions are

$$\sigma_{xy} = 0 \quad \text{and} \quad u_y = 0 \tag{4}$$

for $y = 0$ and

$$\sigma_{xy} = -mk \quad \text{and} \quad u_y = -U \tag{5}$$

for $y = H$. The first condition in (5) is the friction law where m is the friction factor. Other boundary conditions are not essential for Orowan's method. The dependence of k on the equivalent strain rate is supposed to be of the form

$$k = k_0\left[1 + \left(\frac{\xi_{eq}}{\xi_0}\right)^n\right] \tag{6}$$

where k_0, ξ_0 and n are constant. An approximate semi-analytic solution to the system of equations comprising (1), (3), (6) and the equilibrium equations has been found in [13]. In particular, in our nomenclature k as a function of y is determined from the following implicit equation:

$$\left(\frac{k}{k_0} - 1\right)^{1/n} = \eta\left(1 - \frac{m^2 k_f^2 y^2}{k^2 H^2}\right), \quad \eta = \frac{2\sqrt{3}U}{H\xi_0} \tag{7}$$

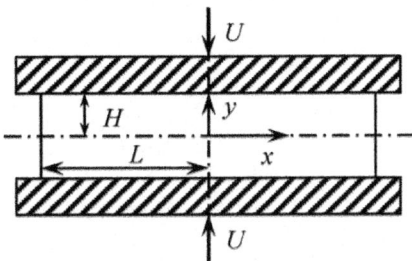

Fig. 1. Schematic diagram of upsetting.

where k_f is the shear yield stress at $y = H$. Hence, it follows from (7) that

$$\frac{k_f}{k_0} = 1 + \eta^n \left(1 - m^2\right)^{-n/2}. \tag{8}$$

The Cartesian stresses are determined as

$$\sigma_{xx} = \frac{mk_f x}{H} + 2\sqrt{k^2 - \frac{m^2 k_f^2 y^2}{H^2}} + C_1, \quad \sigma_{yy} = \frac{mk_f x}{H} + C_1, \quad \sigma_{xy} = -\frac{mk_f y}{H}. \tag{9}$$

3 Rolling of Viscoplastic Sheets

A schematic diagram of the process of flat rolling is shown in Fig. 2. In this figure, φ is the angular coordinate, φ_m is its value corresponding to the plane of entry, $\varphi = 0$ at the plane of exit, R is the radius of the rolls, ω is the angular velocity of the rolls, $2H_0$ is the initial thickness of the strip, and $2H_f$ is the final thickness of the strip. The dependence of the thickness of the strip on φ in the range $0 \leq \varphi \leq \varphi_m$ is

$$2H = 2R(1 - \cos \varphi) + 2H_f. \tag{10}$$

Let $f(\varphi)$ be the resulting horizontal force acting across any vertical plane between the plates of entry and exit. In [3], the following equation for $f(\varphi)$ has been derived:

$$df/d\varphi = 2R(s \sin \varphi + \tau \cos \varphi). \tag{11}$$

Note that $f(\varphi)$ is taken as positive if it is a compressive force. In (11), s is the normal pressure on the roll surface and τ is the friction stress. It is assumed that τ is positive for the exit side and τ is negative for the entry side. Equation (7) is also valid for viscoplastic sheets. In order to connect s and τ to the solution given in the previous section, it is necessary to consider an infinitesimal element near the friction surface. Since s and τ act on the side of this element inclined at angle φ to the x-axis, equilibrium requires

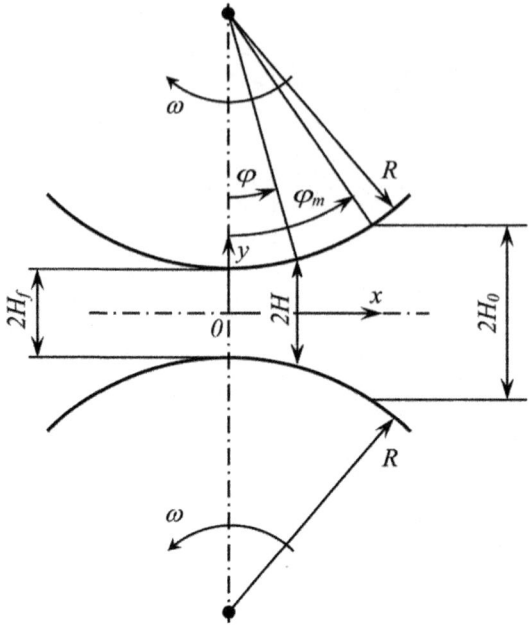

Fig. 2. Schematic diagram of rolling.

$$2\tau = -\left(\sigma_{xx} - \sigma_{yy}\right)\sin 2\varphi + 2\sigma_{xy}\cos 2\varphi,$$
$$2s = -\left(\sigma_{xx} + \sigma_{yy}\right) + \left(\sigma_{xx} - \sigma_{yy}\right)\cos 2\varphi + 2\sigma_{xy}\sin 2\varphi. \tag{12}$$

Substituting (9) into (12) yields

$$\tau = -k_f \sqrt{1 - m^2}\sin 2\varphi - mk_f \cos 2\varphi,$$
$$s = -\sigma_{yy} + k_f\left[\sqrt{1 - m^2}(1 + \cos 2\varphi) - m\sin 2\varphi\right]. \tag{13}$$

The value of τ is prescribed by a friction law. Assuming that the $\tau = \mu k_f$ the first equation in (13) can be solved for m. Thus m is a function of φ. This value of m should be used in the solution given in the previous section. By definition,

$$\frac{f}{2} = -\int_0^H \sigma_{xx} dy. \tag{14}$$

It follows from (9) that

$$\sigma_{xx} = \sigma_{yy} + 2\sqrt{k^2 - \frac{m^2 k_f^2 y^2}{H^2}}. \tag{15}$$

Equations (14) and (15) combine to give

$$\frac{f}{2} = -\sigma_{yy}H - 2\int_0^H \sqrt{k^2 - \frac{m^2 k_f^2 y^2}{H^2}} dy. \tag{16}$$

In this equation k can be eliminated by means of the solution of Eq. (7) and k_f can be eliminated by means of (8). Therefore, the integral in (16) can be evaluated. Then, σ_{yy} in the second equation in (13) can be eliminated by means of (16) giving s as a function of f and φ. Substituting this function for s in (11) and eliminating in the resulting equation τ by means of (13) yield the differential equation for f. This equation should be solved numerically.

4 Conclusion

The approach proposed in [3] for calculating the roll pressure in flat rolling of rigid perfectly plastic strips has been extended to rolling of viscoplastic strips. The major difference between the approaches is that a velocity solution is necessary to find the through thickness distribution of the yield stress, whereas the original approach [3] does not involve any velocity solution. The differential equation for the total horizontal force should be solved numerically.

Acknowledgements. The author acknowledges financial support of this research through the grant RFBR-17-08-01063.

References

1. Montmitonnet, P.: Hot and cold strip rolling processes. Comput. Methods Appl. Mech. Eng. **195**, 6604–6625 (2006)
2. Cawthorn, C.J., Loukaides, E.G., Allwood, J.M.: Comparison of analytical models for sheet rolling. Procedia Eng. **81**, 2451–2456 (2014)
3. Orowan, E.: The calculation of roll pressure in hot and cold flat rolling. Proc. Inst. Mech. Eng. **150**, 140–167 (1943)
4. Kimura, H.: Application of Orowan theory to hot rolling of aluminum. J. Jpn. Inst. Light Met. **35**, 222–227 (1985)
5. Kimura, H.: Application of Orowan theory to hot rolling of aluminum: computer control of hot rolling of aluminum. Sumitomo Light Metal Techn. Rep. **26**, 189–194 (1985)
6. Lenard, J.G., Wang, F., Nadkarni, G.: Role of constitutive formulation in the analysis of hot rolling. Trans. ASME J. Eng. Mater. Technol. **109**, 343–349 (1987)
7. Atreya, A., Lenard, J.G.: Study of cold strip rolling. Trans. ASME J. Eng. Mater. Technol. **101**, 129–134 (1979)

8. Domanti, S., McElwain, D.L.S.: Two-dimensional plane strain rolling: an asymptotic approach to the estimation of inhomogeneous effects. Int. J. Mech. Sci. **37**, 175–196 (1995)
9. Owen, A.G., Griffin, A.W.J.: Rapid solution of Orowan's equations using a hybrid computer. Proc. Inst. Electric. Eng. **119**, 1510–1516 (1972)
10. Rusia, D.: Improvements to Alexander's computer model for force and torque calculations in strip rolling processes. J. Mater. Shap. Technol. **8**, 167–177 (1990)
11. El-Bitar, T.A.: Computer program for the calculation of roll force and torque with strip tension in cold rolling. Iron Steelmak. **20**, 87–96 (1993)
12. Nadai, A.: Plasticity: Mechanics of the plastic state of matter. McGraw-Hill, New York (1931)
13. Alexandrov, S., Lyamina, E.: Extension of Orowan's method to analysis of rolling of three-layer sheets. Proc. Eng. **207**, 1391–1396 (2017)
14. Adams, M.J., Briscoe, B.J., Corfield, G.M., Lawrence, C.J., Papathanasiou, T.D.: An analysis of the plane-strain compression of viscoplastic materials. Trans. ASME J. Appl. Mech. **64**, 420–424 (1997)

Using the Upper Bound Technique
for Calculating the Strain Rate Intensity Factor

Sergei Alexandrov[1(✉)], Dragisa Vilotic[2], and Daria Grabco[3]

[1] Ishlinsky Institute for Problems in Mechanics RAS, 119526 Moscow, Russia
sergei_alexandrov@spartak.ru
[2] University of Novi Sad, 21000 Novi Sad, Serbia
[3] Institute of Applied Physics, MD 2028 Chisinau, Moldova

Abstract. The strain rate intensity factor is the coefficient of the leading singular term in a series expansion of the equivalent strain rate in the vicinity of maximum friction surfaces. This factor can be used to describe the generation of fine grain layers in the vicinity of friction surfaces in metal forming processes. However, a difficulty is that the strain rate intensity factor follows from singular solutions and commercial finite element packages are not capable of finding this factor. In the present paper, the upper bound technique is used for this purpose. The kinematically admissible velocity field chosen accounts for the exact asymptotic expansion of the equivalent strain rate. Therefore, an approximate value of the strain rate intensity factor can be found from this field.

Keywords: Friction · Strain rate intensity factor · Metal forming

1 Introduction

The strain rate intensity factor (SRIF) has been introduced in [1]. It has been shown in this work that the second invariant of the strain rate tensor approaches infinity in the vicinity of maximum friction surface in three-dimensional flow of a rigid perfectly plastic material obeying an arbitrary pressure-independent yield criterion and its associated flow rule. The coefficient of the principal singular term in a series expansion of the second invariant of the strain rate tensor has been named SRIF. The maximum friction law for this material model postulates that the friction stress at sliding is equal to the shear yield stress. The main result reported in [1] has been extended to other rigid plastic models [2, 3]. The technological value of SRIF comes from its potential to provide an approach to describing the generation of fine grain layers in the vicinity of frictional interfaces in metal forming [4]. The main difficulty with the further development of the approach [4] is that commercial FE packages are not capable of determining SRIF [5]. To the best of authors' knowledge, the only available accurate numerical solution for the strain rate intensity factor has been based on the method of characteristics [6]. It is therefore reasonable to develop an approximate method for calculating SRIF.

© Springer International Publishing AG, part of Springer Nature 2019
E. E. Gdoutos (Ed.): ICTAEM 2018, SI 5, pp. 85–90, 2019.
https://doi.org/10.1007/978-3-319-91989-8_15

2 Conceptual Approach

For the sake of simplicity, the subsequent analysis is restricted to the von Mises yield criterion. The yield stress in tension is denoted by σ_0. The equivalent strain rate is

$$\xi_{eq} = \sqrt{(2/3)\xi_{ij}\xi_{ij}} \tag{1}$$

where ξ_{ij} are the components of the strain rate tensor. It has been shown in [1] that

$$\xi_{eq} = D/\sqrt{s} + o\left(1/\sqrt{s}\right) \tag{2}$$

as $s \rightarrow 0$. Here s is the normal distance to the maximum friction surface and D is the strain rate intensity factor. This surface is defined by the condition that the friction stress at sliding is equal to the shear yield stress $\sigma_0/\sqrt{3}$. Let (x, y, z) be a Cartesian coordinate system situated at an arbitrary point, M, of a maximum friction surface, Ω. The z-axis is directed along the normal to Ω, away from the rigid tool and towards the plastic material. Since $s \equiv z$ at M, it follows from (2) that the x – and y – velocity components are expressed as

$$u_x = U_{x0} + U_{x1}\sqrt{z} + o(\sqrt{z}) \text{ and } u_y = U_{y0} + U_{y1}\sqrt{z} + o(\sqrt{z}) \tag{3}$$

as $z \rightarrow 0$. These representations are valid at point M. Also, U_{x0}, U_{x1}, U_{y0}, and U_{y1} are independent of z. Substituting (3) into (2) yields

$$D = \left(\sqrt{3}/4\right)\sqrt{U_{x1}^2 + U_{y1}^2}. \tag{4}$$

According to the upper bound theorem for rigid perfectly plastic material

$$W_{ex} \leq W_p + W_f + W_d \tag{5}$$

where W_{ex} is the real rate of external work, W_p is the power dissipation within plastic regions, W_f is the power dissipation due to friction at sliding contacts between the workpiece and the tooling and W_d is the power dissipation at velocity discontinuities. The terms on the right hand side of (5) are calculated using a kinematically admissible velocity field. In the case of the von Mises yield criterion

$$W_p = \sigma_0 \iiint\limits_V \xi_{eq} dV. \tag{6}$$

Here V is the volume of the plastic regions. It is seen from (2) and (6) that the integral is improper. However, it is evident that it is convergent.

Choosing a kinematically admissible velocity field such that (3) is satisfied it is possible to find U_{x0}, U_{x1}, U_{y0}, and U_{y1} from (5) using a standard procedure. Then, the strain rate intensity factor is immediately determined from (4).

3   Upsetting with a Rotating Die

3.1   Statement of the Problem

A ring of inner radius r_0, outer radius R and thickness $2h$ is forged between two rotating flat dies moving towards each other with velocity u_0. The angular velocity of each die is ω_0. Cylindrical coordinates (r, θ, z) are taken, with the z – axis taken as the axis of symmetry of the process. Also, the process is symmetric relative to the plane $z = 0$. Therefore, it is sufficient to consider the region $z \geq 0$. The stress boundary conditions consist of the maximum friction law at $z = h$, the symmetry conditions

$$\sigma_{rz} = 0 \text{ and } \sigma_{\theta z} = 0 \tag{7}$$

at $z = 0$, and the condition that the surfaces $r = R$ and $r = r_0$ are traction free. Here σ_{rz} and $\sigma_{\theta z}$ are the shear stresses in the cylindrical coordinate system. The velocity boundary conditions are

$$u_z = -u_0 \tag{8}$$

at $z = h$,

$$u_\theta = 0 \text{ and } u_z = 0 \tag{9}$$

at $z = 0$. Here u_z and u_θ are the axial and circumferential velocities, respectively. Moreover,

$$u_\theta = \omega_0 r \tag{10}$$

at $z = h$ over the sticking region, if any. It is convenient to introduce the dimensionless quantities by

$$\varsigma = z/h, \quad \rho = r/R, \quad \eta = \omega_0 R/u_0, \quad t = R/h, \quad q = r_0/R. \tag{11}$$

Due to the axial symmetry, the solution is independent of θ. In particular, all derivatives with respect to θ vanish.

3.2   Kinematically Admissible Velocity Field

The general structure of the kinematically admissible velocity field proposed in [7] for the process of upsetting without twist is shown in Fig. 1. This general structure is applicable for the process of upsetting with rotating dies. The rigid region moves down and rotates about the z-axis along with the die. Lines ab and ac denote velocity discontinuity surfaces. Depending on process parameters line ab may intersect line $r = r_0$ (or $\rho = q$) and line ac may intersect line $r = R$ (or $\rho = 1$). If both of these conditions are satisfied then the regime of sticking occurs over the entire friction surface. The representation of the velocity components in the form of (3) occurs in the vicinity of lines *db* and *ce*. Since the solution is independent of θ, the incompressibility equation in the cylindrical coordinate system reads

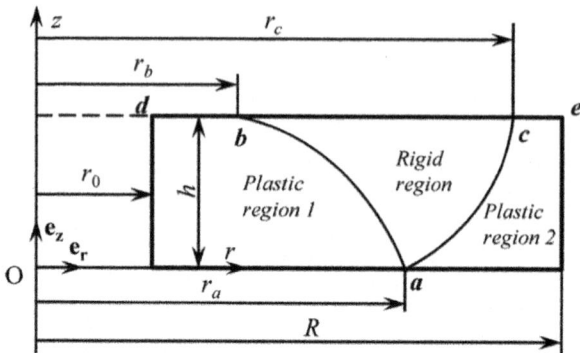

Fig. 1. General structure of the kinematically admissible velocity field.

$$\partial(ru_r)/\partial r + r\partial u_z/\partial z = 0. \tag{12}$$

The circumferential velocity is not involved in this equation. The kinematically admissible velocity field proposed in [7] satisfies Eq. (12) and the boundary conditions (8) and (9) for the axial velocity. Therefore, any combination of the velocity field from [7] and a distribution of the circumferential velocity satisfying the boundary conditions (9) for the circumferential velocity and (10) provides a kinematically admissible velocity field for the problem under consideration. The radial and axial velocities from [7] are represented as

$$u_r/u_0 = \rho/(2t) + f(\zeta)/\rho \text{ and } u_z/u_0 = -\zeta \tag{13}$$

where $f(\zeta)$ is an arbitrary even function of ζ. In the case of shearing without upsetting and radial expansion/contraction the circumferential velocity is given by $u_\theta = \omega_0 r\zeta$. It is therefore reasonable to assume the distribution of this velocity in the case under consideration in the form

$$u_\theta/u_0 = \eta\rho[\zeta - g(\zeta)]. \tag{14}$$

Here g is an arbitrary odd function of ζ and $\zeta \geq g(\zeta) \geq 0$ in the range $0 \leq \zeta \leq 1$.

Consider a generic velocity discontinuity surface. Let $\mathbf{e_r}$, $\mathbf{e_\theta}$ and $\mathbf{e_z}$ be the unit base vectors of the cylindrical coordinate system. Then, the velocity vector in the rigid zone is given by $\mathbf{V_r} = \omega_0 r\mathbf{e_\theta} - u_0\mathbf{e_z}$. The velocity vector in the plastic zone is represented as $\mathbf{V_p} = u_r\mathbf{e_r} + u_\theta\mathbf{e_\theta} + u_z\mathbf{e_z}$. Here u_r, u_θ and u_z are given by (13) and (14). Let $\mathbf{n}$ be the unit normal vector to the velocity discontinuity surface. The normal velocity must be continuous across the velocity discontinuity surface. Therefore, $\mathbf{V_r} \cdot \mathbf{n} = \mathbf{V_p} \cdot \mathbf{n}$ on this surface. Using the expressions for $\mathbf{V_r}$ and $\mathbf{V_p}$ it is possible to get $-u_0\mathbf{n}\cdot\mathbf{e_z} = u_r\mathbf{n}\cdot\mathbf{e_r} + u_z\mathbf{n}\cdot\mathbf{e_z}$ since $\mathbf{n}\cdot\mathbf{e_\theta} = 0$. This equation coincides with that used in [7] to determine the shape of lines ab and ac (Fig. 1). Therefore, the equation for each of these lines is

$$\rho = \rho_d(\zeta) = \sqrt{2tF(\zeta)/(1-\zeta)}, \quad F(\zeta) = \int_1^\zeta f(\chi)d\chi. \tag{15}$$

The amount of velocity jump across the velocity discontinuity surface is given by $[u_\tau] = |\mathbf{V_r} - \mathbf{V_p}|$ and can be found from (13) and (14). Using (15) an infinitesimal surface element of the velocity discontinuity surface is represented as

$$dS = \frac{hR\rho_d(\zeta)}{(1-\zeta)}\sqrt{\left[\frac{\rho_d(\zeta)}{2t} + \frac{f(\zeta)}{\rho_d(\zeta)}\right]^2 + (1-\zeta)^2}d\theta d\zeta. \tag{16}$$

Using (13) and (14) the components of the strain rate tensor in the cylindrical coordinate system can be found with difficulty. Then, ξ_{eq} can be found from (1) as a function of ρ and ζ. Thus the integral in (6) can be evaluated once the functions $f(\zeta)$ and $g(\zeta)$ have been specified. The term W_d involved in (5) is defined as

$$W_d = \frac{\sigma_0}{\sqrt{3}}\iint_S [u_\tau]dS. \tag{17}$$

Since $[u_\tau]$ has been already found, this integral can be evaluated as well. In the case of the maximum friction law the term W_f involved in (5) is defined as

$$W_f = \frac{2\pi\sigma_0 R^2}{\sqrt{3}}\int_{\rho_1}^{\rho_2}\sqrt{u_r^2 + u_\theta^2}\rho d\rho \tag{18}$$

The velocity components are understood to be calculated from (13) and (14) at $\zeta = 1$.

3.3 Strain Rate Intensity Factor

The capacity of the proposed method to evaluate the strain rate intensity factor depends on the choice of the functions $f(\zeta)$ and $g(\zeta)$. One of the simplest choices is

$$f(\zeta) = U_{r0} + U_{r1}\sqrt{1-\zeta^2} \text{ and } u_\theta = U_{\theta0} + U_{\theta1}\zeta\sqrt{1-\zeta^2}. \tag{19}$$

The coefficients U_{r0}, U_{r1}, $U_{\theta0}$, and $U_{\theta1}$ should be found by minimizing the right hand side of (5). Then, the strain rate intensity factor is determined from (4).

4   Conclusions

A general method for determining an approximate value of SRIF by means of the upper bound theorem has been proposed and then adopted to derive the system of equations for calculating the strain rate intensity factor in the process of upsetting between rotating dies. A theoretical method for determining the strain rate intensity factor is required for the development of an approach for predicting the generation of fine grain layer in the vicinity of frictional interfaces in metal forming processes.

Acknowledgment. This work was carried out within the framework of a joint project supported by grants RFBR-18-51-76001 (Russia), 359 (Ministry of education, science and technological development, Serbia) and 18.80013.16.02.01/ERA.Net (Moldova).

References

1. Alexandrov, S., Richmond, O.: Singular plastic flow fields near surfaces of maximum friction stress. Int. J. Non-Linear Mech. **36**(1), 1–11 (2001)
2. Alexandrov, S., Jeng, Y.-R.: Singular rigid/plastic solutions in anisotropic plasticity under plane strain conditions. Cont. Mech. Therm. **25**(5), 685–689 (2013)
3. Alexandrov, S., Mustafa, Y.: Singular solutions in viscoplasticity under plane strain conditions. Meccanica **48**(9), 2203–2208 (2013)
4. Goldstein, R., Alexandrov, S.: An approach to prediction of microstructure formation near friction surfaces at large plastic strains. Phys. Mesomech. **18**(3), 223–227 (2015)
5. Facchinetti, M., Miszuris, W.: Analysis of the maximum friction condition for green body forming in an ANSYS environment. J. Eur. Ceram. Soc. **36**, 2295–2302 (2016)
6. Alexandrov, S., Kuo, C.-Y., Jeng, Y.-R.: A numerical method for determining the strain rate intensity factor under plane strain conditions. Cont. Mech. Therm. **28**(4), 977–992 (2016)
7. Alexandrov, S., Lyamina, E., Jeng, J.-R.: A general kinematically admissible velocity field for axisymmetric forging and its application to hollow disk forging. Int. J. Adv. Manuf. Technol. **88**, 3113–3122 (2017)

Dimension Stability of Thin-Walled Parts from 3D Printed Composite Materials

Zdenek Chval$^{(\boxtimes)}$, Karel Raz, and Frantisek Sedlacek

Faculty of Mechanical Engineering, Regional Technological Institute,
University of West Bohemia, Univerzitni 8, Plzen, Czech Republic
zdchval@kks.zcu.cz

Abstract. 3D print of composite materials is one of the rapidly developing areas of additive manufacturing technology. This technology is using (instead of advanced FDM or SLA technologies) materials reinforced by short or micro fibres (mainly carbon or glass fibres). These fibres are oriented mainly in direction of filament extrusion. This is causing transverse- isotropic mechanical properties of final part. Composite materials with short fibres has higher modulus of elasticity, higher strength in bending, better thermal stability and higher impact resistance. These properties can be defined by change of volume rate between matrix and reinforcement. It is noticed nowadays, that these materials are more used in automotive and aerospace industry.

Composite materials reinforced by continuous long fibres are on the highest level in area of 3D printed composite materials. These fibres (in direction of extrusion) leads to printing of materials, which tensile strength is at higher level comparing to conventional used aluminium alloys. Orthotropic properties of these materials is the biggest technical problem of this technology. Strength of final material is in direction perpendicular to the extrusion equal only to the strength of matrix (polymer material). This value is much lower comparing to strength of fibres. This disadvantage has to be considered as an input parameter during designing process and product has to be loaded with respect to this property. Orthotropic properties of printed composite material with long continuous fibres has to be taking into account during designing process.

Quality of input material and technological parameters has to be carefully checked during 3D printing process and during replacing of parts made from conventional metal materials also. All these parameters (shape of part, combination of thin and thick walled areas, print axis orientation, print layout and defaults layer thickness) are affecting the heat load in printed composite part.

Internal stress and deformation can occur during printing process and after removing part from printer. These deformations have an undesirable influence on the dimensional accuracy and mechanical properties of final composite part.

Most of 3D printers producers are declaring minimal deformation values, but during experiment described in this article was approved, that specific initial conditions (geometrical properties and process values) can lead to deformations up to order of millimetres. These deformations can occur after print but also after several days (relaxing of material). This paper is dealing with dimensional stability and internal stress of 3D printed parts from composite materials.

Keywords: Dimension · Stability · 3D print · Deformation

© Springer International Publishing AG, part of Springer Nature 2019
E. E. Gdoutos (Ed.): ICTAEM 2018, SI 5, pp. 91–92, 2019.
https://doi.org/10.1007/978-3-319-91989-8_16

See Fig. 1.

Fig. 1. Thin- walled part during 3D printing process (left); Printed part with marked area of higher deformation caused by internal stress

References

1. Gibson, I., Rosen, D., Stucker, B.: Introduction and Basic Principles in Additive Manufacturing Technologies, pp. 1–18. Springer, New York (2015)
2. Barnatt, C.: 3D Printing: Third Edition. CreateSpace Independent Publishing Platform (2014)
3. Wimpenny, D.I., Pandey, P.M., Kumar, L.J. (eds.): Advances in 3D Printing & Additive Manufacturing Technologies. Springer, Singapore (2017)
4. Markovicova, L., Zatkalikova, V.: Composite materials based on PA reinforced glass fibers. In: Materials Today: Proceedings, vol. 3, pp. 1056–1059 (2016). E-ISSN 2214-7853

Creep Degradation Processes in Tungsten Modified 9%Cr Martensitic Steel

V. Sklenicka[1,2(✉)], P. Kral[1,2], K. Kucharova[1], M. Kvapilova[1,2], and J. Dvorak[1,2]

[1] Institute of Physics of Materials, Academy of Sciences of the Czech Republic, Zizkova 22, 616 62 Brno, Czech Republic
sklen@ipm.cz
[2] CEITEC IPM, Institute of Physics of Materials, Academy of Sciences of the Czech Republic, Zizkova 22, 616 62 Brno, Czech Republic

Abstract. Advanced creep resistant tungsten modified 9%Cr martensitic steel (ASTM Grade P92) is a promising structural material for the next generation of fossil and nuclear power plant. The P92 steel has been used to construct new coal-fired ultra-supercritical (USC) power plants with higher efficiency. Creep behaviour and fracture processes in creep are phenomena of major practical relevance, often limiting the lives of power plant components and structures designed to operate for long periods under stress at elevated and/or high temperatures. The creep behaviour of P92 steel has widely been reported. Furthermore, in recent years, extensive experimental studies and thermodynamic modelling of the microstructure and its stability during high-temperature creep of P92 steel have been published. Unfortunately, there are rather few published reports on damage processes in P92 steel during high-temperature creep, and the effect of damage evolution on the creep strength is nor fully understood at present. Therefore, it is not surprising that there are different and often controversial opinions about the role of secondary phases resulting from the additions of high concentrations of tungsten and molybdenum in P92 steel. In addition to $M_{23}C_6$ carbides and MX carbonitrides, an intermetallic Laves phase $Fe_2(W,Mo)$ is another dominating precipitating phase.

The possible degradation effects of Laves phase precipitation on the creep rupture processes in P92 steel have not been convincingly proved. There are different opinions about the effect of Laves phase precipitation on the creep resistance. From the creep strengthening point of view the precipitation of Laves phase has two aspects. On the one hand, high amounts of W and Mo content are incorporated in this phase, causing the depletion of these elements from the solid solution and thus a reduction of their contribution to the overall creep resistance (solid solution hardening). On the other hand, the increased volume fraction of Laves phase could lead to higher precipitation hardening during the initial precipitation period so at the beginning the prediction of the fine Laves phase might increase the creep resistance. It cannot be excluded that the reported controversial influence of the Laves phase precipitation on creep strength may be explained by the differences in Laves phase coarsening. Finally, it has been observed that creep cavities were nucleated at coarse precipitates of Laves phase. The formation, growth and coalesce of creep cavities bring about the brittle intergranular creep mode.

E. E. Gdoutos (Ed.): ICTAEM 2018, SI 5, pp. 93–94, 2019.
https://doi.org/10.1007/978-3-319-91989-8_17

In this study, the creep behaviour of P92 steel in the as-received condition and after isothermal stress-free ageing was investigated at 600 °C and 650 °C using uniaxial creep tests in tension. The results clearly show a detrimental effect of isothermal ageing on creep properties even after a few hundred hours. Based on microstructure analysis, a significant decrease in the creep life after ageing resulted from instability (rapid coarsening and clustering) of Laves phase and $M_{23}C_6$ particles. Therefore, Laves phase could lead to higher precipitation hardening only during the initial very short precipitation phase, when the beginning of the precipitation of the fine Laves phase particles might increase the creep resistance. Hence, the precipitation, growth and coarsening of the Laves phase can be accounted for a degradation mechanism of the microstructure. Further, it has been observed that after long-term creep exposure creep cavities are nucleated at coarse and hard precipitates of the Laves phase along grain boundaries. Figure 1 shows creep cavity located at the Laves phase particles. Thus, the decohesion between the metallic matrix and the large Laves phase precipitates could be reason for premature fracture, which occurs at a low value of the fracture strain. The creep damage tolerance factor λ as an important outcome of the continuum damage mechanics approach has been used to assess the possibility of occurrence of creep premature fracture.

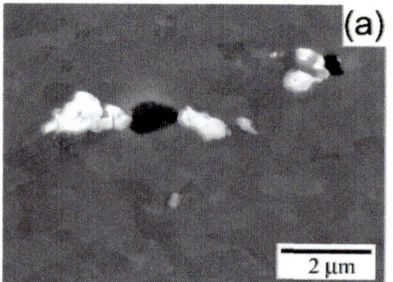 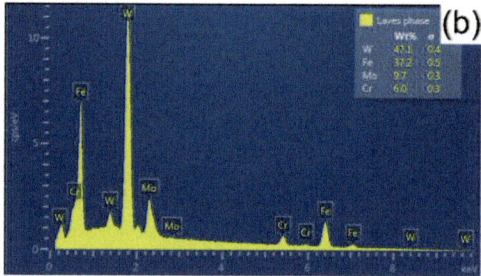

Fig. 1. (a) Creep cavity between two particles of Laves phase, (b) EDX spectra of Laves phase precipitation.

Acknowledgement. The authors acknowledge the financial support for this work provided by the Czech Science Foundation under the Grant Project No. 16-09518S.

Experimental Investigation of Cross-Laminated Timber Shear Wall Under Shear Force by Using Digital Image Correlation Method

Tzu-Yu Kuo[1(✉)], Wei-Chung Wang[1], Chih-Hsien Lin[2], and Te-Hsin Yang[2]

[1] National Tsing Hua University, Hsinchu 30013, Taiwan, Republic of China
`asmit.com@msa.hinet.net`, `wcwang@pme.nthu.edu.tw`
[2] National Chung Hsing University, Taichung 40227, Taiwan, Republic of China
`milkk750i@hotmail.com`, `tehsinyang@dragon.nchu.edu.tw`

Abstract. At the beginning of the 20th century, a new wood manufacturing technology, i.e. cross-laminated timber (CLT), was started. In Taiwan, the manufacturing technology of CLT has just started recently. For the sake of safety, the information of stiffness and strength of the shear wall of the CLT are essential for structural designs. In this paper, by following the method B (ISO 16670 Protocol) of ASTM standard E2126-11, shear test of a real-scale CLT shear wall was performed. The measured shear modulus and cyclic test results of the CLT shear wall were reported in this paper. By using the three-dimensional digital image correlation technique, full-field deformation information of the CLT shear wall were obtained.

Keywords: Cross-laminated timber · Shear wall
Digital image correlation technique

1 Introduction

At the beginning of the 20th century, a new wood manufacturing technology, i.e. cross-laminated timber (CLT), was started. In Taiwan, the manufacturing technology of CLT has just started recently. To improve the resistance against earthquake and gale, information of stiffness and strength of the CLT shear wall are indispensable to a reliable and safe structure [1]. Sebera et al. [2] employed digital image correlation (DIC) method to measure surface of an actual medium-scale CLT panel under torsion. The results were verified by the finite element analysis. Brandner et al. [3] verified a novel test configuration given by Kreuzinger and Seider [4] for in-plane shear properties of CLT diaphragms. Full-field deformation measurement of a real-scale CLT shear wall was performed by using the three-dimensional DIC method.

© Springer International Publishing AG, part of Springer Nature 2019
E. E. Gdoutos (Ed.): ICTAEM 2018, SI 5, pp. 95–97, 2019.
https://doi.org/10.1007/978-3-319-91989-8_18

2 Materials

Japanese cedar (*Cryptomeria japonica*) obtained from Hsinchu of northern Taiwan was used as the experimental specimen. The quality of lumber was selected by following M60A of the standard JAS 3079 [5]. In this paper, two-stage manufacturing technique was adopted to prepare the CLT shear wall: (1) adhering narrow faces of lumber to form single-layer panels and (2) gluing side face of five-laminate panels to build the CLT shear wall. The size of the CLT shear wall is $3,000 \times 850 \times 180$ mm^3. The resorcinol-formaldehyde (RF) resin and bonding pressures of 0.98 MPa were used. The X-RAD connection system (Rothoblaas Co., Ltd.), for CLT structures was employed. The resistances of the CLT structures measured in the range of 0°–315° by X-RAD are from 108.95 to 289.66 kN. The largest resistance is 289.66 kN at 225° [6].

3 Experimental Methods

The experiment of CLT panel under shear force followed the method B (ISO 16670 Protocol) of ASTM standard E2126-11 [7]. To obtain the ultimate displacement as the reference deformation of the cyclic test, a monotonic test was used. The loading schedule of the cyclic test consists two displacement patterns. The first pattern consists five single fully reversed cycles, 1.25%, 2.5%, 5%, 7.5% and 10%, of the reference deformation. The second pattern consists of phases, each containing three fully reversed cycles of equal amplitude, at displacements of 20%, 40%, 60%, 80% 100%, and 120% of the reference deformation. The ratio of displacement was set at 2 mm per second.

The DIC method is a full-field and non-contact measurement method. Commercially available 3D-DIC software Vic-3D V7 (Correlated Solutions, Inc., USA) was utilized in this study. The image was obtained every five frames per second. To increase the surface contrast and characteristics, white paint was lacquered on the CLT shear wall surface first, and black speckle pattern was sprayed on the white surface later. The speckle size was controlled within 2 to 15 pixels [8] so that the error of accuracy is under 0.01 pixel. Littell [9] conducted 6 pixels of one speckle dot to analyze the large field DIC for full-scale aircraft crash test. Therefore, the speckle size in this study was control within 6 to 8 pixels. By using the 3D-DIC technique, full-field deformation information of the CLT shear wall were obtained.

References

1. Chang, W.-S.: Repair and reinforcement of timber columns and shear walls – a review. Constr. Build. Mater. **97**, 14–24 (2015)
2. Sebera, V., Muszyński, L., Tippner, J., Noyel, M., Pisaneschi, T., Sundberg, B.: FE analysis of CLT panel subjected to torsion and verified by DIC. Mater. Struct. **48**(1–2), 451–459 (2013)

3. Brandner, R., Dietsch, P., Dröscher, J., Schulte-Wrede, M., Kreuzinger, H., Sieder, M.: Cross laminated timber (CLT) diaphragms under shear: test configuration, properties and design. Constr. Build. Mater. **147**, 312–327 (2017)
4. Kreuzinger, H., Sieder, M.: Einfaches prüfverfahren zur bewertung der schubfestigkeit von kreuzlagenholz/brettsperrholz. Bautechnik **90**(5), 314–316 (2013)
5. Japanese Agricultural Standard: JAS 3079 Standard for cross laminated timber, Tokyo, Japan (2013)
6. Angeli, A., Polastri, A., Callegari, E., Chiodega, M.: Mechanical characterization of an innovative connection system for CLT structures. In: World Conference on Timber Engineering 2016, Vienna, Austria, pp. 3549–3558 (2016)
7. American Society for Testing Materials: ASTM E2126-11 Standard test methods for cyclic (reversed) load test for shear resistance of vertical elements of the lateral force resisting systems for buildings, PA, USA (2011)
8. Zhou, P., Goodson, K.E.: Subpixel displacement and deformation gradient measurement using digital image/speckle correlation (DISC). Opt. Eng. **40**(8), 1613–1620 (2001)
9. Littell, J.: Large field digital image correlation used for full-scale aircraft crash testing: methods and results. In: International Digital Imaging Correlation Society 2016 Conference and Workshop, Philadelphia, PA, pp. 235–239 (2017)

Reinforcing Concrete with Carbon Nanotubes and Carbon Nanofibers: A Novel Method to Improve the Modulus of Elasticity

P. A. Danoglidis[✉] and M. S. Konsta-Gdoutos

School of Engineering, Democritus University of Thrace, 67100 Xanthi, Greece
pdanogli@civil.duth.gr

Abstract. Modulus of Elasticity (MOE) can be raised by optimizing the mix proportions, increasing the compressive strength and using stiffer aggregates. Unfortunately, such concrete is relatively brittle and has a high propensity of autogenous shrinkage cracking. This study proposes a novel way to improve the modulus of elasticity of concrete without increasing its compressive strength and brittleness, by using carbon nanotubes (CNTs) and carbon nanofibers (CNFs).

Keywords: Modulus of elasticity · Concrete · Carbon nanotubes Carbon nanofibers

1 Introduction

While compressive strength is considered the most important engineering property of concrete, designers of tall building also specify high MOE to limit lateral deformations and ensure occupant comfort [1]. CNTs and CNFs show great promise for enhancing the modulus of concrete, as they exhibit modulus of elasticity values on the order of up to 1 TPa. Studies on cement pastes and mortars reinforced with either CNTs or CNFs reported a dramatic improvement in the mechanical properties i.e. flexural strength, stiffness, energy absorption capability, and fracture toughness [2–4]. This study focuses on improving the flexural performance and elastic response of concrete, without increasing its compressive strength, by using small amounts of CNTs and CNFs.

2 Results

Flexural and compressive strength and modulus of elasticity of the nano-modified concrete were determined through four point bending and uniaxial compression experiments. Figure 1a, shows the flexural and compressive strength of concrete, reinforced with well dispersed CNTs and CNFs at an amount of 0.1 wt%. It is observed that while the flexural strength showed an increase of about 49%, the addition of CNTs and CNFs resulted only in the modest improvement of the compressive strength of 6%. A substantial improvement of the MOE (56%) was observed from the stress strain curves shown in Fig. 1b. Young's modulus was also calculated from flexural tests and the results are in perfect agreement with the values calculated from the compression tests using concrete cylinders (Table 1).

© Springer International Publishing AG, part of Springer Nature 2019
E. E. Gdoutos (Ed.): ICTAEM 2018, SI 5, pp. 98–99, 2019.
https://doi.org/10.1007/978-3-319-91989-8_19

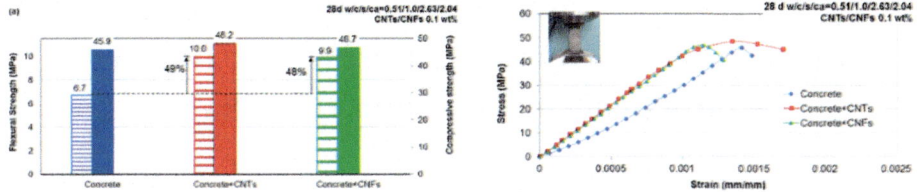

Fig. 1. (a) Flexural and compressive strength and (b) stress-strain curves of 28d concrete and concrete reinforced with CNTs and CNFs at amount of 0.1 wt%

Table 1. Effect of CNT and CNF on the compressive and tensile Young's modulus of concrete

	Young's modulus (GPa) 4 point bending	Young's modulus (GPa) uniaxial compression
Concrete	29.3 ± 0.98	29.0 ± 0.99
Concrete + CNTs 0.1 wt%	45.7 ± 0.83	44.3 ± 0.87
Concrete + CNFs 0.1 wt%	45.8 ± 0.88	44.1 ± 0.84

3   Conclusions

The experimental results of this study show that a small amount of CNTs and CNFs is capable of greatly enhancing the modulus of elasticity of concrete and its flexural strength. The increase of the MOE is not associated with increased compressive strength and decreased ductility. It is possible that CNTs and CNFs provide concrete with a toughening mechanism, (i) by reinforcing the nanostructure of C-S-H [2]; and (ii) by modifying the interface between aggregates and cement matrix at a nanoscale level, similarly to silica fume.

Acknowledgements. The authors would like to kindly acknowledge Glonatech S.A. for supplying the CNTs.

References

1. Shah, S.P., Konsta-Gdoutos, M.S.: Uncoupling Modulus of Elasticity and Strength-Effect of small addition of carbon nanotubes on concrete properties. Concr. Int. **39**(11), 37–42 (2017)
2. Konsta-Gdoutos, M.S., Metaxa, Z.S., Shah, S.P.: Highly dispersed carbon nanotube reinforced cement based materials. Cem. Concr. Res. **40**, 1052–1059 (2010)
3. Gdoutos, E.E., Konsta-Gdoutos, M.S., Danoglidis, P.A.: Portland cement mortar nanocomposites at low carbon nanotube and carbon nanofiber content: a fracture mechanics experimental study. Cem. Concr. Compos. **70**, 110–118 (2016)
4. Danoglidis, P.A., Konsta-Gdoutos, M.S., Gdoutos, E.E., Shah, S.P.: Strength, energy absorption capability and self-sensing properties of multifunctional carbon nanotube reinforced mortars. Constr. Build. Mater. **120**, 265–274 (2016)

Flexural Strength Optimization in CNF Cement Based Nanocomposites with Tailored Electrochemical Impedance Properties

M. G. Falara$^{(\boxtimes)}$, P. A. Danoglidis, M. E. Maglogianni,
and M. S. Konsta-Gdoutos

School of Engineering, Democritus University of Thrace, 671 00 Xanthi, Greece
mfalara@civil.duth.gr

Abstract. To unlock the efficiency of carbon nanofibers (CNFs) in potential applications, it is necessary to take into consideration their distribution in cement-based materials. In this work, it is observed that the relationship between flexural strength and electrochemical impedance properties such as capacitance and resistivity may provide valuable information on the actual CNF dispersion state in the matrix and a method to optimize flexural strength in reinforced mortars.

Keywords: Carbon nanofibers · Flexural strength · Capacitance
Ultrasonication

1 Introduction

Ultrasonic energy is a key parameter of CNFs' dispersion in cementitious materials, affecting the mechanical and electrical properties of the nanoreinforced cement pastes and mortars [1, 2]. In this study, the application of an ultrasonic energy close to 2400 kJ/l was found to be most effective, as the CNF reinforced mortars exhibit the highest flexural strength and simultaneously the lowest resistivity and energy storage ability, expressed by the capacitance values. The observed relationship between capacitance and resistivity values provide valuable information about the optimization of the flexural performance of nanocomposites.

2 Results

CNFs were added in mortar mixtures at an amount of 0.1 wt%. The CNF suspensions were subjected to different ultrasonic energies, ranging from 800 to 4000 kJ/l. The processor was set to deliver constant energy of 1900–2100 J/min [1]. Figure 1 illustrates the flexural strength and capacitance of the nanoreinforced mortars. Compared to the plain mortar, it is observed that the addition of CNFs increases the nanocomposites' flexural strength from 25% to 80%. Nanocomposites with CNFs dispersed at an energy of 2400 kJ/l exhibit the highest flexural strength (10.5 MPa) and the lowest capacitance value, while mortars prepared with CNFs dispersed with energies either lower or higher

© Springer International Publishing AG, part of Springer Nature 2019
E. E. Gdoutos (Ed.): ICTAEM 2018, SI 5, pp. 100–102, 2019.
https://doi.org/10.1007/978-3-319-91989-8_20

exhibit more modest increases in strength and higher energy storage ability. Mortars reinforced with CNFs sonicated at the energy of 2400 kJ/l also exhibit the lowest resistivity (Table 1), indicating the formation of a complete network of individual fibers, allowing current passage.

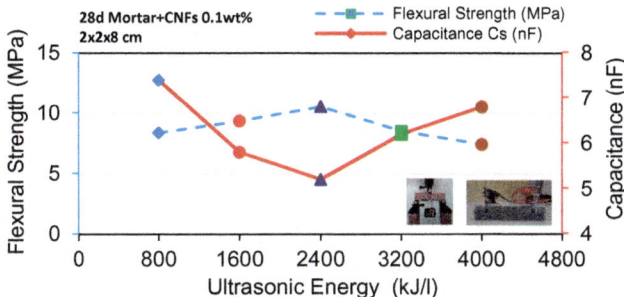

Fig. 1. Flexural strength and capacitance of 28d mortars reinforced with 0.1 wt% CNFs, dispersed using different ultrasonic energies.

Energies lower than 2400 kJ/l result to either more, or bigger in size agglomerates, increasing this way the electrical charge storage ability, and at the same time preventing a successful load transfer and the current passage. Higher energies may cause a breakage of individual nanofibers; also their aspect ratio is reduced. In these cases, it is possible that the electron transport mechanism is accomplished rather by electron hopping paths, than by the formation of a strong and continuous conductive network [3]. Hence, as seen from both Table 1 and Fig. 1, resistivity and capacitance values increase.

3 Conclusions

In this study, a substantial enhancement of the flexural strength, associated with addition of CNFs in mortars, was demonstrated. It was observed that a relationship between the capacitance and resistivity exists that provide valuable information about the optimization of the flexural performance of nanocomposites. The application of an ultrasonic energy close to 2400 kJ/l was found to be most effective, as the nanocomposites exhibit the highest flexural strength and the lowest resistivity and energy storage ability, expressed by the capacitance values. Lower energies lead to the formation of agglomerates, while higher energy breaks and reduces the CNFs' aspect ratio, which in turn weaken their reinforcing efficiency. As a result, flexural strength is modestly enhanced and at the same time the values of both resistivity and capacitance increase.

Table 1. Resistivity of CNF nanocomposites as a function of ultrasonic energy

Ultrasonic energy kJ/l	Resistivity kOhm·cm
800	5.2
1600	4.7
2400	3.9
3200	5.9
4000	7.7

References

1. Gdoutos, E.E., Konsta-Gdoutos, M.S., Danoglidis, P.: Portland cement mortar nanocomposites at low carbon nanotube and carbon nanofiber content: A fracture mechanics experimental study. Cem Concr Compos. **70**, 110–118 (2016)
2. Konsta-Gdoutos, M.S., Aza, C.A.: Self sensing carbon nanotube (CNT) and nanofiber (CNF) cementitious composites for real time damage assessment in smart structures. Cem. Concr. Compos. **53**, 162–169 (2014)
3. Watts, P.C., Hsu, W.K., Kroto, H.W., Walton, D.R.: Are bulk defective carbon nanotubes less electrically conducting. Nano Lett. **3**(4), 549–553 (2003)

Usage of Composite Materials in Design of Recuperative Member with Respect to Buckling Phenomena

Karel Raz$^{(\boxtimes)}$ and Miroslav Kepka

Regional Technological Institute, University of West Bohemia,
Univerzitni 8, Plzen, Czech Republic
kraz@rti.zcu.cz

Abstract. This paper deals with composite materials usage in design of recuperative member. This member is designed as parallelogram and it is used as energy storage in winding device. This member consist of two linear springs, made from fibre-reinforced composite (carbon T700S with epoxy matrix MTM57). These two springs are mounted on both sides in aluminium brackets.

Keywords: First composite leaf · Energy recuperation · Buckling

1 Introduction and State of Art

Usage of composite materials in field of winding technology is very rare and nowadays are recuperation members (springs) made mainly from metals. The usage of composite materials is beneficial because of lower energy consumption of winding device. These members are performing an oscillating motion and therefore exists requirement for lower weight of moving parts [1, 2].

2 Design, Analysis and Testing

A design of recuperating member was created according described scheme. Deflection of moving end is in range 0–58 mm. It is unavoidable, that buckling will appear on upper spring during loading (Fig. 1). This spring is in compression and the lower is in tension. The buckling analysis was performed using Nastran software. This analysis was considering composition of individual layers and it is described in article. The safety value of buckling is only 1.11 and this is not enough for any application. The buckling is resulting in significant stiffness decreasing (Graph 1) [3, 4].

3 Optimization

The design of recuperating member was improved according results from testing and virtual simulation. Main aim was to minimize buckling influence and achieve desired response (reacting force caused by displacement of 58 mm). Special additional

© Springer International Publishing AG, part of Springer Nature 2019
E. E. Gdoutos (Ed.): ICTAEM 2018, SI 5, pp. 103–105, 2019.
https://doi.org/10.1007/978-3-319-91989-8_21

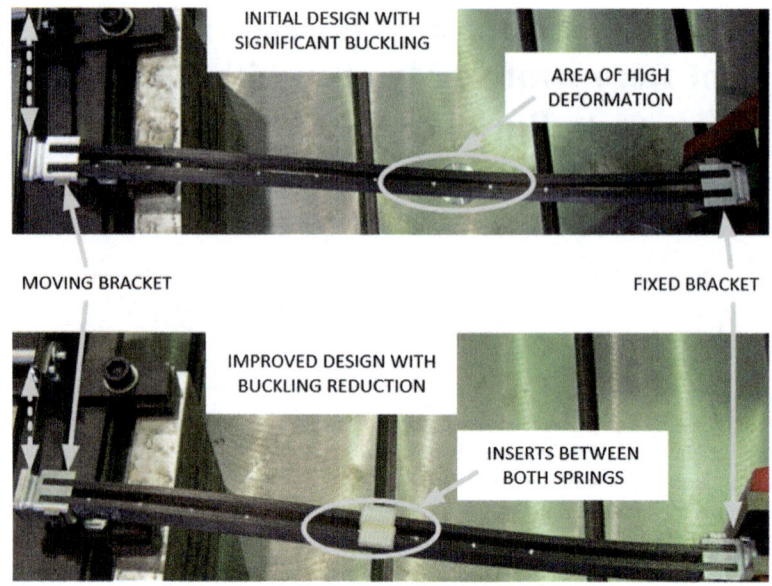

Fig. 1. Both designs of recuperative member under deformation

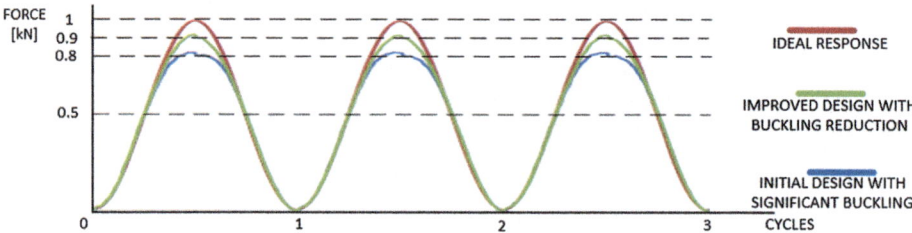

Fig. 2. Comparing of response- ideal, initial and improved design.

"L"-shaped inserts were created by 3D printing technology and they were placed between both springs (Fig. 1). This leads to reduction of buckling by approximately 50% (see Fig. 2).

4   Conclusion

This paper describes usage of two composite linear springs in form of parallelogram. FEM analysis of buckling shows high influence of this phenomenon, which is reducing stiffness. Optimized design is shown and reduction of buckling is shown in graph.

References

1. Mallick, P.K.: Composites Engineering Handbook. M. Dekker, New York (1997)
2. Wahl, A.M.: Mechanical Springs. McGraw-Hill, New York (1963)
3. Morris, C.J.: Composite integrated rear suspension. Compos. Struct. **5**(3), 233–242 (1986). ISSN 0263-8223
4. Beardmore, P., Johnson, C.F.: The potential for composites in structural automotive applications. Compos. Sci. Technol. **26**, 251–281 (1986). ISSN 0266-3538

Optimization of Additive Manufactured Components Using Topology Optimization

Frantisek Sedlacek$^{(\boxtimes)}$ and Vaclava Lasova

Regional Technological Institute, Faculty of Mechanical Engineering,
University of West Bohemia, Univerzitni 8, Plzen, Czech Republic
fsedlace@kks.zcu.cz

Abstract. Additive manufacturing (AM) is today one of the fastest growing industries. Previously, this technology was called Rapid Prototyping. The term 'rapid prototyping' (RP) was used in a variety of industries to describe a process for rapidly production of parts before final release or commercialization [1]. Today, this term is not current, because now 3D printing is not used only for fabricating prototypes, but it is increasingly used for small-series production of final parts. This leads to higher requirements on sufficient strength and stiffness of these parts. This problem can be solved using advanced numerical simulations, which also allows to found appropriate solution in combination with structural optimizations. There are many types of structural optimizations (such as geometry, topography, topometry, shape or topology optimization). Topology optimization is one of the best optimizations for this purpose because it can find the best use of material within a given design space [2]. Its use is often overlooked because the final optimized shape of the part is often too complicated for conventional manufacturing technologies, but it is not for AM technologies.

The goal of the paper was to find the appropriate methodology for optimizing a part for FDM fabrication using numerical simulations and topology optimization. The first stage of the intake system of a Yamaha YZF-R6 engine for a one-seat racing car was used as a typical part (a case study) for optimization. The simulation was done with respect to temperature. The PA6 copolymer with short carbon fibres was used as the material for the printed part. Experimental tensile testing of this material at temperatures up to 160 °C according to ISO standards was done to find the mechanical parameters [3]. The transverse isotropic mechanical properties of the material with high temperature dependency were found (see Figs. 1 and 2).

The optimization was done using the NX Nastran 12 Topology Optimization solver with special manufacturing constrains for AM which allow the creation of a design with respect to minimum thickness of the walls, maximum angles for overhangs and an internal lattice structure.

The optimized structure of the part with respect to FDM AM technology and sufficient stiffness and strength of the design was found using the designed methodology. The mass of the currently manufactured aluminium alloy part was reduced by more than 68%.

Keywords: Topology optimization · FEM · Additive manufacturing

© Springer International Publishing AG, part of Springer Nature 2019
E. E. Gdoutos (Ed.): ICTAEM 2018, SI 5, pp. 106–107, 2019.
https://doi.org/10.1007/978-3-319-91989-8_22

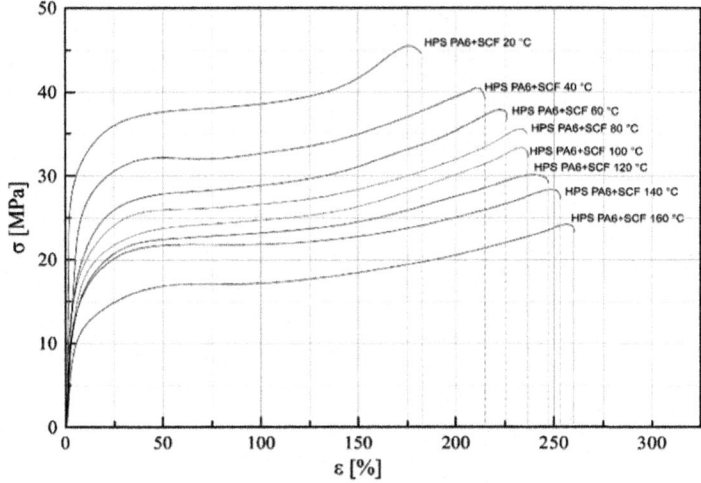

Fig. 1. Measured tensile stress-strain curves for vertically printed specimens of PA6 with short CF.

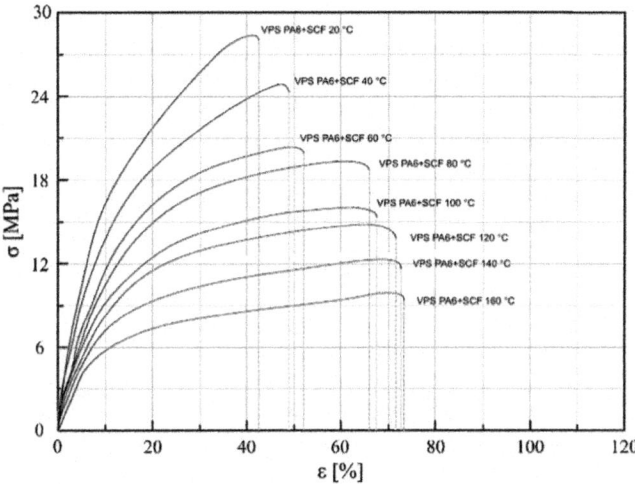

Fig. 2. Measured tensile stress-strain curves for horizontally printed specimens of PA6 with short CF.

References

1. Gibson, I., Rosen, D., Stucker, B.: Introduction and basic principles. In: Additive Manufacturing Technologies, pp. 1–18. Springer, New York (2015)
2. Sigmund, O., Maute, K.: Topology optimization approaches: a comparative review. Struct. Multi. Optim. **48**(6), 1031–1055 (2013)
3. ISO 527-2:2012 - Determination of Tensile Properties - Part 2. International Organization for Standardization, February (2012)

Poro-Hyperelasticity: The Mechanics of Fluid-Saturated Soft Materials Undergoing Large Deformations

A. P. S. Selvadurai[(✉)]

Department of Civil Engineering and Applied Mechanics, McGill University,
Montréal, QC H3A 0C3, Canada
patrick.selvadurai@mcgill.ca
http://www.mcgill.ca/civil/people/selvadurai

Abstract. This paper presents recent advances in the modelling of porous hyperelastic materials the pore space of which is filled with an ideal fluid. The incorporation of hyperelastic behavior of the porous skeleton enable the application of the developments to biological materials such as brain matter and arterial tissues and soft industrial materials that are impregnated with a fluid. The theory accounts for the coupled hyperelastic deformations of the porous skeleton and fluid flow through the pore space. The coupled multiphasic theory is applied to examine certain canonical problems involving one-dimensional compression, radial and spherical expansion of annuli and pure shear of poro-hyperelastic materials. The analytical solutions to these problems are used to benchmark computational approaches.

Keywords: Poro-hyperelasticity · Fluid-saturated porous media
Canonical mathematical solutions

1 Introduction

The classical theory of poroelasticity developed by Biot [1] describes the theory of infinitesimal deformations of a porous elastic solid saturated with a fluid. The mechanical behavior of the porous elastic skeleton is described by Hooke's law and the fluid migration through the pore space is described by Darcy's law. This theory has been used quite successfully to describe the mechanics of a wide range of materials including geomaterials, bone and other engineering materials. The concepts of infinitesimal deformations in classical poroelasticity naturally restricts the applicability of the theory to biological materials that are also fluid-saturated and contains a porous skeleton that can undergo large strains. The theory of poro-hyperelasticity is also an advance that was proposed by Biot [2] but the applications of the theory was not extensive. The research presented in this paper re-visits the theory of poro-hyperelasticity by developing certain canonical mathematical solutions, which

---

http://www.mcgill.ca/civil/people/selvadurai/list-research-publications.

© Springer International Publishing AG, part of Springer Nature 2019
E. E. Gdoutos (Ed.): ICTAEM 2018, SI 5, pp. 108–109, 2019.
https://doi.org/10.1007/978-3-319-91989-8_23

includes the one-dimensional compression of axial and spherically symmetric poro-hyperelastic elements, the pure shear of a cuboidal block of material and the inflation of cylindrical and spherical annular regions. The studies will present the complete mathematical development of the non-linear partial differential equations governing the poro-hyperelasticity problems and their solutions using hyperelastic skeletal properties of the neo-Hookean, Mooney-Rivlin and other strain energy functions commonly used to describe the skeletal response (Selvadurai and Suvorov [3–5], Suvorov and Selvadurai [6]). The availability of complete mathematical solutions also enables their use for the calibration computational approaches for the modelling of poro-hyperelastic materials. In these studies, it is shown that commonly used computational approaches give accurate solutions to poro-hyperelasticity problems and this benchmarking exercise is a validation of their accuracy (Fig. 1).

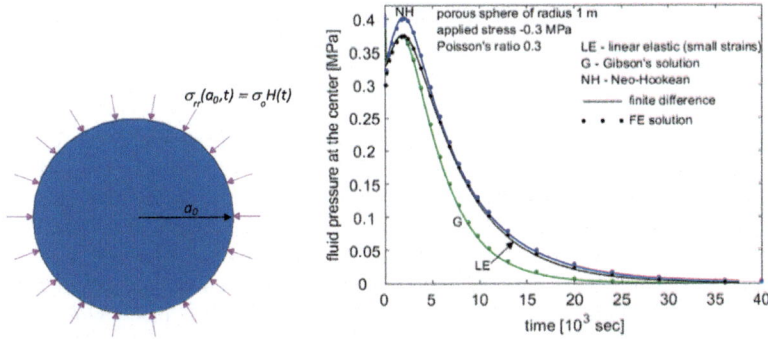

Fig. 1. Development of excess pore fluid pressures at the center of a poro-hyperelastic sphere due to Mandel-Cryer effects

Acknowledgements. The work described in this paper was supported by Natural Sciences and Engineering Research Council of Canada and the James McGill Research chairs program.

References

1. Biot, M.A.: General theory of three-dimensional consolidation. J. Appl. Phys. **12**, 155–164 (1941)
2. Biot, M.A.: Theory of finite deformations of porous solids. Indiana Univ. Math. J. **21**, 597–620 (1972)
3. Selvadurai, A.P.S., Suvorov, A.P.: Coupled hydro-mechanical effects in a poro-hyperelastic material. J. Mech. Phys. Solids **91**, 311–333 (2016)
4. Selvadurai, A.P.S., Suvorov, A.P.: On the inflation of poro-hyperelastic annuli. J. Mech. Phys. Solids **107**, 229–252 (2017)
5. Selvadurai, A.P.S., Suvorov, A.P.: On the development of instabilities in an annulus and shell composed of a poro-hyperelastic material, Unpublished (2018)
6. Suvorov, A.P., Selvadurai, A.P.S.: On poro-hyperelastic shear. J. Mech. Phys. Solids **96**, 445–459 (2016)

Effects of Machining and Heat and Surface Treatments on as Built DMLS Processed Maraging Steel

Dario Croccolo[1], Massimiliano De Agostinis[1], Stefano Fini[1],
Giorgio Olmi[1(✉)], Francesco Robusto[1], Snezana Ciric-Kostic[2],
Aleksandar Vranic[2], Nusret Muharemovic[3], and Nebojsa Bogojevic[2]

[1] Department of Industrial Engineering (DIN),
University of Bologna, Bologna, Italy
giorgio.olmi@unibo.it
[2] Faculty of Mechanical and Civil Engineering in Kraljevo,
University of Kragujevac, Kragujevac, Serbia
[3] Plamingo d.o.o., Gračanica, Bosnia and Herzegovina

Abstract. The main motivations for this study arise from the need for an assessment of the fatigue performance of DMLS produced Maraging Steel MS1, when it is used in the "as fabricated" state. The literature indicates a lack of knowledge from this point of view, moreover the great potentials of the additive process may be more and more incremented, if an easier and cheaper procedure could be used after the building stage. The topic has been tackled experimentally, investigating the impact of heat treatment, machining and micro-shot-peening on the fatigue strength with respect to the "as built" state. The results indicate that heat treatment significantly enhances the fatigue response, probably due to the relaxation of the post-process tensile residual stresses. Machining can also be effective, but it must be followed (not preceded) by micro-shot-peening, to benefit from the compressive residual stress state generated by the latter.

Keywords: DMLS · Maraging steel · "as built" state · Machining
Heat treatment

1 Introduction

Additive Manufacturing (AM) has great potentials, but also drawbacks, mainly arising from the generation of high tensile residual stresses and from the poor quality of the surfaces of the produced parts. Possible solutions to overcome the mentioned issues are performing an aging heat treatment and making parts undergo machining. It is extremely important to assess the mechanical response of the "as built" components, without further heat treatment and/or machining, to determine if it is affordable to avoid these treatments. Therefore, the subject of this study is tackling this point, investigating also the impact of heat treatment, machining and micro-shot-peening on fatigue.

© Springer International Publishing AG, part of Springer Nature 2019
E. E. Gdoutos (Ed.): ICTAEM 2018, SI 5, pp. 110–111, 2019.
https://doi.org/10.1007/978-3-319-91989-8_24

2   Experimental and Conclusions

DMLS Maraging steel parts were tested under rotating bending fatigue. All the samples had vertical orientation with respect to the base plate. The experimental campaign has been arranged as a 2-by-3 factorial plane. In particular, the first two sample sets were in the unmachined untreated and unmachined treated conditions, moreover, both underwent micro-shot-peening just after the building process, as recommended by the material manufacturer. The following two ones, with and without heat treatment, were shot-peened and then machined with 0.5 mm allowance (like in [1]). Finally the last two ones, again with and without heat treatment had shot-peening and machining swapped with respect to the previous case, i.e. micro-shot-peening was performed as the last stage.

The results (some of them are shown in Fig. 1 (a)), in terms of fatigue limits and S-N curves in the finite life domain, were processed and compared by analysis of variance. They indicate that heat treatment, probably due to the relaxation of the post-process tensile residual stresses, has the greatest beneficial impact on the fatigue response. Conversely, operating the micro-shot-peening just after the stacking process seems to have a not positive effect, if followed by machining. Swapping these two treatments is the most proper approach, to benefit of the better surface finishing and of the peening induced compressive residual stress state. Moreover, based on micrographies (an example is provided in Fig. 1 (b)), it can be observed that after the heat treatment the material structure is made more uniform than in the untreated state, however the patterned structure is still undoubtedly present.

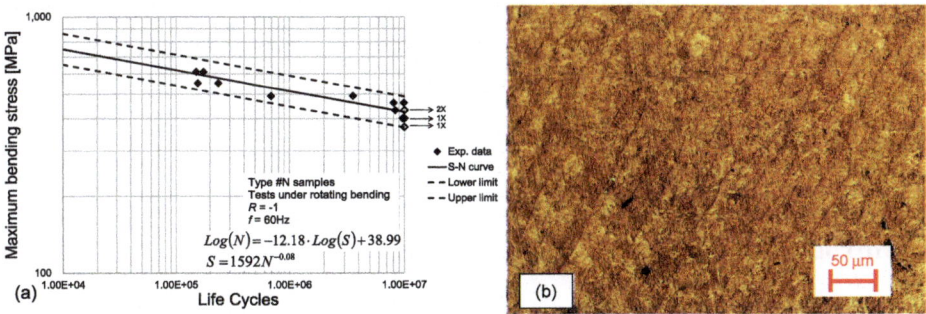

Fig. 1. (a) S-N curve and (b) structure on the build plane for unmachined treated samples.

Acknowledgements. The research presented in this paper has received funding from the European Union's Horizon 2020 research and innovation programme under the Marie Skło-dowska-Curie grant agreement No. 734455.

Reference

1. Croccolo, D., De Agostinis, M., Fini, S., Olmi, G., Vranic, A., Ciric-Kostic, S.: Influence of the build orientation on the fatigue strength of EOS maraging steel produced by additive metal machine. Fatigue Fract. Eng. Mater. Struct. **39**, 637–647 (2016)

DMLS Built Maraging Steel Fatigue Response Investigated for Different Build Orientations and Allowance for Machining

Dario Croccolo[1(✉)], Massimiliano De Agostinis[1], Stefano Fini[1],
Giorgio Olmi[1], Francesco Robusto[1], Nusret Muharemovic[2],
Nebojsa Bogojevic[3], Aleksandar Vranic[3], and Snezana Ciric-Kostic[3]

[1] Department of Industrial Engineering (DIN),
University of Bologna, Bologna, Italy
dario.croccolo@unibo.it
[2] Plamingo d.o.o., Gračanica, Bosnia and Herzegovina
[3] Faculty of Mechanical and Civil Engineering in Kraljevo,
University of Kragujevac, Kragujevac, Serbia

Abstract. This work derives its motivations from the increasing interest towards Additive Manufacturing and the lack of studies, mainly in the field of fatigue. The effect of build orientation and of allowance for machining on DMLS produced Maraging Steel MS1 has been assessed. The experimental results, properly set up by tools of Design of Experiment, have been statistically processed and compared. The outcomes were that, probably due to effect of the thermal treatment, machining and material properties, the aforementioned factors do not have a significant impact on the fatigue response. This made it possible to work out a global curve, accounting for all the result. Fracture surfaces have been carefully studied as well.

Keywords: Additive Manufacturing · DMLS · Maraging steel
Rotating bending fatigue

1 Introduction

A previous study [1] dealt with the effect of build orientation on DMLS produced Maraging steel MS1, following machining and heat treatment, with no significant effect being observed. Conversely, research [2] on Stainless Steel PH1 indicated that the build orientation turns to be effective, when the slanted orientation is introduced in the experiment. Moreover, a higher allowance for machining significantly enhances the fatigue response. The subject of the present study consists in an extension of the outcomes of [1]: it aims at investigating the build orientation effect for incremented allowance and then at deepening the study on the effect of allowance.

© Springer International Publishing AG, part of Springer Nature 2019
E. E. Gdoutos (Ed.): ICTAEM 2018, SI 5, pp. 112–113, 2019.
https://doi.org/10.1007/978-3-319-91989-8_25

2   Experimental and Conclusions

Fatigue tests under rotating bending have been performed on DMLS produced heat-treated Maraging steel MS1 samples. The tests aimed at investigating the effect of build orientation have been arranged, considering three levels of build orientation (horizontal, vertical and slanted) for fixed (3 mm) allowance, thus completing the campaign in [1]. The effect of allowance has been investigated, comparing the fatigue strengths over five levels (allowance of 0.5; 1; 2; 3; 4 mm) for fixed (vertical) build orientation. The results have been statistically processed and compared by an original methodology and the study has then been completed by fractographic and micrographic analyses.

The results indicate that the aforementioned factors do not significantly affect the fatigue response of Maraging steel MS1. Possible reasons may be due to heat treatments and machining removing the possible sources of anisotropy, arising from the different build orientations. Allowance for machining has also a negligible effect unlike for Stainless Steel PH1 [2], due to a more reduced residual stress state, following the stacking process, depending on material properties and on the higher layer thickness. Following the statistical proof of the negligibility of the differences among the results, all the retrieved S-N curves have been merged into a global curve that takes all the fatigue data into account. Based on the fractures surface analysis, it can be emphasized that cracks generally started just beneath the surface (around 80 μm) from porosities having approximately a 40 μm size (Fig. 1).

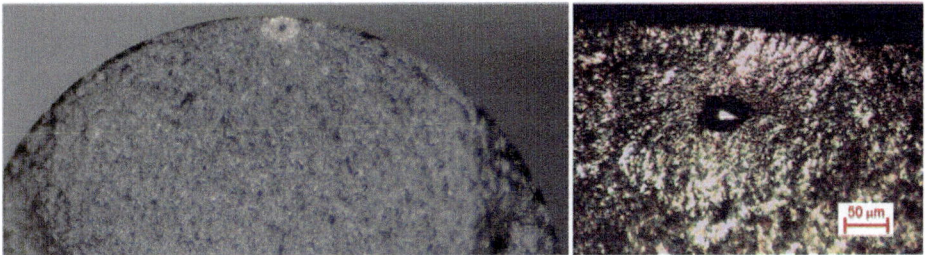

Fig. 1. Crack initiation site (enlarged view on the right).

Acknowledgments. The research presented in this paper has received funding from the European Union's Horizon 2020 research and innovation programme under the Marie Skłodowska-Curie grant agreement No. 734455.

References

1. Croccolo, D., De Agostinis, M., Fini, S., Olmi, G., Vranic, A., Ciric-Kostic, S.: Influence of the build orientation on the fatigue strength of EOS maraging steel produced by additive metal machine. Fatigue Fract. Eng. Mater. Struct. **39**, 637–647 (2016)
2. Croccolo, D., De Agostinis, M., Fini, S., Olmi, G., Bogojevic, N., Ciric-Kostic, S.: Effects of build orientation and thickness of allowance on the fatigue behaviour of 15–5 PH stainless steel manufactured by DMLS. Fatigue Fract. Eng. Mater. Struct. (in Press). https://doi.org/10.1111/ffe.12737

Designing Future Materials with Desired Properties Using Numerical Analysis

Constantine G. Fountzoulas$^{(\boxtimes)}$ and Jian H. Yu

U.S. Army Research Laboratory, WMRD, 6300 Rodman Road,
Aberdeen Proving Ground, Baltimore, MD 21005-5069, USA
{Constantine.fountzoulas.civ,jian.h.yu.civ}@mail.mil

Abstract. The swift advancement of the computer power and the recent advances in the numerical techniques, and improved strength and failure material models, resulted in accurate simulation of manufacturing processes and events, such as stress optimization during lamination process of polymers and impact into multi-layer opaque and transparent armor configurations. Parametric analysis of materials of known material models can contribute to the development of future materials. The systematic numerical analysis of the effect of material properties, such as modulus of elasticity, yield strength and ultimate tensile strength, on the performance of various systems can provide researchers and manufacturers crucial design information for the technologies of the future. This paper presents the modeling efforts at U.S. Army Research Laboratory (ARL) to develop a correlation between failure mechanisms and the material properties obtained experimentally leading to future materials technologies for personnel protection.

Keywords: Future materials design · Modeling · Properties · Technologies

1 Background

1.1 Software

To study in depth the response of materials in diverse loading environments, various commercial and specialized software for numerical analysis is used at ARL, such as ANSYS/WORKBENCH, which includes AUTODYN, COMSOL, LSDYNA and THERMOCALC. In particular, LSDYNA, AUTODYN and EPIC are used for the dynamic characterization of materials, while THERMOCALC is used for construction of phase diagrams and sintering processing. The effect of defects, such as cracks, and the crack propagation in the failure mechanism of ceramics has being studied for the past few decades. For example, the effect of defect clustering, almost impossible to introduce in a transparent material experimentally for validation of the modeling results, with respect to distance from the line of impact, defect density, and defect shape on the damage propagation in the ceramic hard face of the laminate was studied (Fig. 1) [1].

© Springer International Publishing AG, part of Springer Nature 2019
E. E. Gdoutos (Ed.): ICTAEM 2018, SI 5, pp. 114–115, 2019.
https://doi.org/10.1007/978-3-319-91989-8_26

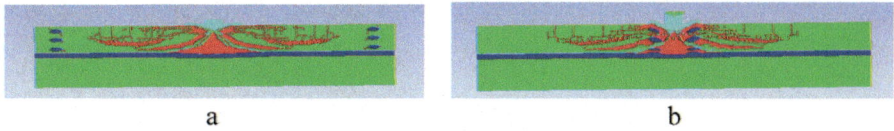

Fig. 1. Damage propagation after 30 μs for elliptical defects with the minor axis parallel to line of impact; (a) 5 mm and (b) 65 mm from the target edge

1.2 Experimentation

X-ray computed tomography

The need for production of material strength and failure models that can be used at the multiscale modeling requires advanced nondestructive characterization techniques, such as Digital Image Correlation (DIC) and X-ray Computed Tomography (XCT) [2]. Currently, the presence of cracks and porosity, in materials such aluminum and high melting temperature polymer with ceramic particles as processing aid in selective laser sintering [2] was studied by XCT using the in-house ZEISS Xradia 520 Versa microscope [3], a nondestructive technique for visualizing interior features within solid objects, and for obtaining digital information on their 3-D geometries and properties (Fig. 2). The findings of these studies will be used for the production of new materials strength and failure models or for the optimization of existing ones.

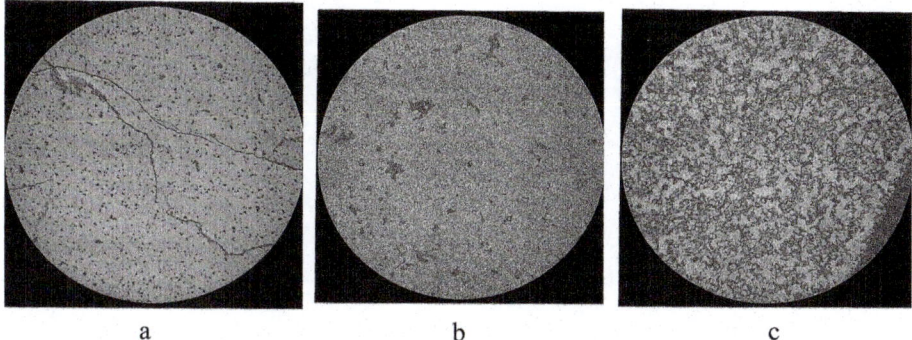

Fig. 2. XCT slices: (a) cracks in aluminum (resolution 2 μm); (b) and (c) void density in the polymer with different stoichiometry of ceramic particles (resolution, 4.7 μm)

References

1. Fountzoulas, C.G., Sands, J.M.: Proceeding of 26th International Symposium of Ballistics, Miami, FL, USA, 12–16 September 2011
2. Yu, J.H, et al.: Private communication, 6 March 2018
3. Carl ZEISS X-Ray Microscopy, Inc, Pleasanton, CA

Experimental Mechanics

Prediction of Interlaminar Shear Strength of Uni-Laminates Based on Fiber Bundle Composites

Li Wen[(⊠)], Jiang Zhenyu, Chen Wei, Huang Peiyan, and Yang Yi

School of Civil Engineering and Transportation,
South China University of Technology, Guangzhou 510641, China
liiween@qq.com

Abstract. Compare to unidirectional laminates, the preparation of fiber bundle composites is relatively simple. Previous studies have shown that fiber bundle composites and uni-laminates demonstrate similar failure features in the standard Iosipescu shear tests. But a considerable different strength value is observed between the two kinds of specimens. In this paper, three types of fiber bundle composites and uni-laminates were prepared using two kinds of carbon fibers and epoxy resins. The shear strength of fiber bundle composites and uni-laminates were compared to explore the underlying mechanisms resulting in the different shear strength. It is found through the analysis of the interfacial stress field between fiber bundle and matrix using the interface element method that the interfacial stress state in fiber bundle composite specimen is a coupling of tensile stress and shear stress, whereas the interface in unidirectional laminate specimen is in an ideal shear stress state. This difference makes the fiber bundle composites fail at a lower shear stress level. In this paper, a strength model was proposed to bridge the shear strength of fiber bundle composites and uni-laminates using Yamada-Sun strength theory. The experimental verification shows that the shear strengths of the uni-laminate predicted by the proposed model reach a good agreement with the measured values, with the relative deviation of that about 10%.

Keywords: Fiber bundle composites · Uni-laminates
Interface element method · Shear strength · Strength model

1 Experiment

1.1 Experimental Methods

The carbon fiber used in this experiment is Toray T700 and Tairyfil TC35. The Bisphenol A (Bisphenol A, BPA) was E51. The model of Bisphenol F (Bisphenol A, BPF) epoxy resin is NPEF-170. The curing agent is methyl hexahydrobenzoic anhydride (MeHHPA), and the promoter was N, N- dimethyl benzylamine (BDMA).

© Springer International Publishing AG, part of Springer Nature 2019
E. E. Gdoutos (Ed.): ICTAEM 2018, SI 5, pp. 119–123, 2019.
https://doi.org/10.1007/978-3-319-91989-8_27

1.2   Experimental Methods

The fiber bundle Iosipescu shear (FBIS) was prepared as shown in Fig. 1(a). The specific size of the specimen is referred to ASTM D5379 [1], and the Transverse fiber bundle tension (TFBT) is prepared as shown in Fig. 1(b). Figure 1(c) and (d) show that the carbon fiber is evenly distributed in the epoxy resin, and the average volume fraction of the fiber is about 40%.

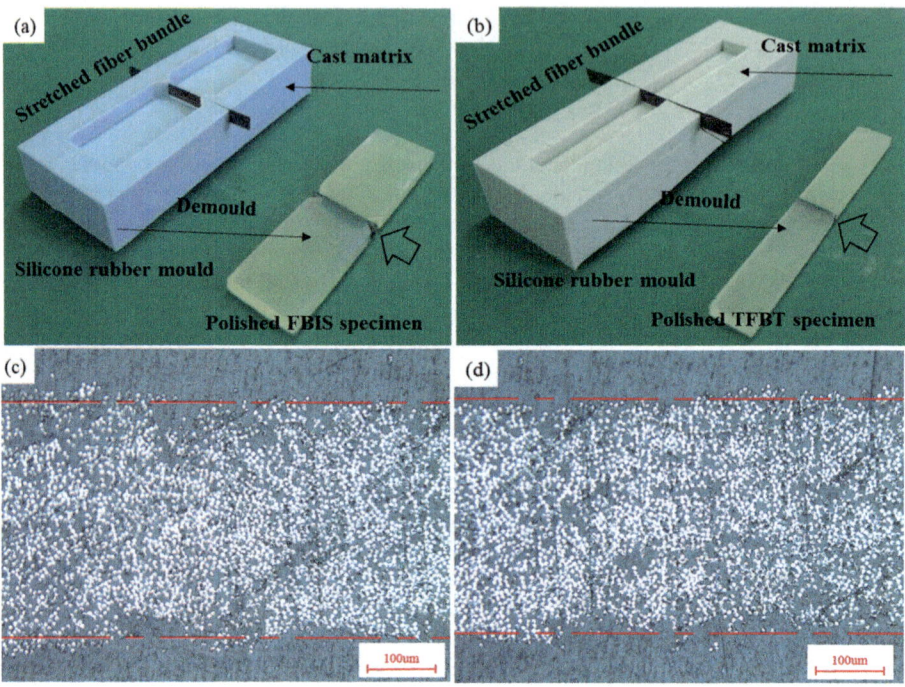

Fig. 1. (a) and (b) Illustration of SFBC specimen and TFBT specimen preparation, (c) and (d) optical microscopic photo on fiber distribution in the TFBT specimen and SFBC specimen taken in the direction of the arrow displayed in (a) and (b)

1.3   Experimental Results

Figures 2(a) and (b) show the fracture of FBIS specimen and TFBT specimen respectively. The fracture of both specimens is covered with carbon fibers on one side and only a few of carbon fibers on the other side (Fig. 3).

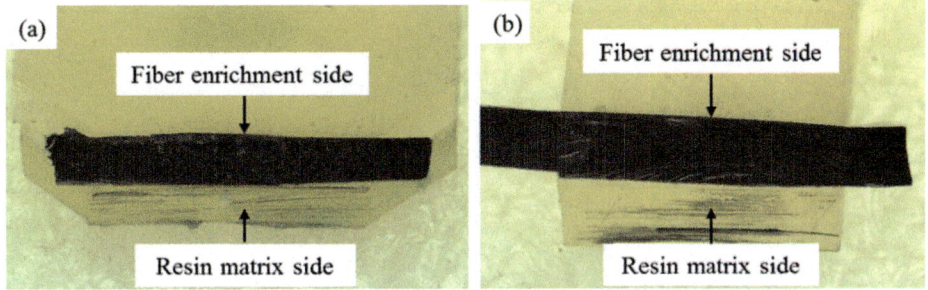

Fig. 2. Fracture surfaces of FBIS specimen (a) and TFBT specimen (b)

The strength of FBIS specimens and TFBT specimens and the shear strength of uni-laminates are shown in Table 1. The test scheme of FBIS specimen and uni-laminar shear specimen is based on the ASTM D5379 standard [1]. The test scheme of TFBT specimen is based on ISO 527-2 standard [2].

Table 1. Test results of FBIS specimens, TFBT specimens and unidirectional laminates shear specimens

	FBIS strength (95% Confidence limit)/MPa	TFBT strength (95% Confidence limit)/MPa	Uni-laminate shear strength (95% Confidence limit)/MPa
T700/BPA	23.7 ± 9.8 (7.87)	7.6 ± 0.5(0.44)	45.2 ± 4.8 (3.54)
TC35/BPA	27 ± 6.5 (4.81)	9.6 ± 2.2(2.17)	49.7 ± 6.1 (4.85)
TC35/BPF	27 ± 6.5 (5.71)	11.1 ± 2.5(1.65)	40.9 ± 5.4 (4.71)

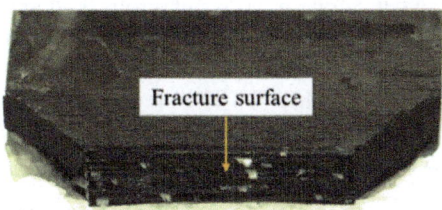

Fig. 3. Fracture surface of interlaminar shear specimens of unidirectional laminate

2 Numerical Analysis

The interface of fiber bundle composites and matrix is a typical heterogeneous interface. With the traditional finite element algorithm often can't get the continuous interface stress field. The usual method is to embed an interface element between the heterogeneous interface [3]. In this study, the precise interface stress field (Fig. 4) is calculated by using the interface element method.

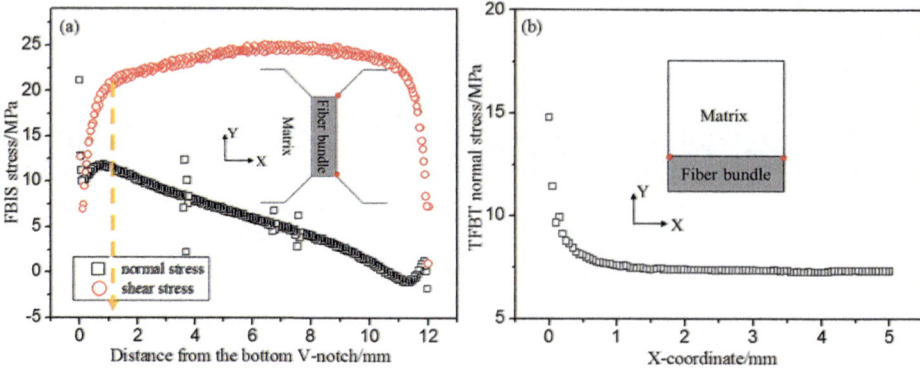

Fig. 4. Interfacial normal and shear stress field of FBIS specimen (a), interfacial normal stress field of TFBT specimen (b)

3   Prediction Model and Prediction Results of Uni-Laminated Shear Strength

In this paper, Yamada-Sun's tensile and shear coupled strength model is introduced, and the prediction model of shear strength of uni-laminate is as follows:

$$\tau_{Lam} \geq \frac{f_2(y)}{\sqrt{1 - \left(\dfrac{f_1(y)}{\sigma_{yy}\big|_{X=0,Y=0}}\right)^2}} \tag{1}$$

$f_1(y)$ and $f_2(y)$ were normal stress and shear stress in Fig. 4(a), respectively. Unidirectional laminate shear strength τ_{Lam}, therefore, can be predicted through the most easily damage point on the interface in the fiber bundle composite shear specimens (as shown in Fig. 4(a) in the orange dotted arrow), the same as the maximum of the stress state in the right side of in Eq. (1). The predicted results are shown in Table 2. It can be found that the predicted value of the unidirectional laminate is more consistent with the measured value, and the relative deviation is about 10%.

Table 2. Comparison of measured strength and predicted strength of unidirectional laminates

	Measured laminate shear strength/MPa	Predicted laminate shear strength/MPa
T700/BPA	45.2 ± 4.8	36.4
TC35/BPA	49.7 ± 6.1	54.7
TC35/BPF	40.9 ± 5.4	37.3

References

1. American Society of Testing Materials. ASTM-D5379-Standard Test Method for Shear Properties of Composite Materials by the V-Notched Beam Method. American: ASTM (2013)
2. International Organization for Standardization. ISO-527-2 Plastics - Determination of tensile properties - Part 2: Test conditions for moulding and extrusion plastics. Switzerland (1993)
3. Lei, X.: Contact friction analysis with a simple interface element. Comput. Methods Appl. Mech. Eng. **190**, 1955–1965 (2001)
4. Yamada, S.E., Sun, C.T.: Analysis of laminate strength and its distribution. J. Compos. Mater. **12**(3), 275–284 (1978)

An Experimental Research of High Temperature Strain Localization and a Method for Non-touch Measurements at a High Temperature Experiments

W. V. Teraud[(✉)] [iD]

Lomonosov Moscow State University, Moscow, Russia
ldrnww@gmail.com

Abstract. There are present high temperature experiments of Al and Ti specimens in a high temperature creep up to fracture. The arm of the research is strain localization in the stretched specimens at various initial tensile stresses.

Keywords: Experiment · Creep · Strain localization

1 Introduction

Fracture propagation in metals and alloys under tension is, as a rule, accompanied by necking, i.e. localization of strains, which arises from losing material strength and stability to further reduce its load-bearing capacity. A moment where uniform straining becomes non-uniform one is of utmost interest.

Systematic experimental researches were first conducted by Davidenkov and Spiridonova [1] and Bridgman [2].

As noted in [3], in 1885, V. Considerant proposed a kinematic approach to research materials under tension loads, according to which thinning of a test specimen occurs when a value of tensile load reaches a maximum. At a moment of strain localization (time τ), strain hardening $d\sigma/d\varepsilon$ ceases to compensate for an increasing stress σ due to a cross-sectional area of the sample to become less.

A strain criterion in the analysis of fracture under tension is studied in [4], where it was concluded that necking takes place upon straining of a test specimen to come to 12%. It was Malygin [5] who proposed to develop that approach and related uniform straining ε_0 with a function n in the power law of creep ($\sigma = \chi\varepsilon'^n$): $\varepsilon_0 < n$.

When considering a time criterion, it is assumed that a necking time τ is of the order of $0{,}9t^*$ (t^* is a time from loading start to fracture). This approach is true and applicable to high-strength metals, for example, Cr-Mo-V of steel [6] or steel grade 12X18H10T at $T = 850\ ^\circ\mathrm{C}$ [7]. The time criterion is also used in many similar studies [8, 9].

To determine a necking time, one uses other approaches, for example, those based on learning algorithms [9].

In general, the known criteria of localization of strains do not consider a shape of a test specimen. Creep tests are usually carried out in a closed furnace, therefore, as a rule, the only indicator of a strained condition to be measured is a rate of elongation

© Springer International Publishing AG, part of Springer Nature 2019
E. E. Gdoutos (Ed.): ICTAEM 2018, SI 5, pp. 124–129, 2019.
https://doi.org/10.1007/978-3-319-91989-8_28

Δl of a test specimen in dependence to t. The tests may at times be interrupted so that parameters of a test specimen can be measured [10], but this considerably affects accuracy of the test results. These experimental difficulties lead to development of generalized criteria. The method of non-touch measurement developed by the author of this paper, to be applied to the parameters of a test specimen during its testing enabled to investigate occurrence and development of necking at high temperatures in real time, with its first version to be presented in [11]. Preliminary studies were carried out in [12]. In this paper, we use a flat cross-section hypothesis. It is assumed that the stresses throughout a cross section are equal, with no structural conditions of various cross-sections of test specimens to be considered during occurrence of necking and its further progress. The purpose of the article is to study localization of strains and to describe a measurement method developed.

2   A Strain Localization

A test specimens were made from aluminum alloy D16T (Russian GOST 4784-97) and titanium alloy VT-6 (Russian GOST 19807-91), similar to an american Grade5 alloy. All the test specimens had a fillet-like shape and were cut out from the same rolled sheet longitudinally. Before testing, the surface of a test specimen was gridded. A test specimen heated to an operating temperature (400 °C for aluminum and 500 °C for titanium) was stretched under a constant load at various initial stresses up to a moment of fracture. As a measurement system, the non-touch method described in Sect. 3 was used.

2.1   A Strain Localization in a Titanium Alloy

As an example, Fig. 1 presents three pictures of the same test specimen of the titanium alloy VT6 at a temperature of 500 °C, taken at an initial time, at a time of localization and shortly before a fracture. A well-formed neck is clearly visible in the last picture, whereas the picture in the middle shows no localized strain visible. At an initial time, test specimen sizes were 30 * 6 * 1 mm. A moment of localization was found by means of criterion Ω_5 from [12] with a sensitivity value of the criterion $k_5^1 = 0.3$ MPa. It is pointed on a creep curve, at a time $\tau = 4.57$ h, whereas a time to fracture was $t^* = 5.34$ h. Elongation of a test specimen was $\Delta l^* = 18.8$ mm. In the pictures, vertical lines above the test specimen show boundaries of a working area. It is seen that a moment of localization approximately corresponds to half the elongation of a test specimen, relative to elongation at a fracture time.

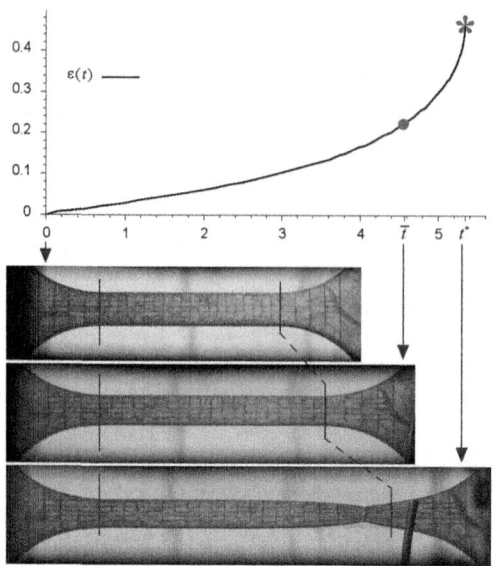

Fig. 1. Logarithmic longitudinal strain ε(t) from time t in hours and pictures taken: right after loading, at a moment of localization and shortly before fracture.

2.2   Checking a Time Criterion

Time criteria [6–9] are based on an assumption that there is a linear regression [13] between values of t^* and τ. Let us verify this assumption using experimental data [12].

Let us express a linear approximation as follows:

$$\hat{\tau} = \hat{\beta} \cdot t^* + \hat{\alpha} \tag{1}$$

$\hat{\tau}$ means predicted by the linear model a values τ, $\hat{\alpha}$ и $\hat{\beta}$ estimate of a accurate values α и β. Based on a localization time τ_5 at $k_5^1 = 0.3$ MPa and a fracture time t^* from the paper [12], we obtain: $\hat{\alpha} = -0.306$, $\hat{\beta} = 0.838$. Note that the line (1) must pass through the origin of coordinates, so $\hat{\alpha} = 0$, then $\hat{\beta} = 0.773$. We assume that the quantities τ are normally distributed. Calculating corresponding scatters of s_ε^2, s_α^2 and s_β^2, and using Student's t-distribution t_{n-2}^*, we can get a confidence interval $\beta \in \left[\hat{\beta} - s_\beta t_{n-2}^*, \hat{\beta} + s_\beta t_{n-2}^*\right]$. So, for the 0.975 quantile of Student's t-distribution with 21 degrees of freedom is $t_{n-2}^* = 2.0796$, and thus the 95% confidence intervals for β are:

$$\beta \in [0.617, 0.929] \tag{2}$$

Note, that β must be less one ($\beta < 1$). The product-moment correlation coefficient might also be calculated $\hat{r} = 0.99963$.

A cross-section of test specimens used was 8.9 * 1.3 mm, so the coefficient k_5^1 = 0.3 MPa approximately corresponds to narrowing of the specimen by 0.087 mm at a moment of localization. Therefore, if we assume that localization of creep strains takes place with a coefficient of width variation of (0.087/8.9) * 100% ~ 1%, then a linear relationship between a localization time and a time to fracture exists, with a probability of 95%. In the example considered, an average localization time is $\hat{\beta}$ · 100% = 77% of a time to fracture.

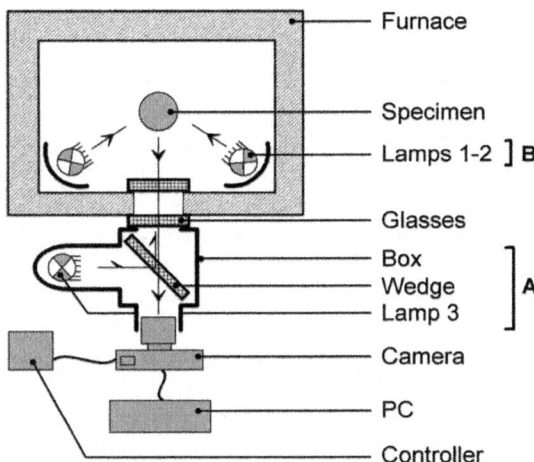

Fig. 2. The non-touch measuring system diagram.

3   A Non-touch Measurement Method

The measurement system was based on a non-touch principle to measure geometric parameters. The geometry of a test specimen and coordinates of reference points were calculated with use of pictures taken during high-temperature strains.

To provide visual access to the Specimen during the experiment, a hole was made in the furnace wall, covered on both sides with an optical quartz Glass to avoid penetration of cold air into the furnace space (see Fig. 2).

In order to obtain an optimal number of frames of one or different samples at various strain rates, a Controller was developed, which enabled to take pictures at predetermined intervals of test specimens' strains; in the tests, 1 picture was taken every 0.1 mm of elongation under tensile load. A Nikkor d300s camera, installed far enough from the high temperature furnace, connected to a high-power computer (PC) was used.

One of the key features of the system was a lighting-related problem. For good pictures of test specimens being strained, the heater-based lighting was not sufficient. The first lighting option (Fig. 2, Option **A**, without **B**) consisted of as follows: Box 230 * 200 * 120 mm with three holes and a semitransparent optical wedge; at an angle of 45° to the Box, a direct light (Lamp) was placed. The light falling from the

Lamp turned on the wedge at 90° and stroke the test specimen. This made it possible to have a direct visual access to them, being lit. This lighting system had a number of drawbacks, therefore another one was developed (Option **B**, without **A**). The lamp was placed inside the furnace. Initially, only one lamp was installed, but to achieve better reliability in long-term experiments, a second lamp was used to double the effect. To increase a lamp service life operating close to a titanium alloy (operating temperature of 600 °C), the Controller was modified in such a way that shortly before sending a signal to the camera, it also turned on the lamp. The quality of lighting and reliability of the system were considered sufficient for long experiments to take days or even weeks.

To process pictures taken, a software package was written (Recognizer, Solver and Post-processor), which enables to automatically or half-automatically recognize a test specimen on a picture taken, measure geometry of its working part and position of the reference points on each frame. Solver converts data obtained into metric geometry, from which, during post-processing, it is possible to build graphs of values that are of interest. In addition to the standard data (elongation $\Delta l(t)$ and strain $\varepsilon(t)$), the system also provides distribution of the true stress $\sigma(t, x)$ along the working part and the radius of a tangent circle $R(t)$ in the neck (the most important parameter for the theory [1, 2], etc.).

Financial support was provided by the Russian Foundation for Basic Research (grants 16-38-60200).

4   Conclusions

If we assume that strain localization takes place with a coefficient of width variation of a test specimen to come to 1%, then there is a linear relationship between a localization time and a time to fracture with a probability of 95%. On average, a localization time is 77% of a time to fracture.

The developed system enables to obtain unique experimental data, making it possible to more accurately check theoretical developments available. In 2017, the system was patented, and the software package was certified.

References

1. Davidenkov, N., Spiridonova, N.: An analysis of the state of stress in the neck of the stretched sample. Industr. Lab. Mater. Diagn. **6**, 583–593 (1945). In Russian
2. Bridgman, P.: Studies in Large Plastic Flow and Fracture with Special Emphasis on the Effects of Hydrostatic Pressure. McGraw-Hill, New York (1952)
3. Altmeyer, G.: Mod´elisation th´eoretique et num´erique des crit`eres d'instabilit´e plastique: Application la pr´ediction des ph´enom`enes de striction et de localisation lors d'operation demise en forme par emboutissage. PhD thesis, L' ´Ecole Nationale Sup´erieure d'Arts et M´ etiers, Fran,cai (2012)
4. Nirmal, K.: Constant-load tertiary creep in nickel-base single crystal superalloys. Mater. Sci. Eng. **432**, 129–141 (2006)

5. Malygin, G.: Influence of the grain size on the resistance of micro and nanocrystalline metals against the neck like localization of plastic deformation. Phys. Solid State **53**(2), 363–368 (2011)
6. Wilshire, B., Burt, H.: Long-term creep design data for forged 1Cr–1Mo–0.25 V steel. Strength Fract. Complex. **4**, 65–73 (2006)
7. Radchenko, V., Nebogina, E., Andreeva, E.: Structural model of material softening at creepunder complex stress conditions. J. Samara State Tech. Univ. Ser. Fiz.-Mat. Nauki **19** (1), 75–84 (2009). In Russian
8. Srinivas, B., Janaki, P., Ganesh, R.: Application of a few necking criteria in predicting the forming limit of unwelded and tailor-welded blanks. Strain Anal. **45**, 79–96 (2009)
9. Volk, W., Suh, J.: Reliable and robust evaluation of local necking with a generalized thinning limit diagram. Forum Forming Technology, 115–120 (2012)
10. Nirmal, K.: Constant-load tertiary creep in nickel-base single crystal superalloys. Mater. Sci. Eng. **432**, 129–141 (2006)
11. Lokoshchenko, M., Teraud, W.: Nontouch experimental method for recording and measuring of fracture under high temperature. In: 19th European Conference on Fracture: Fracture Mechanics for Durability, Reliability and Safety 2012, p. 268. Resh. Cent. for Power Engineering Problems of the Russia Academy of Sciences, Kazan, Russia (2012)
12. Teraud, W.: Localization of the creep in rectangular samples at high temperature. Russ. Eng. Res. **37**(10), 850–856 (2017)
13. Simple linear regression. https://en.wikipedia.org/wiki/Simple_linear_regression. Accessed 14 Feb 2018

DIC Assisted FCG Testing for Materials Used in Shale Gas Mining

Sandra Musial[1], Tomasz Brynk[2(✉)], and Zbigniew Pakiela[2]

[1] Institute of Fundamental Technological Research Polish Academy of Sciences,
Pawinskiego 5B str., 02-106 Warsaw, Poland
[2] Faculty of Materials Science and Technology,
Warsaw University of Technology, Woloska 141 str., 02-507 Warsaw, Poland
tbrynk@inmat.pw.edu.pl

Abstract. Limited mineral resources amount induce research in natural gas sequestration from non-conventional reservoirs with shale rock being a prominent example. CO_2 based fracturing is promising technique for both shale rock fracturing and greenhouse effect causing gas underground storage, however CO_2 in the presence of water might be corrosive for traditional materials used in natural gas wells limiting their performance in static and dynamic loading conditions. The aim of this work is to develop minisamples based technique for Fatigue Crack Growth (FCG) rate testing on casing pipe material used in conventional gas mining. 3-point bending type samples of three different W dimension has been selected for testing with FCG standard based approach as well as with optical, noncontact displacement measurements made with Digital Image Correlation (DIC) near the propagating crack tip. Results of different testing techniques for P110 steel have been discussed and conclusion drawn out giving general guidelines for proposed method usage in pipeline materials investigation. Additionally scale effect on FCG results has been revealed.

Keywords: Fatigue Crack Growth (FCG) · Digital Image Correlation (DIC)
Minisample

1 Motivation and Methods

1.1 CO_2 Based Shale Rock Fracturing

CO_2 based technique developed at Military University of Technology under patent No. P.398228 is a promising method for shale gas sequestration. It allows not only for efficient shale rock fracturing but also underground deposition of greenhouse effect causing gas. The method utilizes thermodynamic transformations of CO_2 from liquid to supercritical state for effective generation of pressure larger than tensile strength of shale rock. The main drawback of the method is the risk of well material corrosion (usually carbon steels) in the case of moisture presence in CO_2, therefore materials properties changes have to been investigated for appropriate well assessment. However, mechanical testing of pipeline materials is related with some difficulties and novel experimental techniques have to be developed.

© Springer International Publishing AG, part of Springer Nature 2019
E. E. Gdoutos (Ed.): ICTAEM 2018, SI 5, pp. 130–135, 2019.
https://doi.org/10.1007/978-3-319-91989-8_29

1.2   FCG Tests of Pipeline Material

One of the most important materials properties for pipeline materials working in dynamic loading conditions is fatigue crack growth (FCG) rate. Pipe's geometry (insufficient wall thickness) implies mini-samples usage instead of standardized samples. Under this research FCG rate of P110 steel commonly used for gas transmission has been investigated.

The single edge notched bending specimens (SENB) were designed and cut out from pipes according to scheme presented in Fig. 1. Three types of geometrically similar samples were prepared with W dimension equal 18, 7.6 and 3.8 mm. Orientation B gave an opportunity for the largest samples preparation and comparing the results of optical displacement fields based technique with the results from ASTM E 1820 standard procedure. Samples cut in orientation A had the most dangerous crack orientation from the point of view of expected pipeline loading conditions.

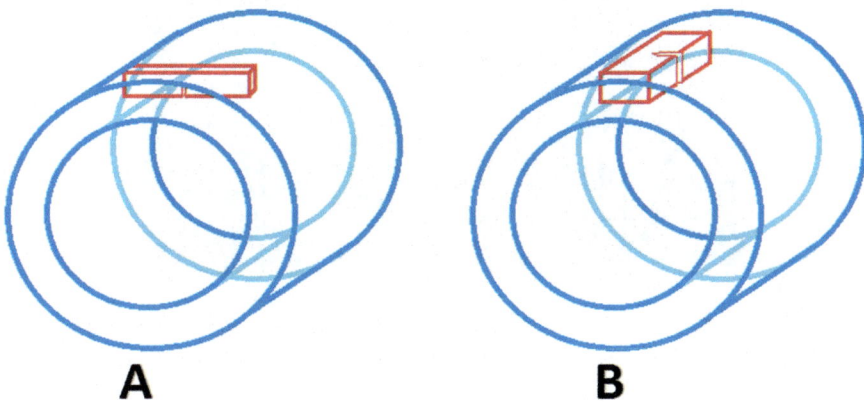

Fig. 1. The scheme of SENB samples cutting from P110 casing pipe.

For in-plane displacement measurement in smaller samples two-dimensional Digital Image Correlation (2D DIC) [1] technique was used, which requires a random grey intensity distribution on the specimen surface. The speckle pattern was made by spraying black and white paints onto the specimen surface.

All of the specimens were precracked prior testing. After that fatigue crack propagation test were made. In order to calculate crack length and stress intensity factor two methods were used: compliance based on DIC extensometers and inverse method based on displacement field measurements.

The first proposed method utilized three virtual extensometers of the same length placed in the known distance from sample edge measuring displacement related with crack advance. The results of this measurements has been interpolated to the sample edge allowing further stress intensity calculations and crack growth rate according do ASTM 1820 standard.

In the second method DIC displacement field measurements near the crack tip has been used as an input data for inverse method algorithm described in [2, 3]. Here, stress

intensity factors and crack tip positions were delivered directly from iterative calculations aimed to determine optimized set of several parameters of analytical equations
describing displacements fields near the crack tip in the way minimizing difference
between the model and DIC measurements results. Calculations were done with different size and positioning of the area from which DIC measurement data were taken as
the input data. There were examined square fitting areas with 0.3 W to 0.7 W width
and height located having the crack tip in its centre. Additionally, for 0.5 W case
calculations for three different positioning of crack tip were made, non-symmetric in
horizontal direction (with crack tip located closer to left edge of fitting area, see Fig. 2).

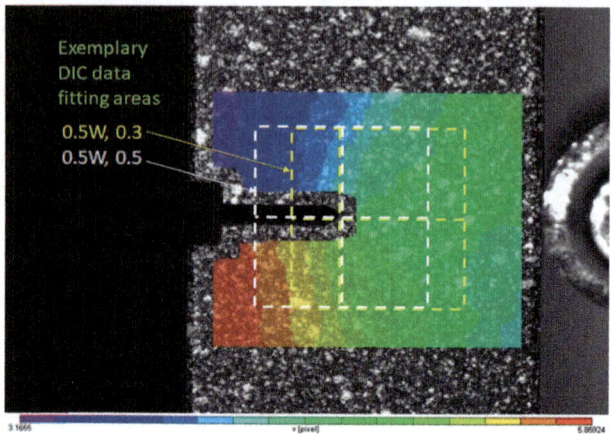

Fig. 2. Exemplary DIC fitting area positioning for inverse method calculations with 0.5 W size
and horizontal positioning ratio 0.3 and 0.5 on the background of DIC vertical displacement field
measurements

Digital images required for DIC measurements in both proposed method were
acquired automatically with camera trigger synchronized with cyclic loading.

2   Results and Discussion

2.1   DIC Based Compliance vs. Standardized Procedure

Figure 3 presents comparison of results for the largest samples (W = 18 mm) cut from
pipe material in B type orientation obtained from standardized procedure utilizing
clip-on-gage extensometer and the method based on virtual DIC extensometers. Results
from both methods are in good accordance, showing usability of the later to smaller
samples where extensometer cannot be applied.

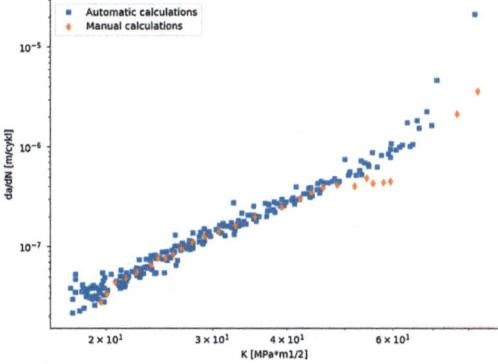

Fig. 3. Comparison of FCG results for DIC extensometer compliance method measurements (manual) and standard based procedure (automatic) for W = 18 mm sample

2.2 DIC Based Compliance vs. Inverse Method

Inverse method based FCG test results showed the influence of fitting area size and positioning on the obtained results (see Fig. 4). Generally, crack growth rates are in the same range, however stress intensity factors are translated to larger values with the growing fitting area size and with crack tip positioning closer to the right side according to the orientation presented in Fig. 2 (with larger horizontal positioning ratio value in Fig. 3b). This effect is related with large plastic zone occurrence in the investigated material, having influence on displacement values measured on areas positioned far from the crack tip. In cases of larger fitting area and higher horizontal positioning ratio values larger fraction of large displacement are taken into account during calculations enlarging obtained stress intensity values. Thus, stress intensity factors are larger when delivered from virtual extensometers based compliance even when comparing to delivered from inverse method with smaller horizontal positioning ratio (see Fig. 5).

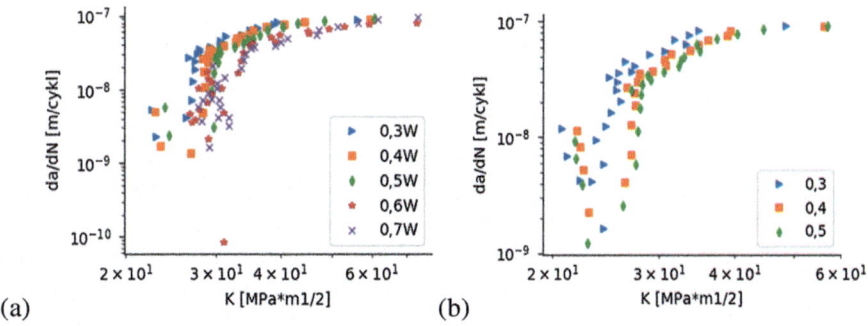

Fig. 4. Results of inverse method FCG rate test for W = 7.6 mm sample; (a) fitting area size impact, (b) crack tip positioning impact

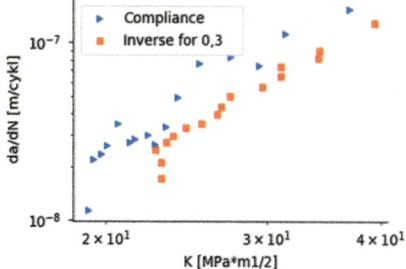

Fig. 5. Comparison of FCG rate results obtained from compliance and inverse based method (W = 0.5, horizontal positioning ratio = 0.3)

2.3   Scale Effect

Comparison of FCG growth results delivered from DIC virtual extensometers measurements based method is presented in Fig. 6. It is clearly visible decrease of crack grow rates with decrease of samples size.

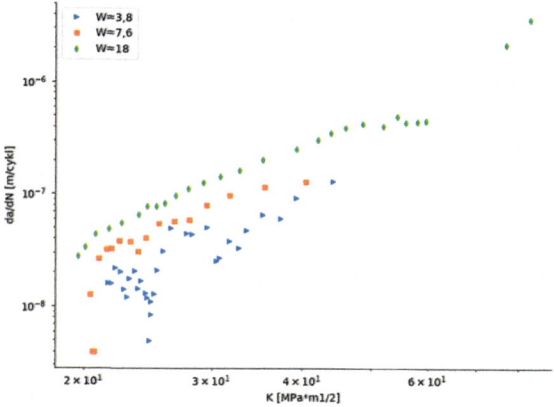

Fig. 6. FCG rate results for geometrically similar samples of three sizes

3   Conclusions

DIC based measurements techniques has been successfully applied in FCG growth rate test of minisamples cut from P110 casing pipe used in natural gas mining. More automated, inverse based method with displacement fields measurements utilization requires preliminary testing for fitting area positioning determination while more laborious approach based on DIC extensometers gave reliable results confirmed in standardized test. The later method has been used to demonstrate scale effect in the investigated material.

Acknowledgements. The Project is financed by The National Centre for Research and Development within Blue Gas II Programme, Contract no. BG2/DIOX4SHELL/14.

References

1. Sutton, M.A., Wolters, W.J., Peters, W.H., Ranson, W.F., McNeill, S.R.: Determination of displacements using an improved digital correlation method. Image Vis. Comput. **1**, 133–139 (1983)
2. Brynk, T., Rasinski, M., Pakiela, Z., Olejnik, L., Kurzydlowski, K.J.: Investigation of fatique crack growth rate of Al 5484 ultrafine grained alloy after ECAP process. Phy. Status Solidi (A) Appl. Mater. Sci. **207**(5), 1132–1135 (2010)
3. Brynk, T., Pakiela, Z., Kulczyk, M., Kurzydlowski, K.J.: Fatigue crack growth rate in ultrafine-grained Al 5483 and 7475 alloys processed by hydro-extrusion. Mech. Mater. **67**, 46–52 (2013)

Micro-scale Non-destructive Stress Measurement for Ultra-thin Glass Plates

Po-Chi Sung[1], Wei-Chung Wang[1(✉)], Mao-Chi Lin[2], Yu-Wei Kuo[1], and Tzu-Hsuan Hsu[2]

[1] Department of Power Mechanical Engineering, National Tsing Hua University, Hsinchu, Taiwan 30013, Republic of China
wcwang@pme.nthu.edu.tw
[2] Laser and Additive Manufacturing Technology Center, Industrial Technology Research Institute, Tainan, Taiwan 73445, Republic of China

Abstract. In this paper, a micro-scale stress measurement system was developed to non-destructively measure the laser cutting residual stresses on the edges of ultra-thin glass plates. Three 50 µm thickness ultra-thin glass plates with different laser cutting processing procedures were measured. Experimental results showed that the difference between these three ultra-thin glass plates not only in stress value but also in stress distribution can be clearly identified by the micro-scale stress measurement system.

Keywords: Ultra-thin glass plate · Laser cutting · Residual stress
Micro-scale · Photoelasticity

1 Introduction

In comparison with the mechanical cutting, the laser cutting is more suitable for the precision cutting of ultra-thin glass plates. However, the magnitudes of the residual stresses by laser cutting on the edges of ultra-thin glass plate may be a critical factor affecting the quality of ultra-thin glass plate. Therefore, non-destructive inspection for laser cutting stress in ultra-thin glass plate has been imperatively needed.

The recent study proposed by the authors [1] demonstrated that the laser cutting stress measurement of ultra-thin glass plates can be achieved by the low-level stress measurement method - enhanced exposure theory of photoelasticity (EEToP) [2]. Besides, by using the measurement system of EEToP established in [1], the difference between different glass plates in stress distribution with the same laser cutting processing procedure can be detected. However, per the stress measurement results shown in [1], measurement system of EEToP may not be capable of distinguishing the dissimilarity between the glass plates with different laser cutting processing procedures. This might be because the laser cutting is one kind of micro-scale processing procedures.

To observe the detail of the edges of the ultra-thin glass plates in micro-scale and measure the corresponding stresses, a micro-scale stress measurement system was established in this paper. The micro-scale stress measurement system is based on the macrophotography and EEToP. Three ultra-thin glass plates with different laser cutting processing procedures were measured. Their laser cutting stress measurement results were then compared to each other to analyze the difference between them.

© Springer International Publishing AG, part of Springer Nature 2019
E. E. Gdoutos (Ed.): ICTAEM 2018, SI 5, pp. 136–138, 2019.
https://doi.org/10.1007/978-3-319-91989-8_30

2   Experimental Setup

In the micro-scale stress measurement system, a plane polariscope was employed to implement the measurement procedure of EEToP. The image acquisition equipment is composed of an advanced CMOS camera (with 4 million spatial resolution, 16 bit grayscale resolution, and 6.45 μm pixel size) as well as a macro lens (with 4X magnification and 6.5 mm focal length). Thus, the visible area in a circle with 3 mm diameter and the spatial resolution in 1.61 μm/pixel in object space can be achieved. Three 50 μm thickness ultra-thin glass plates cut by different laser cutting processing procedures labeled as S1, S2, and S3 were measured to analyze their difference in edge stress to each other. The laser cutting processing procedures of these three specimens were performed by Laser and Additive Manufacturing Technology Center, Industrial Technology Research Institute (ITRI). Especially, the specimen S2 was cut with the advanced laser processing technique - gLaserTrim [3] developed by ITRI.

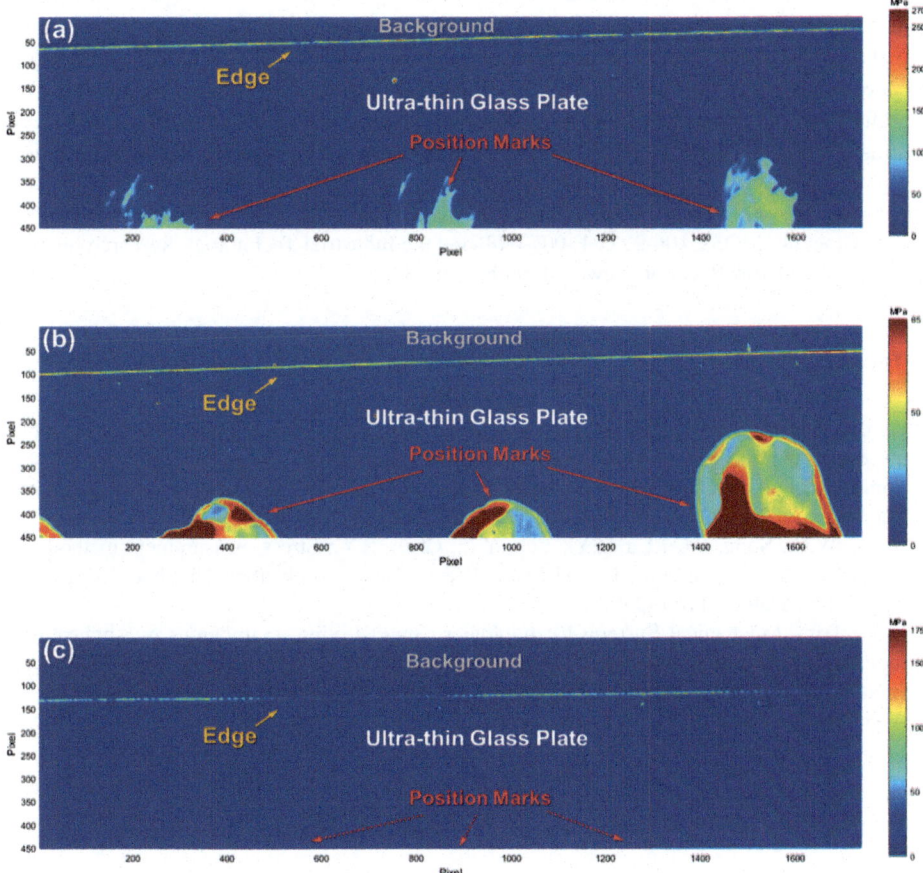

Fig. 1. Stress measurement results several locations of edges of ultra-thin glass plates (a) S1 (b) S2 and (c) S3.

3   Experimental Results

Figures 1(a), (b), and (c) show the stress measurement results of several locations of edges of specimens S1, S2 and S3, respectively. In Fig. 1, the position marks indicated on the specimens are used to conveniently identify which location of the edge is under measurement. The distance between two neighboring position marks is approximately 1 mm. Apparently, measurement results of these three specimens are different to each other not only in stress value but also in stress distribution. The values of the maximum stress of specimens S1, S2 and S3 are 270 MPa, 85 MPa and 175 MPa, respectively. It shows the superiority of the laser processing technique of gLaserTrim.

4   Conclusions

In this paper, the micro-scale stress measurement system based on macrophotography and EEToP was established so that the laser cutting stresses on the edges of ultra-thin glass plates can be measured. Through the analysis of experimental results, it is confirmed that the micro-scale stress measurement system is able to distinguish the difference between the glass plates with different laser cutting processing procedures. Therefore, the micro-scale stress measurement system is full of potential to be developed as the non-destructive inspection system used in laser cutting and glass industries.

Acknowledgement. This paper was supported in part by the Ministry of Science and Technology (Grant no. MOST 106-2221-E-007-048) and the Industrial Technology Research Institute (Grant no. A200-106 BA1) of Taiwan, Republic of China.

References

1. Kuo, Y.W., Sung, P.C., Wang, W.C., Lin, M.C., Hsu, T.H.: Investigation of non-destructive inspection method for ultra-thin glass plates. In: International Symposium on Optomechatronic Technology 2017, pp. 1–2. Shangri-La's Far Eastern Plaza Hotel, Tainan, Taiwan, Republic of China (2017)
2. Wang, W.C., Sung, P.C., Lu, Z.Y., Yeh, Y.L., Chen, P.Y.: Stress measurement method and system for optical materials. United States Patent, Patent Application Number: 15/618,145, Application Date 2017/06/09
3. gLaserTrim, Mechanical Systems Technologies, Smart Living, Innovations & Applications, Industrial Technology Research Institute, https://www.itri.org.tw/eng/Content/MsgPic01/Contents.aspx?SiteID=1&MmmID=620651706136357202&MSid=655414016464677501. Accessed 30 Jan 2018

Measurement of Displacement and Its Derivatives from a Phase Fringe Pattern

Rishikesh Kulkarni[1($\boxtimes$)] and Pramod Rastogi[2]

[1] Department of Electronics and Electrical Engineering,
Indian Institute of Technology, Guwahati, India
rishi.k@iitg.ernet.in
[2] Applied Computing and Mechanics Laboratory,
Ecole Polytechnique Federale de Lausanne, Lausanne, Switzerland

Abstract. Noting the fact that the phase and phase derivatives carry the information on the mechanical deformation in optical interferometric measurements techniques, we propose a noise robust fringe analysis technique for the simultaneous estimation of unwrapped phase and arbitrary order phase derivative from a phase fringe pattern. The continuous phase along row or column of the phase fringe pattern is approximated as a weighted linear combination of linearly independent basis functions such as Fourier basis. The weights are accurately estimated based on a state space analysis using the linear Kalman filter. Simulation and experimental results are provided to substantiate the applicability of the proposed method.

Keywords: Phase estimation · Phase derivative estimation
Linearly independent basis functions · State space analysis · Kalman filter

1 Introduction

Optical measurement techniques such as electronic speckle pattern interferometry (ESPI), shearography and digital holographic interferometry (DHI) have been widely used in the areas of experimental mechanics such as measurement of deformations on rough object surfaces, non-destructive testing and fracture mechanics [1–3]. In the ESPI setup, the information on the object surface displacement is recorded in the form of real interferogram of the form [4]

$$I(x, y) = a(x, y) + b(x, y) \cos \psi(x, y)$$

where, $x \in [1, N]$ and $y \in [1, M]$ represent the spatial coordinates of the interferogram $I(x, y)$ of size $M \times N$. $a(x, y)$ and $b(x, y)$ represent background intensity and amplitude of the interferogram, respectively; The displacement and its spatial derivatives are directly proportional to the phase $\psi(x, y)$ and its derivatives, respectively. As a result, accurate estimation of phase and its derivatives is of vital importance [5]. Typically, the unambiguous phase estimate is derived from a stack of phase shifted interferograms [4, 5]. However, the estimated phase in a wrapped form, i.e., the phase values lies in the

© Springer International Publishing AG, part of Springer Nature 2019
E. E. Gdoutos (Ed.): ICTAEM 2018, SI 5, pp. 139–144, 2019.
https://doi.org/10.1007/978-3-319-91989-8_31

range of $[-\pi, \pi)$. Such a phase map is usually termed as *phase fringe pattern*, $\phi(x, y)$. The relationship between $\psi(x, y)$ and $\phi(x, y)$ can be given as

$$\phi(x, y) = W\{\psi(x, y)\} = mod(\psi(x, y) + \pi, 2\pi) - \pi,$$

where, $W\{\cdot\}$ represents the *wrap* operator. In the case of DHI setup, the displacement information is recorded in the form of a complex interferogram as $I(x, y) = \exp[j\phi(x, y)]$, where, $j = \sqrt{-1}$. It can be observed that the phase estimate $\phi(x, y)$ can be derived directly with the application of *angle* function to the interferogram without any need of phase shifting mechanism. In this paper, we propose a technique for the estimation of unwrapped (true) phase $\psi(x, y)$ and its derivatives from the measurement of $\phi(x, y)$.

2 Phase Approximation with Weighted Linear Combination of Linearly Independent Basis Functions

In the proposed method, the phase and phase derivative estimation is performed either in a row-wise or column-wise manner. The true phase $\psi(x, y)$ is assumed to be spatially continuous. Considering a row-wise estimation procedure, the phase $\psi(x)$ along a specific row y can be represented as a weighted linear combination of linearly independent basis functions as,

$$\psi(x) = \sum_{k=0}^{K-1} C_k \beta_k(x), \tag{1}$$

where, C_k represents the weight of k th basis function $\beta_k(x)$ and K represents the *basis dimension*. Different types of basis functions are reported in the literature such as polynomial, Fourier, Legendre, Chebyshev and radial basis. In accordance with Eq. (1), once basis functions are defined, the computation of appropriate weights can provide the required phase estimate. In addition, Eq. (1) indicates that the pth derivative of phase $\psi(x)$ can be obtained as

$$\psi^{(p)}(x) = \sum_{k=0}^{K-1} C_k \frac{d^p \beta_k(x)}{dx^p}. \tag{2}$$

Note that the computed weights simultaneously provide the phase estimate and the estimate of phase derivative of arbitrary order p.

3 Estimation of Weights of Basis Functions Based on State Space Analysis

The weights of basis functions are defined to be elements of a state vector C as $C = [C_0, C_1, C_2, \cdots, C_{K-1}]^T$. The superscript T indicates the transpose operation. The state space formulation can be given as

$$C_x = FC_{x-1}, Process\ update\ equation$$

where, F represents the state transition matrix of size $K \times K$. Since the state vector is constant in a given row, the state transition matrix is an identity matrix. Based on Eq. (1), the measurement model can be given as

$$\Gamma_x = \phi(x) = W\{C_x^T \beta_x\},$$

where, $\beta_x = [\beta_0(x), \beta_1(x), \beta_2(x), \cdots, \beta_{K-1}(x)]^T$. It is important to note that the term $C_x^T \beta_x$ basically represents the true phase $\psi(x)$. The measurement Γ_x, however, is a wrapped its value. We propose to use a wrapped Kalman filter [6] based on circular data statistics to the estimate of state vector. The algorithm of state vector estimation is given as follows:

1. Initialize the state vector estimate at $x = 0$ and its error covariance matrix as

$$\hat{C}_0^+ = E[C_0],$$

$$\hat{P}_0^+ = E[(C_0 - \hat{C}_0^+)(C_0 - \hat{C}_0^+)^T],$$

where, $E[\cdot]$ is the expectation operator. The superscripts $-$ and $+$ indicate the a priori and a posteriori estimates of the associated variables, respectively.
2. For $x = 1, 2, \cdots, N$, since F is an identity matrix, set

$$\hat{C}_x^- = \hat{C}_{x-1}^+, \hat{P}_x^- = \hat{P}_{x-1}^+.$$

3. Perform the measurement update of the state vector as

$$K_x = \hat{P}_x^- \beta_x^T (\beta_x \hat{P}_x^- \beta_x^T + R_x),^{-1}$$

$$\hat{C}_x^+ = \hat{C}_x^- + K_x W(\Gamma_x - \hat{C}_x^{-T} \beta_x),$$

$$\hat{P}_x^+ = (I - K_x \beta_x)\hat{P}_x^- (I - K_x \beta_x)^T + K_x R_x K_x^T,$$

where, K_x is the Kalman gain, R_x is the noise variance and I is an identity matrix.

The estimated weights in the state vector at $x = N$ are substituted in Eqs. (1) and (2) to obtain the estimates of phase and phase derivatives, respectively. The weight estimation is performed in each row to obtain the complete two dimensional estimate of $\psi(x, y)$.

4   Simulation and Experimental Results

Figure 1(a) shows a phase fringe pattern simulated with noise variance = 0.4. The proposed method is implemented using the Fourier basis functions with the basis dimension of $K = 25$. The state vector estimation is performed in a row-wise manner. Figures 1(b), (c) and (d) show the estimated phase, first order and second order phase derivatives with respect to x, respectively. The peak valley error in the phase and phase derivative estimation are found to be 0.78 rad, 0.0360 rad/pixel and 0.0028 rad/pixel^2, respectively.

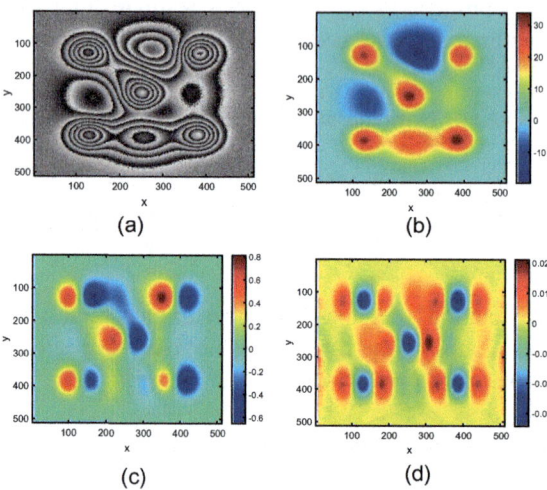

Fig. 1.   (a) Simulated noisy phase fringe pattern. Estimated (b) phase $\psi(x, y)$ in radians (c) first order phase derivative in rad/pixel with respect to x and (d) second order phase derivative in rad/pixel^2 with respect to x.

The experimental validation of the proposed method is performed with a phase fringe pattern shown in Fig. 2(a) recorded in a DHI setup corresponding to out-of-plane displacement of a square aluminum plate subjected to load. The proposed method is implemented using the Fourier basis functions with the basis dimension of $K = 15$. Figures 2(b), (c) and (d) show the estimated phase, first order and second order phase derivatives with respect to x, respectively. The point of application of load is predominantly visible in the second order derivative map.

The basis dimension is selected empirically depending on the data length (N or M).

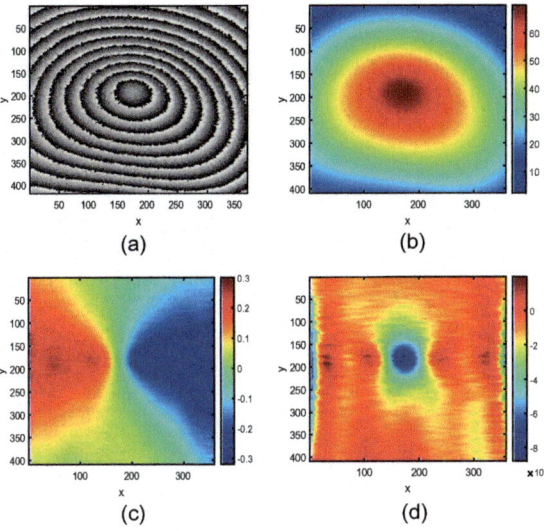

Fig. 2. (a) Experimentally recorded phase fringe pattern in a digital holographic interferometry setup corresponding to the out-of-plane displacement of the aluminum plate. Estimated (b) phase $\psi(x, y)$ in radians (c) first order phase derivative in rad/pixel with respect to x and (d) second order phase derivative in rad/pixel^2 with respect to x.

5 Conclusion

A novel method is proposed for the estimation of unwrapped phase and phase derivatives from a phase fringe pattern. The continuous phase assumption allows us to consider the phase along a row or column as a weighted linear combination of linearly independent basis functions and a pre-defined basis dimension. Considering the weights to be elements of a state vector, the linear Kalman filter based state vector estimation provides robust estimates of weights from a noise corrupted phase fringe pattern. The estimated weights subsequently provide simultaneous estimation of unwrapped phase and phase derivative of any arbitrary order.

References

1. Jacquot, P.: Speckle interferometry: a review of the principal methods in use for experimental mechanics applications. Strain **44**, 57–69 (2008)
2. Sciammarella, C.A., Kim, T.: Frequency modulation interpretation of fringes and computation of strains. Exp. Mech. **45**, 393–403 (2005)
3. Rastogi, P.K.: Techniques to measure displacements, derivatives and surface shapes. extension to comparative holography. In: Rastogi, P.K. (ed.) Holographic Interferometry. Springer Series in Optical Sciences, vol. 68. Springer, Heidelberg (1994)
4. Servin, M., Quiroga, J.A., Padilla, J.M.: Fringe Pattern Analysis for Optical Metrology: Theory, Algorithms, and Applications. Wiley-VCH, Weinheim (2014)

5. Kulkarni, R.D., Rastogi, P.: Single and Multicomponent Digital Optical Signal Analysis: Estimation of Phase and Its Derivatives. Institute of Physics Publishing (2017)
6. Traa, J., Smaragdis, P.: A wrapped Kalman filter for azimuthal speaker tracking. IEEE Sig. Process. Lett. **20**, 1257–1260 (2013)

Evaluation of Calibration Performance by Conical Targets

Chi-Hung Hwang[2]([✉]), Wei-Chung Wang[1], and Yung-Hsiang Chen[1,2]

[1] Department of Power Mechanical Engineering, National Tsing Hua University,
No. 101, Sec. 2, Kuang Fu Rd, Hsinchu 30013, Taiwan, Republic of China
wcwang@pme.nthu.edu.tw
[2] Instrument Technology Research Center,
National Applied Research Laboratories, No. 20, Yanfa 6th Rd, Hsinchu 30076,
Taiwan, Republic of China
{chhwang,yschen}@narlabs.org.tw

Abstract. Three conic calibration targets (CCTs) were designed and implemented for three-dimensional (3D) digital image correlation (DIC) camera calibration. To understand the associated performance of CCTs and the potential problems while using CCTs to determine the camera parameters, in this paper, the CCTs image data and the corresponding spatial distance were first utilized to determine the camera parameters, and then the 3D CCT object was reconstructed with the same images and the determined camera parameters of different camera-pairs from different viewing directions. Based on the calculated deviations of the overlapping regions of the reconstructed 3D CCTs, it is clear that the proposed CCTs can be used for determining the camera parameters.

Keywords: Digital image correlation · Camera calibration · 3D reconstruction
Conic calibration targets

1 Introduction

In computer vision applications, system calibration procedure is essential for determining intrinsic and extrinsic camera parameters for a pair of cameras. Geometrical mapping relationship between pixels of images and the spatial set data points of an imaging system can be established by those calibration-determined parameters. In this paper, three conical calibration targets (CCTs) [1, 2], i.e. a conical, a hemispherical, and a hollow cylindrical were adopted. To identify the spatial data points on the surface of CCTs, the surface of CCTs was anodized into black with white circular taggers attached to the intersection of latitude and longitude lines. A semi-circular frame was designed to mount three CCD cameras at different positions. Three different camera-pairs with different viewing angles with respect to the CCT can be formed. In this paper, the first CCD camera named as camera C2 was located at the center of the semi-circular frame, the other two CCD cameras C1 and C3 were arranged symmetrically with respect to the camera C2.

All three CCTs were designed and manufactured so that the distance between attached circular taggers on the circumference of the CCTs are equal in distance or in

© Springer International Publishing AG, part of Springer Nature 2019
E. E. Gdoutos (Ed.): ICTAEM 2018, SI 5, pp. 145–148, 2019.
https://doi.org/10.1007/978-3-319-91989-8_32

angle according to the geometrical shape of the CCT. When the geometrical symmetrical center of a CCT is directly viewed by a camera, the taggers on the captured image should be symmetric with respect to the symmetrical axis of the CCT projected on the image. Then camera view-angle and the angle between line-of-sight and conical symmetrical axis can be determined by the offset between the real symmetrical axis of the CCT projected on the image with respect to the ideal one. Thanks to the unique geometrical characteristics of CCTs, the intrinsic and extrinsic camera parameters can be directly determined. In this paper, the camera parameters were determined by iteration from the CCTs' spatial data and the corresponding image points.

2   Experiment Results

As a typical example, Fig. 1 shows the semi-circular multi-camera imaging system [3] with a hemispherical calibration target placed on the circle center of the semi-circular frame. From left to right, three cameras mounted on the semi-circular frame are named as C1, C2 and C3, respectively. By selecting any two of the three cameras, 3 different camera-pairs can be obtained, i.e. C1C2, C1C3 and C2C3, respectively. In this paper, conical, hemispherical, hollow cylindrical CCTs were respectively placed to the circle center of the semi-circular frame. Taking hemispherical calibration target as a typical example, the images of CCT obtained by cameras C1, C2 and C3 are shown in Fig. 2. The region of interest (ROI) is confined by yellow lines. All points inside the selected ROI were used for camera calibration. The center of the circular taggers on the circumference was extracted. As shown in Fig. 2, for better observation, the center of the

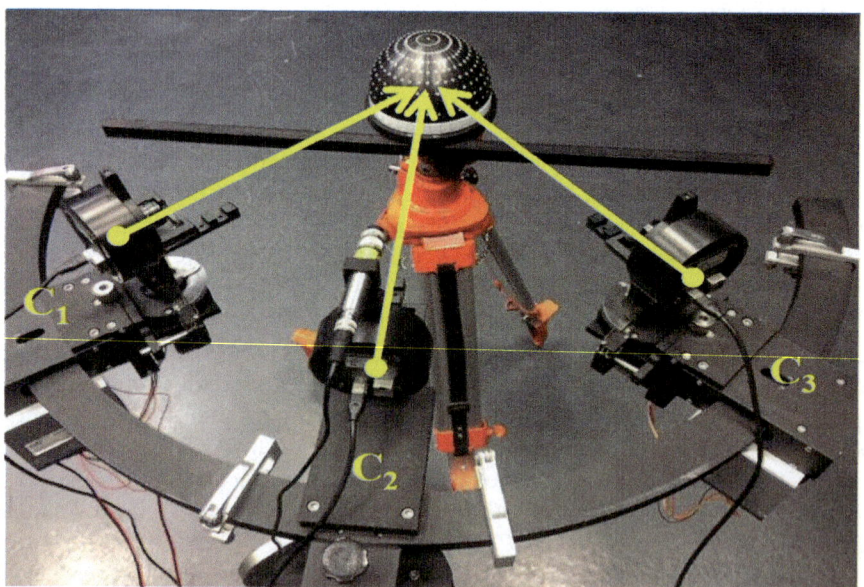

Fig. 1. The semi-circular multi-camera imaging system.

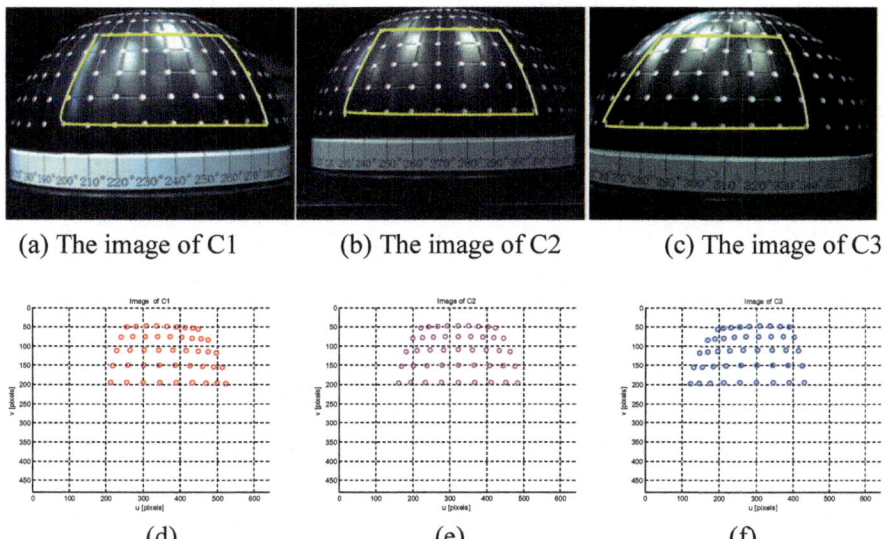

(a) The image of C1   (b) The image of C2   (c) The image of C3

(d)   (e)   (f)

Fig. 2. The hemispherical calibration target

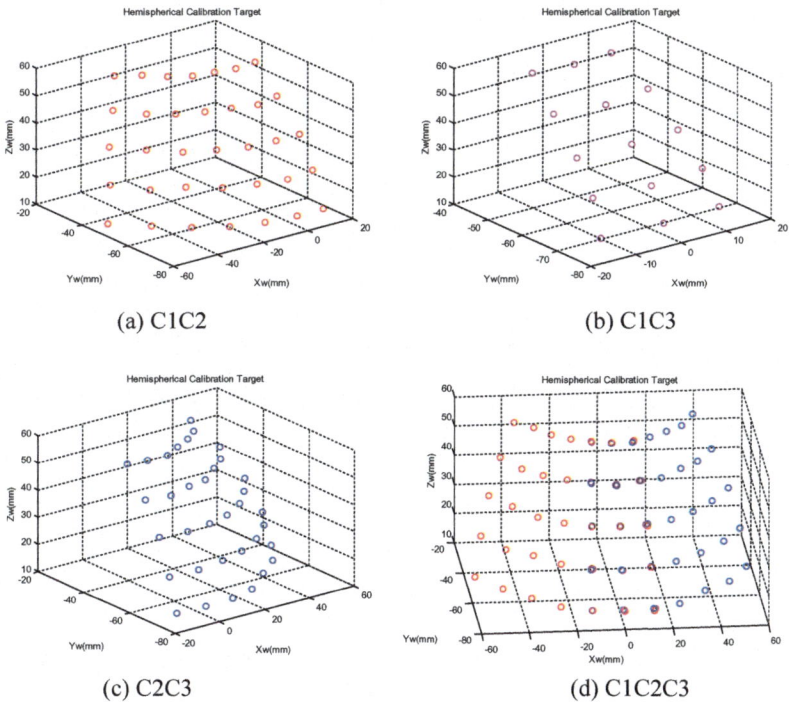

(a) C1C2   (b) C1C3

(c) C2C3   (d) C1C2C3

Fig. 3. The 3D reconstruction for hemispherical calibration target

circle is labelled with a circle. The reconstructed 3D spatial data set of the hemi-spherical calibration target from camera-pairs C1C2, C1C3 and C2C3 are shown in Figs. 3(a), (b) and (c), respectively. Since the depth is determined by the associated disparity of the camera-pair, therefore, the reconstructed spatial data set is only determined within the overlapped region of two cameras. The three reconstructed 3D spatial data set were then stitched as shown in Fig. 3(d). Based on the residuals of the overlapped region and errors of reconstructed CCTs, the calibration performance can be evaluated.

3 Conclusion

In this paper, to remove drawbacks and limitations of currently commercially available calibration targets, three new CCTs were proposed for determining intrinsic and extrinsic calibration parameters of camera for stereo-vision applications. The results shows that the camera parameters can be well determined from different viewing angles. The performance of three different CCTs, the hemispherical, cylindrical and conical ones, can also be determined by reconstructing the CCTs spatial data set with three camera-pairs. It can be concluded that the proposed CCTs can be used to replace the widely used planar calibration target.

Acknowledgement. This paper was supported in part by the Office of Naval Research (ONR), United States (Award Number: N62909-16-1-2064) and Ministry of Science and Technology, Taiwan (Grand no. MOST-102-2221-E-492-014, MOST-106-2221-E-492 -013).

References

1. Huang, C.H., Chen, Y.H., Wang, W.C.: Conical Calibration Target Used for Calibrating Image Acquisition Device. TWI 576652 (2017)
2. Huang, C.H., Chen, Y.H., Wang, W.C.: Conic section calibration object for calibrating image capturing device. US9762897 B2 (2017)
3. Hwang, C.-H., Wang, W.-C., Chen, Y.-H.: Camera calibration and 3D surface reconstruction for multi-camera semi-circular DIC system. In: International Conference on Optics in Precision Engineering and Nanotechnology (icOPEN2013), vol. 8769. International Society for Optics and Photonics (2013)

Optical Method Applying to the Measure of Transverse Deformation

C. Huang$^{(\boxtimes)}$ and Xu Liu

Wuhan Textile University, Wuhan 430074, People's Republic of China
ab6688us@yahoo.com

Abstract. An optical measure system is developed to measure transverse deformation of stressed materials. The paper discusses two problems to be solved in the measure system. Experimental results show that the constructed system works well.

Keywords: Transverse deformation · Homogenization of light intensity
White noise

1 Introduction

Theoretical simulation indicates that differences in dynamic transverse deformations of brittle material can lead to the similar responses as the ones that transverse confinement pressures are applied to the material increasingly when it is loaded dynamically in axial direction. To explain the observed results, one may need to explore the feature of transverse inertia induced by the dynamic loading, because transverse inertia may play the same role as the transverse confinement pressure does. Studying transverse deformation of brittle materials will be able to further the understanding of the effect of transverse inertia on dynamic responses.

Deformation measurement can be generally categorized as the contact measure and the noncontact measure based on the gauging manner. The contact measure mounts gauging instrument on the surface of deforming specimen and forces the instrument to deform with the deformation of the specimen simultaneously. The contact measure can now be further divided into the conventional strain gage measurement and optical fiber measurement based on the type of the mounted gage. The conventional strain gage measurement uses a metallic foil to gauge the deformation of specimen; The other one is the newly developed optical fiber measurement which uses optical fiber as the gauging instrument to deform with the specimen; Since the measure instrument either the strain gage or the optical fiber has the stiffness which is larger than the one of some materials, soft materials, for example rubber and soft goods are unable to match the stiffness of the measure instruments. It is hard to use the above measure methods to measure the deformation of the soft materials. The noncontact measurement adopts laser based gauging and machine vision to measure deformation of stressed materials. With the development in the measure technique, the laser based optical method can now provide 3-D solution for various deformation tests. Due to the huge calculation

© Springer International Publishing AG, part of Springer Nature 2019
E. E. Gdoutos (Ed.): ICTAEM 2018, SI 5, pp. 149–151, 2019.
https://doi.org/10.1007/978-3-319-91989-8_33

work resulting from the machine vision, however, it is difficult for the measure to provide instantaneous testing results when the measure is performed.

2 Construction of the Measure System

A new optical method is developed to measure transverse deformation. The optical measure system is composed of: semiconductor laser diode that can emit laser beam gauging transverse deformation; high speed photodetector that is used to convert collected light signal into voltage signal; high speed digital oscilloscope that is the device to capture and store the electrical signal from the photodetector and in the mean time the oscilloscope can also present the captured signal on its screen as the instantaneous testing results. The following is the working mechanism of the laser based optical method. Semiconductor laser diode emits a rectangular laser beam as shown in Fig. 1. The light arms at a side surface of the tested specimen. Part of the light is projected on to the side surface of the specimen and the rest light is let pass through the specimen and is collected by the high speed photodetector placed right behind the specimen. As the specimen is loaded in uniaxial compression, expansion deformation in transverse direction due to the effect of Poisson's ratio can occlude some of the passing light and the photodetector collects less light accordingly. Reduction of the light intensity leads to a decrease of the converted voltage. If the decreasing voltage curve can be correlated with the transverse deformation, the oscilloscope will be able to present the deformation on its screen instantaneously. If the specimen is loaded in uniaxial tension, the voltage signal captured by the oscilloscope will show an increasing curve in comparison with the one in uniaxial compression.

There are two problems to be solved in the optical measure system. Laser light emitted from semiconductor laser diode usually has the light intensity in Gaussian distribution, i.e., light intensity is the highest in center of the light beam. With the increase of distance away from the center, the intensity decreases very fast. The non-uniformity of light intensity in laser light beam undoubtedly brings in a significant error in the measurement of the transverse deformation, since the relationship between the light intensity and transverse deformation is not in an expected linear manner. To eliminate the nonlinearity of the relationship, one has to take advantage of some optical technique to homogenize the light intensity, so as to obtain a linear relationship of the light intensity with transverse deformation. Therefore, the homogenization of laser light has to be done before taking any measurement to ensure the measure accuracy of transverse deformation. High frequency electrical noise (white noise) is another primary problem to be solved in the measure system. Generated by electronic devices in electric circuits, the white noise is another important error source for the optical measure system and is an undesired random disturbance of the experimental signal. The noise can distort the real signal seriously and sometimes it makes the oscilloscope failed in the capturing of real signal. Filter circuits have to be built up in the measure system to eliminate the effect of noise and to ensure the accuracy of experimental results.

3 Experimental Results

Based on the above discussion, the optical deformation measure system is constructed and is applied to the measure of rubber tension deformation. Figure 2 shows the measure setup for rubber tension testing and Fig. 3 shows the measure results of the optical measure system. With the tension deformation, transverse size of the rubber is reduced. The reduction of the transverse size makes the collected light intensity increased in the photodetector and an instantaneously increasing curve is exhibited on the screen of the oscilloscope as Fig. 3 shows. The system works well for the measure of transverse deformation.

Fig. 1. Laser light beam

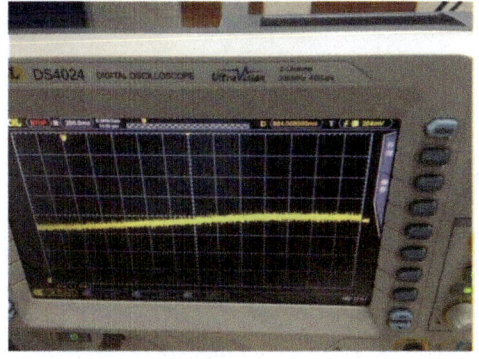

Fig. 3. Measure results of the transverse defor-   **Fig. 2.** Setup for the optical measure system
mation of the rubber

Experimental Study on Fatigue Performance of CFRP-RC Beams Under Variable Amplitude Overloads

Zhanbiao Chen[1] and Peiyan Huang[1,2(✉)]

[1] School of Civil Engineering and Transportation,
South China University of Technology, Guangzhou, China
pyhuang@scut.edu.cn
[2] State Key Laboratory of Subtropical Building Science,
South China University of Technology, Guangzhou, China

Abstract. In the past two decades, external bonding of carbon fiber reinforced polymer (CFRP) composites is widely used for strengthening reinforced concrete (RC) structures of highway bridges [1]. Researchers have demonstrated that external bonding of CFRP can obviously improve the fatigue performance of RC structures [2]. However, although the super measures have been taken to limit overload from truck, the overload phenomenon is still widespread for highway transportation in China.

Therefore, in order to prove the fatigue property and reliability of the RC members strengthened with CFRP in highway bridges under the vehicle overload, experimental study on fatigue performance of RC beams strengthened with CFRP under variable amplitude overloads is carried out in this paper.

Keywords: Fatigue performance · CFRP · RC beams · Overload

1 Experimental Program

Fatigue testing specimens were RC beams strengthened with carbon fiber laminate (CFL). The size of RC beam was 1850 mm length × 100 mm width × 200 mm height, with the span length of L = 1600 mm. The strengthened beam mainly consists of concrete, steel and CFL. The main steel bars were Grade II Φ10 and the other steel bars were Grade I Φ8, the reinforcement ratio was 0.981%. The mechanical properties of main composition materials of the specimen as shown in Tables 1, 2 and 3 of reference [3].

The definition method of overload was carried out in this paper. Basing on the vehicle loading history of a typical highway bridge's superstructure in Guangzhou, the variable amplitude overloading spectrum was generated, as shown in Fig. 1. This provides an experimental method for studying the fatigue performance of bridge structures strengthened with CFRP under overloads. Fatigue tests were conducted to study the fatigue failure mechanism and analyze the fatigue lives of RC beams strengthened with CFL under variable amplitude overloads.

© Springer International Publishing AG, part of Springer Nature 2019
E. E. Gdoutos (Ed.): ICTAEM 2018, SI 5, pp. 152–154, 2019.
https://doi.org/10.1007/978-3-319-91989-8_34

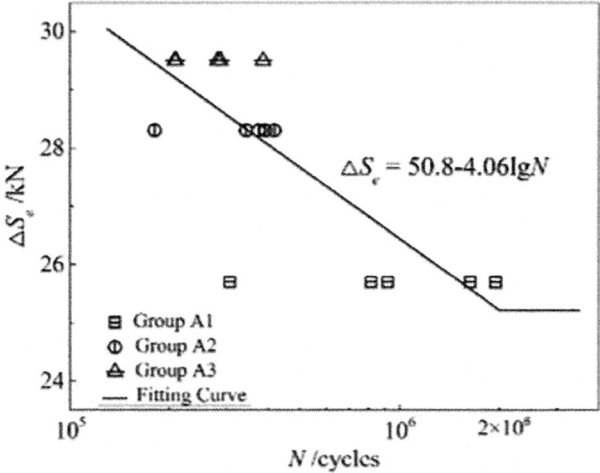

Fig. 1. $\Delta S_e - N$ curve under variable amplitude overloads

Totally 15 specimens, divided into 3 groups (A1–A3) were tested under 3 different overload levels with proved MTS-810 system. The overload level of group A1 was same as the spectrum mentioned above. Overload levels of group A2 and A3 were 10% and 15% higher than that of group A1 respectively.

2 Experimental Results and Analyses

In order to compare the fatigue lives of the strengthened RC beams under overloads and constant amplitude [3], the 3 steps overload amplitude can be conversed to equivalent load amplitude by Eq. 1.

$$\Delta S_e = [\frac{\sum_1^n n_i \Delta S_i^m}{ND}]^{1/m} \tag{1}$$

where, ΔS_e is equivalent load amplitude.

$\Delta S_e - N$ curve was shown in Fig. 1. As shown in Fig. 1, with the increase of equivalent load amplitude, the fatigue lives decrease rapidly. This proves that the overload has a significant effect on fatigue performance of RC beam strengthened with CFL.

Semi-empirical formula of fatigue lives of RC beams strengthened with CFL under variable amplitude overloads was proposed:

$$\Delta S_e = 50.8 - 4.06 \lg N \tag{2}$$

Compare the results with previous studies under constant amplitude [3], it can be seen that the fatigue limit of the strengthened RC beams under overloads is lower. This indicates that the fatigue lives of RC beams strengthened with CFRP would significantly reduce under variable amplitude overloads.

Acknowledgments. The project is supported by National Natural Science Foundation of China (Nos. 11627802, 51678249, 11132004, 51508202).

References

1. Dong, J.F., Wang, Q.Y., Guan, Z.W.: Structural behavior of RC beams externally strengthened with FRP sheets under fatigue and monotonic loading. Eng. Struct. **41**, 24–33 (2012)
2. Meneghetti, L.C., Garcez, M.R., Da Silva Filho, L.C.P.: Fatigue life of RC beams strengthened with FRP systems. Struct. Concrete **15**, 219–228 (2014)
3. Huang, P.Y., Zhou, H., Wang, H.Y.: Fatigue lives of RC beams strengthened with CFRP at different temperatures under cyclic bending loads. Fatigue Fract. Eng. Mater. Struct. **34**(9), 708–716 (2011)

Fatigue Crack Propagation Behavior of RC Beams Strengthened with Prestressed CFL

Huang Kainan, Huang Peiyan[(✉)], Guo Xinyan, Liu Dong,
and Li Wen

School of Civil Engineering and Transportation,
South China University of Technology, Guangzhou 510641, China
pyhuang@scut.edu.cn

Abstract. Compared with the reinforcement technique with non-prestressed Carbon Fiber Laminate (CFL) for reinforced concrete (RC) structures, prestressed CFL can Increase the cracking load of the strengthened members and make the existed crack closure. In this paper, numerical and experimental methods were applied to investigate fatigue crack propagation behavior of reinforced concrete (RC) beams strengthened with prestressed carbon fiber laminate (CFL).

Keywords: Prestress · Carbon fiber laminate (CFL)
Fatigue crack propagation · Reinforced concrete (RC) beam · Life prediction

Fatigue crack propagation tests were carried out to obtain the crack propagation rate of the RC beams strengthened by CFL with prestressing levels of 8%. According to the existing experimental results of prestress loss (Fig. 1), the method of tensioning was improved. The prestress losses of CFL in the five stages of tensile, pasting, curing, releasing and long-term loss were tested and the results show that the prestress loss can be reduced effectively by improving the tensioning method.

All of the beams were submitted to three-point bending. The fatigue crack propagation rate, da/dN can be calculated based on a-N curves obtained by fatigue crack propagation tests. The specimens with 8% prestressing levels and 30KN loads was shown in Fig. 2 which shows that the main crack propagation behavior on the strengthened beams can be summarized into three stages: (1) fast propagation stage; (2) stable propagation stage; (3) unstable propagation stage.

According to the semi-empirical formula of the second expansion stage, the stress intensity factor (SIF) amplitude ΔK was calculated by finite element method (FEM) which was used to predict fatigue life of reinforced beams. The results of life prediction are verified by experiments (Figs. 3 and 4).

© Springer International Publishing AG, part of Springer Nature 2019
E. E. Gdoutos (Ed.): ICTAEM 2018, SI 5, pp. 155–157, 2019.
https://doi.org/10.1007/978-3-319-91989-8_35

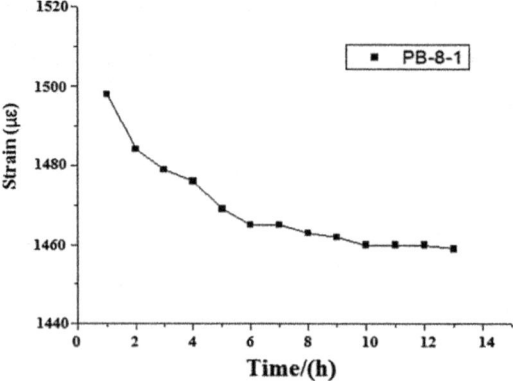

Fig. 1. Change of strain during solidify (14 h).

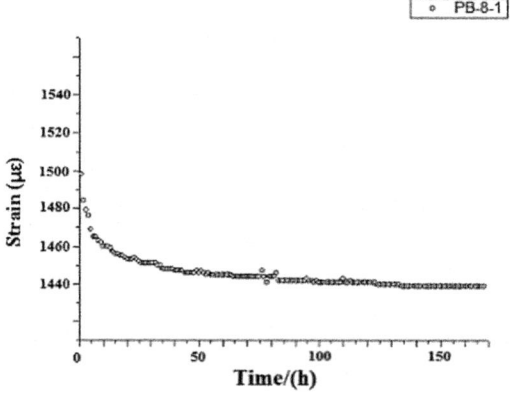

Fig. 2. Change of strain during solidify (7day).

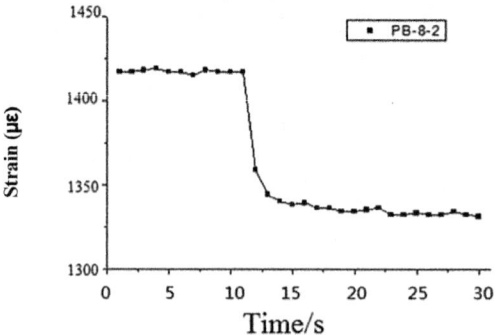

Fig. 3. CFL strain changes during release.

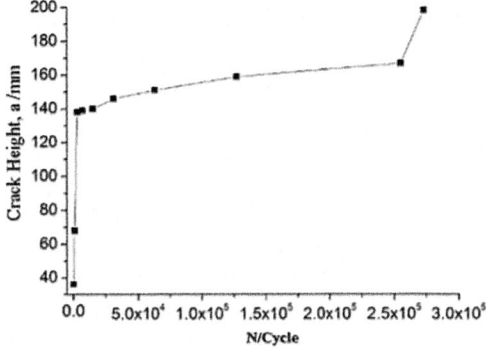

Fig. 4. PB-8-2 a - N curve

References

1. Kim, Y.J., Wight, R.G., Green, M.F.: Flexural strengthening of RC beams with prestressed CFRP sheets: development of nonmetallic anchor systems. J. Compos. Constr. **12**(1), 35–43 (2008)
2. Bymaster, J.C., Dang, C.N., Floyd, R.W., et al.: Prestress losses in pretensioned concrete beams cast with lightweight self-consolidating concrete. Structures **2**, 50–57 (2015)
3. Yang, D.-S., Park, S.-K., Neale, K.W.: Flexural behaviour of reinforce concrete beams strengthened with prestressed carbon composite. Compos. Struct. **88**, 497–508 (2009)
4. Huang, P., Liu, G., Guo, X.: Fatigue life prediction of RC beams strengthened with prestressed CFRP under cyclic bending loads. Acta Mech. Solida Sin. **26**(1), 46–52 (2013)

On the Fracture Mechanics of Prince Rupert's Drops

Hillar Aben[1]([⊠]), Johan Anton[1], Pearu Peterson[1], and Marella Õis[2]

[1] Faculty of Science, Tallinn University of Technology,
21 Akadeemia Tee, 12618 Tallinn, Estonia
aben@cs.ioc.ee
[2] Ericsson Estonia AS, 9 Järvevana Tee, 11314 Tallinn, Estonia

Abstract. Stresses in several Prince Rupert's drops (PRDs) were measured with integrated photoelasticity. Measurements show that the surface layer of thickness about 0.5 mm of the head of a PRD is under high compression stresses reaching at the surface the value of about 500 MPa. That explains the extraordinary strength of the heads of PRD's. The external part of the tails of PRD's is also under high compressive stress reaching at the surface the value of 600 MPa. Internal part of the tail is under high tensile stresses. Stress distribution in the tail is used to explain catastrophic disintegration of the tail when it is cut.

Keywords: Prince Ruperet's drops · Photoelasticity · Glass
Fracture mechanics

1 Introduction

Prince Rupert's drops (PRDs) are made by dropping red hot blobs of molten soda-lime glass into cold water. On cooling down and solidifying the drops are of a tadpole shape (Fig. 1a). They have a bulbous head 5–15 mm in diameter and a tail with diameter in the range of 0.5 to 3 mm. PRD's have exceptional strength properties: the head of a PRD can withstand impact with a small hammer, but the tail can be broken with just finger pressure leading to catastrophic disintegration of the PRD. Disintegration of a PRD is accopanied by centrifugal scattering of the pieces of broken glass.

2 Experimental

Stresses in PRD's were measured with integrated photoelasticity [1] similarly to [2]. The PRD's were embedded in an immersion tank with immersion liquid, the refraction index of which matched that of the glass. On the basis of measurement data the stresses were determined using algorithms of integrated photoelasticity. Figure 1b shows integrated fringe pattern of the head of a PRD. Figure 1c shows radial distribution of the stresses both in the middle section of the head of a PRD and in the tail. Figure shows that the head of the PRD is surrounded by a layer of high compressive

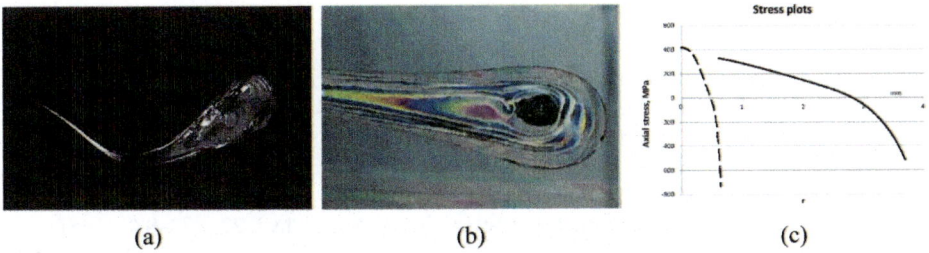

Fig. 1. (a) - Prince Rupert's drop; (b) - Fringe pattern of the head of the PRD; (c) - axial stress distribution in the cross section of the head (solid curve) and in the tail (dashed curve).

stresses. Thickness of this layer is about 0.7 mm. This compression stress layer explains the extraordinary strength of the head of a PRD [2].

3   Disintegration Mechanism of the Tail of the PRD

Chandrasekar and Chaudhri have carried out experimental investigation of the disintegration of the PRD's by using high-speed photography [3]. This paper explains several features of the disintegration mechanism. However, it does not explain why the pieces of broken glass are scattered centrifugally around the tail during the disintegration process.

Distribution of the longitudinal stress σ_z in the tail, shown in Fig. 1c is determined directly from the photo-elastic measurement data. It is known that in cylindrical glass objects with residual stress the so-called sum rule [4] is valid.

$$\sigma_z = \sigma_r + \sigma_\theta$$

Here σ_θ denotes the circumferential stress. Since on the surface of the tail $\sigma_r = 0$, at the tail surface we have $\sigma_\theta = \sigma_z$. Thus the external part of the tail is under high compression both in the longitudinal and circumferential directions.

Cherepanov and Esparragoza have [5] shown that if a spot of a material under high compression is freed from pressure, a selfsustaining fracture wave is created. In our case the fracture wave has two tensile stress components – the longitudinal stress σ_z and the circumferential stress σ_θ. The centrifugal scattering of the pieces of broken glass by disintegration of the tail of a PRD is evidently caused by the circumferential stresses. Thus the strange behaviour of PRD's is completely explained.

Acknowledgement. This work was supported by Estonian Research Council Grant No. IUT33/24.

References

1. Aben, H.: Integrated Photoelasticity. McGraw-Hill, New York (1979)
2. Aben, H., Anton, J., Õis, M., Viswanathan, K., Chandrasekar, S., Chaudhri, M.M.: On the extraoerdinary strength of Prince Rupert's drops 109 (2016)
3. Chandrasekar, S., Chaudhri, M.M.: The explosive disintegration of Prince Rupert's drops. Philos. Mag. **70**(6), 1195–1218 (1994)
4. O'Rourke, R.C.: Three-dimensional photoelasticity. J. Appl. Phys. **22**, 872–878 (1951)
5. Cherepanov, P.G., Esparragoza, I.E.: On self-sustaining fracture waves. Int. J. Fract. **144**, 197–202 (2007)

A Further Exploration on Loading Strain Rate

C. Huang[(⊠)]

School of ME&A, Wuhan Textile University,
Wuhan, People's Republic of China
ab6688us@yahoo.com

Abstract. The paper performs a series of SHPB experiments using the material of polymethyl methacrylate (PMMA). 8 lengths of cylindrical specimens are designed. The 8 length specimens can be divided into two groups based on two different diameters. Under the identical loading, which means the specimens are all subjected to the identical incident pulse, the failure strengths of the specimens do not show the well acknowledged strain rate sensitivity as the strain rates increase with the decrease of specimen lengths. It is concluded that the strain rate sensitivity of failure strength has to be further clarified as the sensitivity of loading strain rate that one can define it as the strain acceleration.

Keywords: SHPB experiment · Loading strain rate · Strain acceleration

1 Introduction

In order to explore the strain rate sensitivity of PMMA material, a series of SHPB experiments are carried out to examine the effect of strain rate on failure strength. In SHPB experiment, since SHPB loading strain rate is inversely proportional to specimen length, cylindrical specimens are designed with 8 lengths and with two diameters and can be divided into two groups based on the two different diameters. The first group has the length dimensions of 8 mm, 10 mm, 12 mm and 14 mm with the diameter of 8 mm and the second group has the lengths of 15 mm, 19 mm, 23 mm and 27 mm with the identical diameter of 15 mm as shown in Fig. 1. To ensure the accuracy of the SHPB experiments, each designed length is tested with ten specimens to obtain a more significant statistic mean value.

2 SHPB Experiment Results

The cylindrical specimens are made of quasi-brittle material PMMA with the characteristics of low density and transparency. The specimens are all set to be tested under identical SHPB loading condition. Dynamic responses of the designed specimens are investigated and compared to explore the effects of loading strain rates. Figure 2 shows the distribution of failure strengths and maximum loading strain rates for the specimens with the 8 lengths. In Fig. 2, comparing the 8 mm diameter specimens with the 15 mm diameter ones, the mean strengths with the increase of specimen lengths show the identical trend in the two diameter groups respectively and in general, the failure

© Springer International Publishing AG, part of Springer Nature 2019
E. E. Gdoutos (Ed.): ICTAEM 2018, SI 5, pp. 161–163, 2019.
https://doi.org/10.1007/978-3-319-91989-8_37

Fig. 1. Designed 8-length PMMA Specimens with two different diameters

strengths of the 15 mm diameter specimens are slightly larger than the corresponding ones in the 8 mm diameter. The experiment results indicate the transverse size effects of PMMA specimen. Another noticeable phenomenon shown in Fig. 2 is that the shortest specimens in the two diameter groups (i.e., the 8 mm and 15 mm length specimens) do not show the feature of apparent rate sensitivity of SHPB failure strengths in comparison with the other three length specimens. In terms of the SHPB experimental theory, SHPB loading strain rate is inversely proportional to specimen length. Under the action of identical incident pulse, short specimen subjects to higher strain rate loading than any other longer specimens (Fig. 2). The loading strain rates acting on the two shortest specimens should bring in the highest strengths in the two diameter groups, respectively. However, the specimens do not show the rate sensitivity of SHPB failure strengths, even the shortest 8 mm length specimens in the two groups are tested that they do not show the relatively higher failure strength in comparison with the other length specimens. On the other hand, the mean maximum strain rates of the 8 length specimens do show a clear increase trend with the reduction of specimen length in Fig. 2. The difference between the highest and lowest strain rates can reach more than 2.5 times. The failure strengths presented by the different length specimens can not show any features of the strain rate sensitivity. The concept of the strain rate sensitivity must be explored more again.

It is generally accepted that PMMA is a sensitive material to loading strain rate. As reported by many research papers, the failure strength of PMMA increases with the strain rate that the material is subjected to. However, in Fig. 2, under the identical loading condition the specimens do not show the rate sensitivity to the failure strengths as the rates increase with the decrease of specimen lengths. The observation indicates that under the identical loading, the specimens do not agree with the sensitivity of failure strength. There must be something misunderstood here. PMMA material can present the sensitivity only in one condition that as the loading rate is raised, the material can show the strong rate sensitivity of failure strength. The loading rate can affect the strain rate of deformation, but to more accurately distinguish that it directly results in the response of strain acceleration. Therefore, the concept of rate sensitivity should be clarified as the acceleration sensitivity to reflect the characteristic of dynamic response.

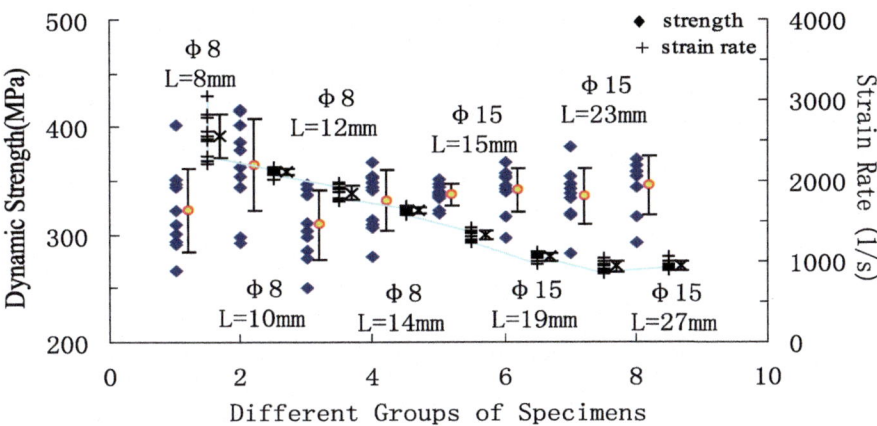

Fig. 2. Scatter strengths and max strain of 8 length PMMA specimens, the mean values, and the corresponding standard deviations

3   Conclusion

Based on the experimental investigation and analysis, it is observed that larger diameter specimens lead to relatively higher failure strength. The concept of rate sensitivity should be clarified as the acceleration sensitivity, so as to help more accurately characterize the dynamic fracture of brittle materials.

Acknowledgement. The research of the paper is under the support of National Scientific Foundation of China (Grant no. 11072181). The author is grateful for the financial support. The author is also particularly grateful for the support received from the Education Bureau of Hubei province (Grant no. D20101607).

Fracture

Determination of Mode I and II Adhesion Toughness of Monolayer Thin Films by Circular Blister Tests

Christopher M. Harvey[1], Simon Wang[1,2(✉)], Bo Yuan[1],
Rachel C. Thomson[3], and Gary W. Critchlow[3]

[1] Department of Aeronautical and Automotive Engineering,
Loughborough University, Loughborough, Leicestershire LE11 3TU, UK
[2] College of Mechanical and Equipment Engineering,
Hebei University of Engineering, Handan 056038, China
[3] Department of Materials Engineering, Loughborough University,
Loughborough, Leicestershire LE11 3TU, UK

Abstract. Mechanical models are developed to determine the mode I and II adhesion toughness of monolayer thin films using circular blister tests under the pressure load. The interface fracture of monolayer thin film blisters is mode I dominant for linear bending with small deflection while it is mode II dominant for membrane stretching with large deflection. By taking the advantage of the large mode mixity difference between these two limiting cases, the mode I and II adhesion toughness are determined in conjunction with a linear failure criterion. Thin films under membrane stretching have larger adhesion toughness than thicker films under bending. Experimental results demonstrate the validity of the method.

Keywords: Adhesion toughness · Circular blisters · Energy release rate
Interface fracture · Thin films

1 Introduction

Thin films can detach the substrates by buckling [1–5], pockets of energy concentration [6–8], etc.; therefore, the adhesion toughness of thin films is a major concern in engineering applications. Several experimental techniques have been developed to determine the film's adhesion toughness, such as peeling [9], scratching [10] and blister tests [11–13]. The blister tests are widely used in microelectronics and coating fields. The first blister test was reported by Dannenberg in 1961 [11], which was further developed by Jensen [12, 13]. Also, multiple theoretical models have been developed to correlate the adhesion toughness of thin films with the blister morphology [14, 15] that is induced by either a pressure load or a point load. Films, such as graphene, and substrates with various material properties and thickness are employed in the blister tests, hence the limits of membrane stretching with large deflections and linear bending with small deflections, and even transition characteristics are necessary in deriving the film's adhesion toughness. Furthermore, the adhesion toughness is influenced by the

© Springer International Publishing AG, part of Springer Nature 2019
E. E. Gdoutos (Ed.): ICTAEM 2018, SI 5, pp. 167–173, 2019.
https://doi.org/10.1007/978-3-319-91989-8_38

through-thickness shearing and film sliding [16]. The present work aims to develop mechanical models to determine the mode I and II adhesion toughness of thin films by using circular blister tests under a pressure load, then the mechanical models are validated with experimental results [17]. Also, the mechanical models for the circular blister test under a point load are developed in Ref. [18].

2   Analytical Mechanical Model for the Circular Blister Test with a Pressure Load

Figure 1 shows a circular blister under a pressure load p. The blister radius is R_B with B denoting the blister tip, and the central deflection is δ. The film thickness is h, which is assumed much smaller than the substrate thickness. Hence, the global deformation of the substrate is negligible. The film's Young's modulus is E and the Poisson's ratio is v.

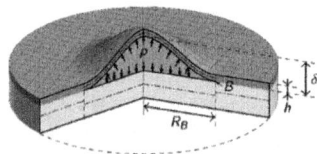

Fig. 1.   A circular blister test with a thin film under a pressure load on a thick substrate.

2.1   Linear Bending Mechanical Model for Small Deflection

For linear bending limit with small deflections, denoted by the subscript b, the crack tip loads [3–5], i.e. bending moment M_B (Nm m^{-1}), in-plane force N_B (N m^{-1}), and through-thickness shear force P_B (N m^{-1}) are

$$M_{Bb} = \frac{1}{8}pR_B^2, \; N_{Bb} = 0 \text{ and } P_{Bb} = \frac{1}{2}pR_B \qquad (1)$$

The mode I and II energy release rates (ERRs) based on 2D partition theories [16, 19–21] are,

$$G_I = 0.6227 \times \frac{1}{2}p\delta_b(1+\lambda)^2 \qquad (2)$$

$$G_{II} = 0.3773 \times \frac{1}{2}p\delta_b \qquad (3)$$

where δ_b is the blister central deflection,

$$\delta_b = \frac{3}{16}\frac{(1-v^2)pR_B^4}{Eh^3} \qquad (4)$$

and λ represents the through-thickness shear effect,

$$\lambda = \frac{P_{\text{Bb}}h}{1.0063 M_{\text{Bb}}} = \frac{4h}{1.0063 R_{\text{B}}} \tag{5}$$

The mode mixity ρ is readily obtained by Eqs. (2) and (3),

$$\rho = G_{\text{II}}/G_{\text{I}} = 0.6059 \big/ (1 + \lambda)^2 \tag{6}$$

Based on Eqs. (5) and (6), it is found the mode mixity approaches to pure mode I for large λ (or small R_{B}/h), and approaches to 0.6059 for very small λ. In addition, the total ERR can be given by combining Eqs. (2) and (3) and expressed by

$$G = G_{\text{J}} + G_{\text{S}} = G_{\text{J}}(1 + G_{\text{S}}/G_{\text{J}}) = G_{\text{J}}[1 + 0.6227\lambda(2 + \lambda)] \tag{7}$$

where $G_{\text{J}} = 1/(2p\delta_{\text{b}})$ is the ERR component from Jensen's work [3–5], which does not account for through-thickness shear; but G_{S} demonstrates the ERR component due to the crack tip through-thickness shear force P_{Bb}. It is seen that the through-thickness shear tends to decrease the mode mixity shown by Eq. (6) and consequently to reduce the adhesion toughness, as per Eq. (7).

2.2 Membrane Stretching Mechanical Model for Large Deflection

For membrane stretching limit with large deflections, denoted by the subscript m, the crack tip loads [3–5] in Eq. (1) become

$$M_{\text{Bm}} = \frac{h}{4} \frac{\left(Ehp^2 R_{\text{B}}^2\right)^{1/3}}{[3(1 - v^2)\phi(v)]^{1/2}}, \ N_{\text{Bm}} = \left(Ehp^2 R_{\text{B}}^2\right)^{1/3}\phi(v) \text{ and } P_{\text{Bm}} = \frac{1}{2}pR_{\text{B}} \tag{8}$$

where the parameter $\phi(v)$ is

$$\phi(v) = \frac{(1.078 + 0.636v)^{2/3}}{2[6(1 - v^2)]^{1/3}} \tag{9}$$

The blister central deflection becomes

$$\delta_{\text{m}} = 0.9635 \left[\frac{3(1 - v)}{7 - v}\right]^{1/3} \left(\frac{pR_{\text{B}}^4}{Eh}\right)^{1/3} \tag{10}$$

The mode I and II ERRs [16, 19–21] are

$$G_{\text{I}} = 0.6227 \times \frac{p\delta_{\text{m}}}{8} \frac{(0.7578 - 0.1429v)^2}{\phi(v)f(v)} \tag{11}$$

$$G_{\mathrm{II}} = 0.3773 \times \frac{p\delta_{\mathrm{m}}}{8} \frac{(1.400 + 0.2358v)^2}{\phi(v)f(v)} \tag{12}$$

Hence, the total ERR is

$$G = \left[\frac{1}{8\phi(v)} + \frac{(1 - v^2)\phi(v)^2}{2} \right] \frac{p\delta_{\mathrm{m}}}{f(v)} \tag{13}$$

and the mode mixity for membrane stretching with large deflection is

$$\rho = 0.6059 \left(\frac{1.400 + 0.2358v}{0.7578 - 0.1429v} \right)^2 \tag{14}$$

Equation (14) shows that ρ varies from 2.0680 for $v = 0$ to 2.9634 for $v = 0.5$, and remains constant during blister radial growth. The adhesion toughness G_{c} therefore remains constant with mode II dominant, and consequently it is expected to be larger than the adhesion toughness of films under linear bending.

Now, the linear failure criterion is used to derive the mode I and II adhesion toughness, which is considered as an accurate failure criterion for interfaces with low adhesion toughness [6, 7, 16, 22, 23]. For any given mode mixity ρ, the corresponding adhesion toughness G_{c} is

$$G_{\mathrm{c}} = \frac{(1 + \rho)G_{\mathrm{Ic}}G_{\mathrm{IIc}}}{\rho G_{\mathrm{Ic}} + G_{\mathrm{IIc}}} \tag{15}$$

Based on Eq. (15), by choosing two values of mode mixities, with their corresponding adhesion toughness, the mode I and II adhesion toughness can be readily obtained. But, when used in conjunction with linear bending model, more accurate predictions for G_{Ic} and G_{IIc} can be determined than from using just the linear bending model alone. Hence, by choosing ρ_{b} and ρ_{m}, with G_{cb} and G_{cm}, the mode I and II adhesion toughness can be obtained as

$$G_{\mathrm{Ic}} = \frac{G_{\mathrm{cb}}G_{\mathrm{cm}}(\rho_{\mathrm{b}} - \rho_{\mathrm{m}})}{\rho_{\mathrm{b}}G_{\mathrm{cb}}(1 + \rho_{\mathrm{m}}) - \rho_{\mathrm{m}}G_{\mathrm{cm}}(1 + \rho_{\mathrm{b}})} \tag{16}$$

$$G_{\mathrm{IIc}} = \frac{G_{\mathrm{cb}}G_{\mathrm{cm}}(\rho_{\mathrm{b}} - \rho_{\mathrm{m}})}{G_{\mathrm{cm}}(1 + \rho_{\mathrm{b}}) - G_{\mathrm{cb}}(1 + \rho_{\mathrm{m}})} \tag{17}$$

3 Experimental Validation

The mechanical models developed above are validated using the experimental results as presented in Ref. [17]. In the first experiment group, photoresist films with three different thickness h are blistered from the copper substrate with a thickness of 80 μm, under a pressure load. The film's Young's modulus is 3.6 GPa and the Poisson's ratio is 0.35. The Young's modulus of copper is about 128 GPa, which is much larger than that of the photoresist films. Therefore, the present thin film models are still applicable. Based on the comparison between the predicted and experimental adhesion toughness, it is found that $h = 10 \, \mu m$ corresponds to membrane stretching limit, and both $h = 31 \, \mu m$ and $h = 60 \, \mu m$ correlates to linear bending limit. The predictions of the adhesion toughness based on the analytical model are summarised in Table 1.

Table 1. Analytical predictions of the adhesion toughness for various film thickness.

Thickness (µm)	Mode mixity ($\rho = G_{II}/G_{I}$)	Measured adhesion Toughness (J m^{-2})	Mode I Toughness (J m^{-2})	Mode II Toughness (J m^{-2})
Photoresist/copper				
10	2.6583 Eq. (14)	0.3487 Eq. (13)		
31	0.5189 Eq. (6)	0.2827 Eq. (7) [0.2845 Eq. (15)]	0.2446 Eq. (16)	0.4152 Eq. (17)
60	0.4535 Eq. (6)	0.2805 Eq. (7)		
Photoresist-graphene/copper				
10	2.6583 Eq. (14)	0.4435 Eq. (13)		
31	0.5189 Eq. (6)	0.3711 Eq. (7) [0.3710 Eq. (15)]	0.3240 Eq. (16)	0.5149 Eq. (17)
60	0.4535 Eq. (6)	0.3664 Eq. (7)		

First, the mode mixities and the corresponding adhesion toughness at $R_{B} = 1530 \, \mu m$ for $h = 10 \, \mu m$ and $h = 60 \, \mu m$ are determined by Eqs. (13) and (14), and Eqs. (6) and (7), respectively. Then, the film's mode I and mode II adhesion toughness can be determined with the mode mixities and the corresponding adhesion toughness at for thickness 10 µm and 60 µm by Eqs. (16) and (17). Next, substituting G_{Ic}, G_{IIc} and ρ for thickness 31 µm into Eq. (15) gives the analytical $G_{c} = 0.2845$ J m^{-2}, which is in excellent agreement with the experimental result shown in Table 1.

In the second experiment group, a monolayer graphene is sandwiched between the photoresist films and the copper substrate. The thickness of the monolayer graphene is about 0.347 nm [16]. Even considering the Young's modulus of the graphene is about 1000 GPa, its effective thickness is still much smaller than the thickness of the photoresist films and it is therefore ignored in the present work. The addition of the graphene layer, however, changes the adhesion toughness at the interface. Following the same analytical procedure as that for the first group, the analytical predictions are summarised in Table 1. It is seen again the analytical adhesion toughness $G_{c} = 0.3710$ J m^{-2} for $h = 31 \, \mu m$ is in excellent agreement with the experimental result shown in Table 1.

4 Conclusions

This work shows the mechanical models for circular blister tests under a pressure load. The large mode mixity difference between the limits of membrane stretching and linear bending enables the mode I and II adhesion toughness of thin films to be accurately determined, which can be well validated with experimental results. For linear bending with small deflection, the interface fracture of thin film blisters is mode I dominant. The through-thickness shear force makes an extra contribution to the mode I ERR and decreases the mode mixity. The thicker the film is, the smaller the adhesion toughness is. For membrane stretching with large deflection, the interface fracture is mode II dominant. Membrane films consequently have larger adhesion toughness. Furthermore, the through-thickness shear force has no effect on the mode mixity which is only dependent on the Poisson's ratio. In addition, the mechanical models for circular blister tests under a point load are developed in Ref. [18].

References

1. Chai, H.: Three-dimensional fracture analysis of thin-film debonding. Int. J. Fract. **46**(4), 237–256 (1990)
2. Hutchinson, J.W., Thouless, M.D., Liniger, E.G.: Growth and configurational stability of circular, buckling-driven film delaminations. Acta Metall. Mater. **40**(2), 295–308 (1992)
3. Jensen, H.M., Sheinman, I.: Straight-sided, buckling-driven delamination of thin films at high stress levels. Int. J. Fract. **110**(4), 371–385 (2001)
4. Moon, M.W., Jensen, H.M., Hutchinson, J.W., Oh, K.H., Evans, A.G.: The characterization of telephone cord buckling of compressed thin films on substrates. J. Mech. Phys. Solids **50** (11), 2355–2377 (2002)
5. Jensen, H.M., Sheinman, I.: Numerical analysis of buckling-driven delamination. Int. J. Solids Struct. **39**(13–14), 3373–3386 (2002)
6. Wang, S., Harvey, C.M., Wang, B.: Room temperature spallation of a-alumina films grown by oxidation. Eng. Fract. Mech. **178**, 401–415 (2017)
7. Harvey, C.M., Wang, B., Wang, S.: Spallation of thin films driven by pockets of energy concentration. Theor. Appl. Fract. Mech. **92**, 1–12 (2017)
8. Yuan, B., Harvey, C.M., Thomson, R.C., Critchlow, G.W., Wang, S.: Telephone cord blisters of thin films driven by pockets of energy concentration. In review
9. Kendall, K.: Thin-film peeling - the elastic term. J. Phys. D Appl. Phys. **8**(13), 1449–1452 (1975)
10. Akono, A.T., Ulm, F.J.: An improved technique for characterizing the fracture toughness via scratch test experiments. Wear **313**(1–2), 117–124 (2014)
11. Dannenberg, H.: Measurement of adhesion by a blister method. J. Appl. Polym. Sci. **5**(14), 125–134 (1961)
12. Jensen, H.M.: Analysis of mode mixity in blister tests. Int. J. Fract. **94**(1), 79–88 (1998)
13. Jensen, H.M.: The blister test for interface toughness measurement. Eng. Fract. Mech. **40**(3), 475–486 (1991)
14. Malyshev, B.M., Salganik, R.L.: The strength of adhesive joints using the theory of cracks. Int. J. Fract. Mech. **1**(2), 114–128 (1965)
15. Wang, Y., Tong, L.: Closed-form formulas for adhesion energy of blister tests under pressure and point load. J. Adhes. **92**(3), 171–193 (2016)

16. Wood, J.D., Harvey, C.M., Wang, S.: Adhesion toughness of multilayer graphene films. Nat. Commun. **8**(1), 1952 (2017)
17. Cao, Z., Tao, L., Akinwande, D., Huang, R., Liechti, K.M.: Mixed-mode traction-separation relations between graphene and copper by blister tests. Int. J. Solids Struct. **84**, 147–159 (2016)
18. Harvey, C.M., Wang, S., Yuan, B., Thomson, R.C., Critchlow, G.W.: Determination of mode I and II adhesion toughness of monolayer thin films by circular blister tests. Theor. Appl. Fract. Mech. (In Press)
19. Wood, J.D., Harvey, C.M., Wang, S.: Partition of mixed-mode fractures in 2D elastic orthotropic laminated beams under general loading. Compos. Struct. **149**, 239–246 (2016)
20. Harvey, C.M., Wood, J.D., Wang, S., Watson, A.: A novel method for the partition of mixed-mode fractures in 2D elastic laminated unidirectional composite beams. Compos. Struct. **116**(1), 589–594 (2014)
21. Hutchinson, J.W., Suo, Z.: Mixed mode cracking in layered materials. Adv. Appl. Mech. **29**, 63–191 (1991)
22. Harvey, C.M., Wang, S.: Experimental assessment of mixed-mode partition theories. Compos. Struct. **94**(6), 2057–2067 (2012)
23. Harvey, C.M., Eplett, M.R., Wang, S.: Experimental assessment of mixed-mode partition theories for generally laminated composite beams. Compos. Struct. **124**(C), 10–18 (2015)

Semi-infinite Crack in Piece-Homogeneous Plane with Non-smooth Interface of Media

V. M. Nazarenko and A. L. Kipnis[(⊠)]

S.P. Timoshenko Institute of Mechanics of the National Academy
of Science of Ukraine, Kyiv, Ukraine
a.l.kipnis@gmail.com

Abstract. An exact solution of symmetric problem on the elastic equilibrium of piece-homogeneous isotropic plane with the interface of media in the form the sides of angle, which contains an interior semi-infinite loaded crack is constructed by the Wiener – Hopf method. The stress behavior near the corner point is investigated.

Keywords: Interface of media · Corner point · Semi-infinite crack
Wiener – Hopf method

1 Introduction

The investigation of the cracks contained in piece-homogeneous bodies in case of the non-smooth interface of media is the actual problem of fracture mechanics of composites. The problem which belongs to this class is solved in present work. The stress behavior near the corner point is studied.

2 The Problem Formulation

Let the piece-homogeneous isotropic elastic plane with the interface of media in the form of side angle which contains symmetric interior semi-infinite crack in one part is considered (see Fig. 1). The crack faces are under the normal pressure action which distributed according the law $-F/r^2$, $r \geq l$ ($F-$ is given force dimensionality positive constant).

The boundary conditions of corresponding symmetric problem of elasticity theory are as follows:

$$\theta = \alpha, \ \langle \sigma_\theta \rangle = \langle \tau_{r\theta} \rangle = 0, \ \langle u_\theta \rangle = \langle u_r \rangle = 0; \tag{1}$$

$$\theta = 0, \ \tau_{r\theta} = 0; \ \theta = \pi, \ \tau_{r\theta} = 0, \ u_\theta = 0;$$

$$\theta = 0, \ r < l, \ u_\theta = 0; \ \theta = 0, \ r > l, \ \sigma_\theta = -F/r^2. \tag{2}$$

$(0 \leq \theta \leq \pi; \ \langle a \rangle$ – jump of a).

The stress behavior analysis near the corner point is the aim of a work.

© Springer International Publishing AG, part of Springer Nature 2019
E. E. Gdoutos (Ed.): ICTAEM 2018, SI 5, pp. 174–177, 2019.
https://doi.org/10.1007/978-3-319-91989-8_39

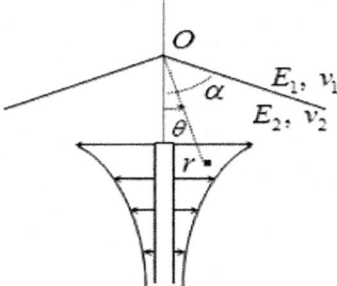

Fig. 1. Piece-homogeneous isotropic plane with symmetric interior semi-infinite crack

3 Wiener – Hopf Equation

For constructing an exact solution of the problem the Wiener – Hopf method combined with Mellin's integral transformation apart was used.

Applying the Mellin's transformation to equilibrium equations, strain compatibility condition, Hooke's law and to conditions (1) and (2) we obtain the Wiener – Hopf functional equation for unknown functions $\Phi^{\pm}(p)$

$$\Phi^-(p) + \frac{\sigma}{p-1} = \operatorname{ctg} p\pi G(p)\Phi^+(p), \tag{3}$$

$$G(p) = \frac{\widetilde{\Delta}(p)\sin p\pi}{2\Delta_0(p)\cos p\pi},$$

$$\Phi^-(p) = \int\limits_0^1 \sigma_\theta(\rho l, 0)\rho^p d\rho,$$

$$\Phi^+(p) = \frac{E_2}{2(1-v_2^2)}\int\limits_1^\infty \left.\frac{\partial u_\theta}{\partial r}\right|_{\substack{r=\rho^l \\ \theta=0}} \rho^p d\rho,$$

$$e = \frac{1+v_2}{1+v_1}e_o, \quad e_0 = \frac{E_1}{E_2}, \quad \ae_{1,2} = 3 - 4v_{1,2}, \quad \sigma = \frac{F}{l^2}$$

$(-\varepsilon_1 < \operatorname{Re} p < \varepsilon_2, \varepsilon_{1,2}$ – sufficiently small positive numbers; $\Delta(p), \Delta(p)$ – are known functions).

The solution of Eq. (3):

$$\Phi^+(p) = \frac{\sigma G^-(1)p}{K^-(1)(p-1)K^+(p)G^+(p)} \quad (\operatorname{Re} p < 0), \tag{4}$$

$$\Phi^-(p) = \frac{\sigma K^-(p)}{(p-1)G^-(p)} \left[\frac{G^-(1)}{K^-(1)} - \frac{G^-(p)}{K^-(p)}\right] (\mathrm{Re}\, p > 0),$$

$$G(p) = \frac{G^+(p)}{G^-(p)} \, (\mathrm{Re}\, p = 0), \quad \exp\left[\frac{1}{2\pi i} \int\limits_{-i\infty}^{i\infty} \frac{\ln G(z)}{z-p} dz\right] = \begin{cases} G^+(p), \, \mathrm{Re}\, p < 0 \\ G^-(p), \, \mathrm{Re}\, p > 0 \end{cases},$$

$$p\, \mathrm{ctg}\, p\pi = K^+(p)K^-(p), \quad K^{\pm}(p) = \frac{\Gamma(1 \mp p)}{\Gamma(1/2 \mp p)}$$

($\Gamma(z)-$ gamma-function).

4   Stress Behavior Analyses Near the Corner Point

Using (3), (4) and inverse Mellin transform we obtain:

$$\sigma_\theta(r,0) = \frac{1}{2\pi i} \int\limits_\gamma \frac{\sigma G^-(1)p\tilde{\Delta}(p)l^{p+1}}{2K^-(1)(p-1)\Delta_0(p)K^+(p)G^+(p)} r^{-p-1} dp, \qquad (5)$$

($\gamma-$ arbitrary straight line lying in the strip $-\varepsilon_1 < \mathrm{Re}\, p < 0$).

Integrand in (5) in the strip $-1 < \mathrm{Re}\, p < 0$ has a unique singular point – simple pole $p = -\lambda_0 - 1$, with λ_0- a unique root in strip $(-1; 0)$ of equation $\Delta_0(-\lambda - 1) = 0$. Values of λ_0 are given in the Table 1.

Table 1. Stress singularity degree ($\nu_1 = \nu_2 = 0,3$)

e_0	α°							
	30	45	60	75	105	120	135	150
2	−0,075	−0,112	−0,112	−0,086	−0,025	−0,054	−0,089	−0,117
3	−0,132	−0,180	−0,184	−0,127	−0,037	−0,081	−0,130	−0,173
5	−0,232	−0,258	−0,248	−0,167	−0,049	−0,104	−0,168	−0,228
10	−0,310	−0,332	−0,308	−0,203	−0,059	−0,124	−0,202	−0,278

Using this information about the singular point in integrand in (5) and applying residue theorem to integral (5) we get the next formula

$$\sigma_\theta(r,0) = \Sigma r^{\lambda_0} + f(r) \, (r \to 0),$$

$$\Sigma = \frac{\sqrt{\pi}\tilde{\Delta}(-\lambda_0 - 1)\Gamma(\lambda_0 + 3/2)G^-(1)}{4(\lambda_0 + 2)s_0\Gamma(\lambda_0 + 1)G^+(-\lambda_0 - 1)} \frac{F}{l^{\lambda_0 + 2}},$$

$$s_0 = \Delta_0'(-\lambda_0 - 1).$$

Here $f(r) \to 0$ when $r \to 0$; $\Delta'_0(p)$ – derivative of function $\Delta_0(p)$.
Similar formulas hold for $\sigma_\theta(r, \theta)$, $\tau_{r\theta}(r, \theta)$, $\sigma_r(r, \theta)$.

5 Results

The analysis of obtained results allows making following conclusions. The corner point O is the singular point of the boundary-value problem of elasticity theory considered and it is a stress-concentrator point with a degree character of singularity. Stresses tend to infinity when the point of domain tens to the point O.

The stress singularity degree at the point O depends on angle, Young's moduli ratio and Poisson's ratios of materials. Corresponding exponent is the unique root in strip (1; 0) of the transcendental equation. Some values of stress singularity degree are given in Table 1.

With increasing of angle α from $\pi/2$ to π stress concentration near the corner point first increases and then decreases. Young's moduli ratio values equals 2; 3; 5; 10 corresponds angles of maximal stresses concentration equals $57, 3°$; $54, 1°$; $46, 7°$; $42, 2°$.

With increasing of Young's moduli ratio, stress concentration near the corner grows and the acute angle of stresses maximal concentration increases and the obtuse angle – decreases.

References

1. Gakhov, F.D.: Boundary-Value Problems. Nauka, Moscow (1977). (in Russian)
2. Lavrentiev, M.A., Shabat, B.V.: The Theory of Complex Variable Methods. Nauka, Moscow (1973). (in Russian)
3. Nobl, B.: Applying the Wiener – Hopf method for solving partial derivatives differential equtions. Inostr. Lit., Moscow (1962). (in Russian)
4. Uflyand, Y.S.: Integral transformations in elasticity theory problems. Nauka, Leningrad (1967). (in Russian)

Probabilistic Assessment of the State of Welded Pipeline Elements and Pressure Vessels with Detected Corrosion-Erosion Defects

Alexey Milenin$^{(\boxtimes)}$, Elena Velikoivanenko, Galina Rozynka, and Nina Pivtorak

E.O. Paton Electric Welding Institute of NAS of Ukraine, 11, K. Malevich Str., Kyiv, Ukraine
asmilenin@ukr.net

Abstract. The main assumptions of the physical and mathematical models of the compatible development of stress-strain state and voids of ductile fracture of welded pipeline elements and pressure vessels with local corrosion-erosion metal losses in weld area have been given to determine the characteristic features of limit state of defective structures. Methods for probabilistic estimation of stressed state of the structures from the point of view of fracture susceptibility have been developed. They are based on the integration of the calculated field of principal stresses within the framework of Weibull statistics. For a correct quantitative assessment of state of critical structures based on complex analysis of the limit state of steel pipelines under internal pressure and different temperatures, functional dependences of Weibull coefficients on metal properties were obtained. Following the typical cases of exploitation damage in the main and technological pipeline elements the specific features of limit state under the different exploitation conditions have been investigated.

Keywords: Mathematical modeling · Probability of fracture · Ductile fracture Welded pipeline · Three-dimensional defect · Limiting state

1 Introduction

Ensuring the industrial safety of critical facilities in a number of cases is associated with the evaluation of the accumulated damage of structures and elements from the point of view of their bearing capacity decrease. In particular, these are typical pressure vessels, main pipelines and technological pipeline systems, for which high internal pressures are combined with the impact of corrosive environments. The resulting defects of local corrosion metal loss reduce load-carrying capacity of the structures because of stress concentration in the region of geometry anomaly and should be taken into account at the assessment of permissibility of their operation. To minimize the conservativeness of assessment of defected critical welded structures workability it is possible to use spatial simulation of their state along with corresponding criteria of limiting state of material and structure element under certain external influence.

Another approach, that allows taking into account the stochastic nature of structural material fracture in a complex stress-strain state, is the probabilistic technique for

© Springer International Publishing AG, part of Springer Nature 2019
E. E. Gdoutos (Ed.): ICTAEM 2018, SI 5, pp. 178–183, 2019.
https://doi.org/10.1007/978-3-319-91989-8_40

integral analysis of current state of specific structure element to fracture. The implementation of such techniques demands both determination of unique constants of material, those establish linkage between stress level and bearing capacity, and prediction of subcritical and critical fracture of material. The features of typical corroded pipeline element limiting state have been investigated within the limits of this work using the corresponding means for numerical simulation of postweld and current state of stresses and strains along with the subcritical and critical damage according to ductile fracture mechanism.

2 Mathematical Model

In general, the structural metal continuity violation under the action of external force loading is initiated in the local geometrical or physicochemical inhomogeneities of the material. Numerical analysis of such a state should take into account the difference in the spatial scales of the fracture processes (microlevel) and the kinetics of stress and strain fields under the action of operational loads (macrolevel). Various approaches to probabilistic analysis could be used depending on the purpose of the study and the type of structural damage. If the geometry of structure under consideration is not characterized with sharp concentrators (that is typical for pipeline elements), the following approach to determination of the probability of fracture could be used, i.e. probabilistic estimation is implemented by integrating the field of main stresses σ_1 within the framework of Weibull statistics for the description of fracture by the mechanism of the "weak link". In this case the probability p_i of fracture in small volume of material (where the stresses are close to homogeneous) can be described by the three-parameter Weibull function

$$p_i = 1 - \exp\left[-\left(\frac{\sigma_1 - A}{B}\right)^{\eta}\right], \ (\sigma_1 > A) \tag{1}$$

where A, B, η are the Weibull parameters.

Integration of stresses over the area of the weakest cross-section S allows estimating the integral fracture probability p:

$$p = 1 - \exp\left[-\int_S \left(\frac{\sigma_1 - A}{B}\right)^{\eta} \frac{dS}{S_0}\right], \ (\sigma_1 > A) \tag{2}$$

where S_0 is the material constant that characterizes the spatial scale of the transition of micro-damage to the macroscopic defect.

The accuracy of the quantitative estimation of probability according to (2) depends on the adequacy of Weibull coefficients used for a particular problem. Accordingly, the determination of the values of A, B, η can be based either on a statistical analysis of data from a large number of experiments on the fracture of an identical material, or on a numerical study of the limiting state of a structure.

The coefficient A characterizes the possibility of fracture nucleation at relatively low stresses. Theoretically, there is a nonzero probability of fracture initiation at stresses close to null ($A = 0$), but this approach is not reasonable for solving applied tasks. To reduce the complexity of the numerical study, it is assumed that the probabilistic nature of failure becomes apparent at stresses exceeding the plastic flow stress σ_{flow}, which depends on the strength properties of a particular material, namely $\sigma_{flow} = (\sigma_Y + \sigma_U)/2$. In turn, η value for the typical structural steel fracture is assumed to be 3–4. Thus, the aim of analyzing the stress fields in the structure in the limiting state is to determine the parameter B of Weibull distribution, which gives an opportunity to carry out the necessary quantitative assessment of the probability of emergency situation in pipeline defective part under exploitation condition.

As the aim of this work is to assess the probability of fracture of welded pipelines elements (pressure vessels) with detected corrosion-erosion wall thinning, the analysis of numerically predicted stress fields at different stages of fracture has been carried out for determination of dependence of Weibull coefficient B on the material properties (such as yield stress or stain hardening coefficient). As it was stated above, due to actual absence of sharp stress concentration (cracks, sharp bends, pipeline branches, etc.) the main mechanism of fracture is ductile one, which consists in nucleation, growth and merging of distributed microscopic voids. And possible presence of voids of ductile fracture with volume concentration f should be considered at development of corresponding mathematical models.

Taking into account major mechanism of deformation, the increment of the strain tensor components could be present as superposition of the next summand:

$$d\varepsilon_{ij} = d\varepsilon_{ij}^e + d\varepsilon_{ij}^p + d\varepsilon_{ij}^c + \delta_{ij}(d\varepsilon_T + df/3) \tag{3}$$

where indexes e, p, c, T corresponds to the components of the strain tensor increment due to elastic deformation mechanism, plastic deformations, creep and the kinetics of temperature field, respectively, $i, j = r, \beta, z$ (see Fig. 1).

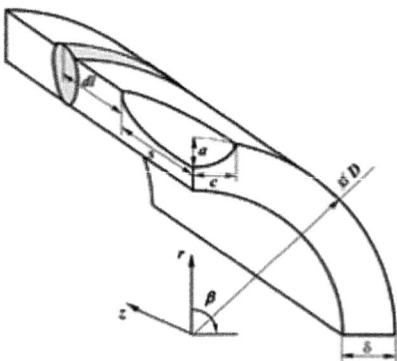

Fig. 1. Scheme of welded pipeline element with corrosion-erosion defect of wall thinning in cylindrical coordinate system.

Analysis of the stress-strain state of the structure with developed ductile fracture porosity was carried out on the basis of numerical solution of the boundary-value problem for continuum with nonuniform bearing cross-section by tracing the elastic-plastic deformations and the ductile fracture void concentration in structure loading to the limit state within the framework of finite element description. The relationship between stresses and deformations was determined by Hooke's law and the associated law of plastic flow, proceeding from the following relationships:

$$\Delta\varepsilon_{ij} = \Psi\left(\sigma_{ij} - \delta_{ij}\sigma_m + \Omega\Delta t\right) + \delta_{ij}\left(K\sigma_m + \Delta\varepsilon_T + \frac{\Delta f}{3}\right) - \frac{1}{2G}\left(\sigma_{ij} - \delta_{ij}\sigma_m\right)^* + (k\sigma_m)^*,$$

(4)

where δ_{ij} is Kronecker symbol, $K = (1-2v)/E$ is the modulus of dilatation, $G = 0.5E/(1 + v)$ is the shear modulus, E is Young modulus, v is Poison's ratio, symbol "*" refers to the variable of the previous tracing step, Ω is the creep function, Ψ is the state function of the material, which is determined with the level of plastic deformation according to plastic flow surface Φ, that was considered in accordance with Gurson-Tvergaard-Needleman model [1, 2]:

$$\Phi = \left(\frac{\sigma_i}{\sigma_Y}\right)^2 - (q_3 f^*)^2 + 2q_1 f^* \cosh\left(q_2 \frac{3\sigma_m}{2\sigma_Y}\right) - 1$$

(5)

where q_1, q_2, q_3 are the constants, f^* is the equivalent concentration of voids, σ_m is the average value of the normal components of the stress tensor σ_{ij}.

As criterion of the local nucleation of ductile fracture in the initially undamaged material the modified Johnson-Cook law [3] has been chosen, the growth of porosity due to plastic flow and creep was determined with of Rice-Tracy law [4]. Numerical criterion of brittle-ductile fracture was used as a condition of macroscopic fracture nucleation (i.e. the loss of bearing capacity of specific finite element). It consists in the fulfillment of one of three conditions [5]:

$$\begin{cases} \left(\Psi - \frac{1}{2G}\right)KP \geq \frac{\varepsilon_f - \varepsilon_p^*}{1.5\sigma_i} \approx \frac{\varepsilon_f - \varepsilon_p^*}{1.5\sigma_s(\varepsilon^p, \varepsilon^c, T)}; \\ f^* \to f_d^* = \frac{1}{q_1}\exp\left(-\frac{3q_2\sigma_m}{2\sigma_Y}\right); \\ \frac{\sigma_1}{1-2f/3} > S_k, \end{cases}$$

(6)

where S_K is the microcleavage stress, ε_f is the critical strain (material deformability).

3   Results and Discussion

As an example of this technique application for the fracture probability analysis, the case of a straight-line welded pipe element (steel X80, $\sigma_Y = 560$ MPa, $\sigma_U = 630$ MPa, $v = 0.3$, $E = 205$ GPa, $\delta = 39$ mm, $D = 800$ mm) with an external semielliptical surface defect has been considered. For correct determination of Weibull parameter B, it has been carried out a complex of calculations of limiting pressure, which the pipe with external surface defects of various sizes can withstand. Limiting state of a defective structure is characterized by two stages of integrity violation, namely nucleation of microvoids with a developed plastic flow of metal (see Fig. 2) and the beginning of finite elements fracture (formation of macrodefects). In the first case, the probability of an emergency situation is assumed to be equal the minimum possible for typical laboratory tests of fracture specimens ($p = 0.01$–0.05), in the second case the condition of the structure is unacceptable ($p = 0.95$). It was shown [6] that first approach is more conservative and less sensitive to inaccuracy of input data, therefore, it could be recommended for application.

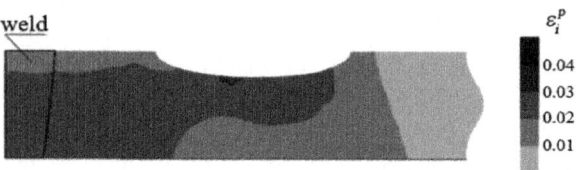

Fig. 2. Distribution of plastic strains in cross-section of pipeline with an external defect of the local wall thinning at origin of macroscopic fracture caused with internal pressure.

Analysis of the results of numerical prediction of limiting state in defected pipeline of different materials allowed deriving the dependences of parameter B on different characteristics of material, such as yield stress σ_Y (a), and stain hardening coefficient m (see Fig. 3).

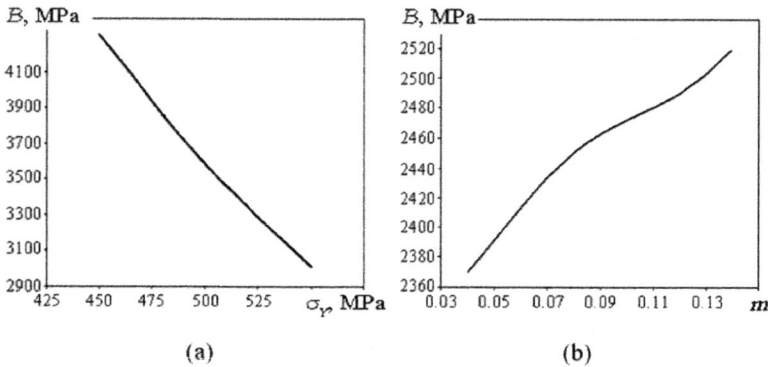

(a)   (b)

Fig. 3. Dependence of conservative value of Weibull parameter B on yield stress σ_Y (a) and stain hardening coefficient m (b) of material of defective pipeline section.

In should be emphasized, that there is no significant effect of size of the defect (or pipe) on the value of this coefficient, so it could be concluded that the parameter B is a characteristic of mechanical properties of the pipeline material, and the geometric and scale factors are correctly described by the integral functional (2). It allows to assess the value of failure probability for different cases of pipeline defectiveness (see Fig. 4), including corresponding stages of risk-analysis of complicated technological and industrial systems.

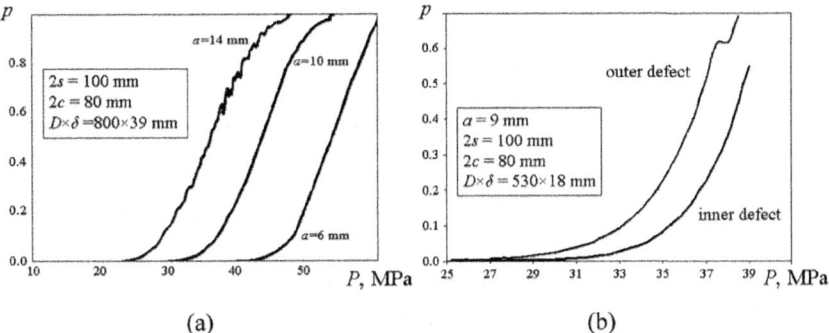

(a)   (b)

Fig. 4. Influence of defect depth a (a) and surface of its formation (b) on fracture probability p of pipeline element under internal pressure P of different values

References

1. Tvergaard, V.: Material failure by void growth to coalescence. Adv. Appl. Mech. **27**, 83–151 (1990)
2. Xue, L.: Damage accumulation and fracture initiation in uncracked ductile solids subject to triaxial loading. Int. J. Solids Struct. **44**(16), 5163–5181 (2007)
3. Velikoivanenko, E., Milenin, A., Rozynka, G., Pivtorak, N.: Evaluation of operability of the main pipeline with local wall thinning at repair by arc surfacing. Paton Weld. J. **1**, 18–23 (2015)
4. Karzov, G., Margolin, B., Shvetsova, V.: Physical-Mechanical Modeling of Fracture Processes. Politekhnika, St. Petersburg (1993)
5. Milenin, A., Velikoivanenko, E., Rozynka, G., Pivtorak, N.: Methodology of numerical prediction of serviceability of pipeline elements with corrosion-erosion defects under conditions of high-temperature operation. Tech. Diagn. Non-Destructive Test. **4**, 7–13 (2017)
6. Velikoivanenko, E., Milenin, A., Rozynka, G., Pivtorak, N.: Probabilistic estimate of the condition of pipeline elements with detected thinning defects. Tech. Diagn. Non-Destructive Test. **2**, 12–18 (2014)

Creep Fracture Ductility of Cobalt-Based Superalloys

Marie Kvapilova[1,2(✉)], Petr Kral[1,2], Jiri Dvorak[1,2],
and Vaclav Sklenicka[1,2]

[1] Institute of Physics of Materials, Academy of Science of the Czech Republic,
Žižkova 22, 61665 Brno, Czech Republic
kvapilova@ipm.cz
[2] CEITEC IPM, Institute of Physics of Materials, Academy of Science
of the Czech Republic, Žižkova 22, 61665 Brno, Czech Republic

Abstract. ·The challenge and demands of environmental protection and energy saving have been more and more serious tasks in some processing technologies of glass industry. Significant effort is being carried out for the development of advanced technology for precision casting of spinner discs for glass industry (glass wool) and high temperature applications up to 1050 °C. In this paper a study creep fracture processes of two different cobalt-based superalloys Co Stelit and Co Ursa Stelit was investigated. Constant load uniaxial stress creep tests in tension were carried out at three different testing temperatures of 900 °C, 950 and 1000 °C and at the applied stress ranging from 40 to 200 MPa. A mutual comparison of the creep characteristics of the investigated alloys under comparable creep loading conditions shows that the cobalt based superalloy Co Stelit exhibits more brittle character of fracture than superalloy Co Ursa Stelit. By contrast, the values of time to fracture (creep life) are longer for the superalloy Co Stelit in comparison to those for superalloy Co Ursa Stelit. The analyses of creep data indicate that creep behaviour of the superalloys under investigation obey both Monkman-Grant and its modified version. The creep fracture modes of these superalloys were determined using empirical formulas based on a continuum damage mechanics approach. The values of damage tolerance factor λ correspond to mixture of transgranular and intergranular type of fracture of both superalloys.

Keywords: Cobalt based superalloys · Creep fracture mode
Damage tolerance factor

1 Introduction

Recently, a new type of Co superalloy named Co-Ursa Stelit has been developed for spinner discs in glass industry with its improved dimensional stability at high temperatures. Therefore, the present study was initiated to provide an experimental investigation of new Co-superalloy enabling a comparative evaluation of its creep properties and fracture modes with an older superalloy Co Stelit. The damage processes in creep often limit the lives of components operated at high temperatures. In this study the creep deformation and fracture characteristic of both superalloys will be analysed

© Springer International Publishing AG, part of Springer Nature 2019
E. E. Gdoutos (Ed.): ICTAEM 2018, SI 5, pp. 184–189, 2019.
https://doi.org/10.1007/978-3-319-91989-8_41

by linking them to the identified creep degradation mechanisms, in terms of empirical formulas for fracture time assessment [1, 2], and by the creep damage tolerance factor λ [3] as an important results of the continuum damage mechanics (CDM) approach.

2 Experimental Materials and Procedures

Both cobalt based superalloys resistant to high-temperature oxidation used for this study were received in the state after casting with no following heat treatment. They were cast by a foundry company PBS Velká Bites a.s., Czech Republic, using conical ingots with the minimum diameter of 13 mm, the maximum diameter of 18 mm and the length of 90 mm. Their chemical compositions (in wt.%) of are shown in Table 1. Both alloys were investigated in their as-cast states. Standard uniaxial tensile constant load creep tests were carried out using cylindrical creep specimens with the gauge diameter of 3.5 mm and the gauge length of 50 mm. The creep tests were performed at applied stresses ranging from 40 to 200 MPa and at three different temperatures of 900 °C, 950 °C and 1000 °C in argon atmosphere. The creep elongations were continuously measured using a linear variable differential transducer, recorded digitally and computer processed. All creep specimens were run up to the final fracture.

Table 1. The chemical composition of studied alloys (in wt.%).

	Composition [wt.%]									
	Ni	Co	C	W	Cr	Fe	Ta	Nb	Si	Mn
Stelit	10.3	Bal.	0.75	8.53	31.1	0.5	x	2.29	0.45	0.33
Ursa	23	Bal	0.6	7	29.3	4.9	2.3	x	1.0	0.22

3 Results and Discussion

The representative standard creep curves for cobalt-based alloys at testing temperature 900 °C are shown in Fig. 1. In Fig. 1a the time dependences of creep strain for both alloys are shown. These standard creep curves can be easily replotted to the time dependences of the creep rate that clearly indicate each stage of creep curves (Fig. 1b). The creep curves for others temperatures and applied stress have the same tendency as those in Fig. 1. From inspection of Fig. 1b ensures that the values of the minimum creep rates are usually smaller for the superalloys Co Stelit. By contrast, the superalloy Co Ursa Stelit exhibits shorter time to fracture and higher creep fracture elongation (Fig. 1a). Figure 2 illustrates creep fracture elongation for both alloys Co Ursa Stelit and Co Stelit at three different temperatures within the selected interval of applied stress. The fracture elongation of crept specimens could be divided in two levels. The level of lower values of creep deformation belongs mostly Co Stelit specimens. The level of higher values of creep elongation involving Co Ursa Stelit alloy specimens only.

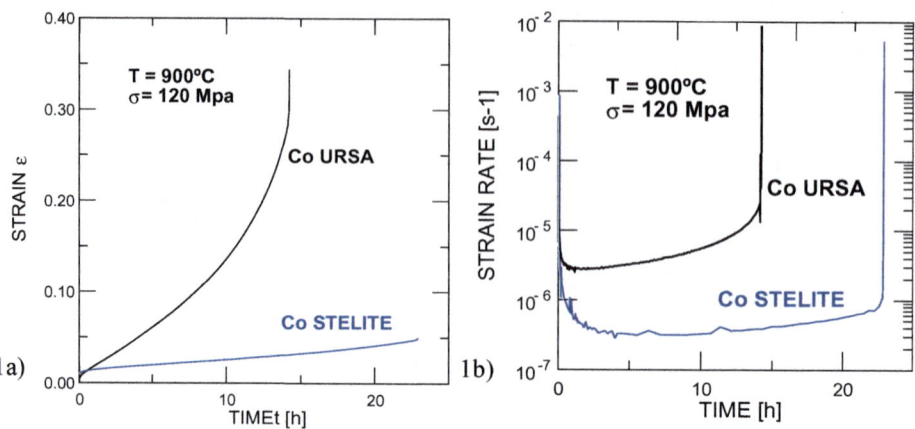

Fig. 1. Time dependences of (a) creep strain, and (b) creep rate for alloys Co Ursa Stelit and Co Stelit

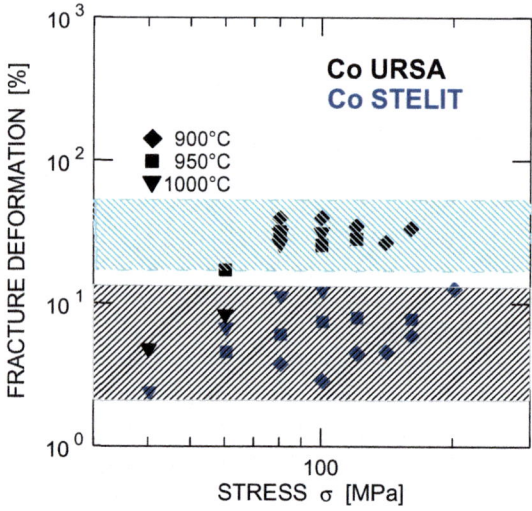

Fig. 2. Creep fracture deformation for both alloys Co Ursa Stelit and Co Stelit.

The relationship between the minimum creep rate $\dot{\varepsilon}_m$ and the time to fracture t_f is given by the well known empirical Monkman-Grant relationship [1]:

$$(\dot{\varepsilon}_m)^{\alpha}t_f = C_{MG}, \tag{1}$$

where α and C_{MG} are constants. The Monkman-Grant (MG) equation was modified by Dobeš and Milička [2] to Eq. (2)

$$(\dot{\varepsilon}_m)^\alpha t_f / \varepsilon_f = C_{MMG}, \tag{2}$$

where C_{MMG} is the modified constant and ε_f is the fracture strain. The MG relationship respective its modified MG version are presented in Figs. 3a and b. The linear dependences of the time to fracture on the minimum creep rate for both superalloys indicate that fracture process is deformation controlled under chosen testing conditions.

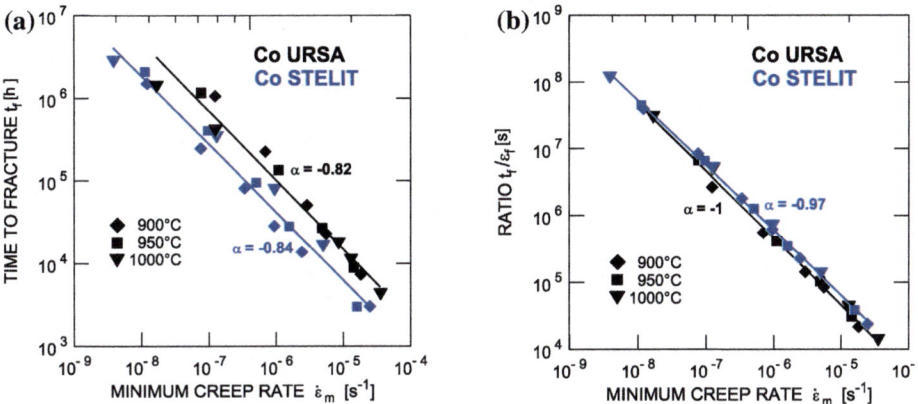

Fig. 3. Double logarithmic plots of (a) Monkmant-Grant relationship (b) modified Monkmant-Grant relationship

Based on a continuum damage mechanics (CDM) approach, the creep damage tolerance factor λ can be defined according Eq. (3)

$$\lambda = \varepsilon_f / \dot{\varepsilon}_m t_f = 1/C_{MMG}. \tag{3}$$

The creep damage tolerance factor λ can be used to assess the creep damage mode [3] and can indicate susceptibility of materials to localized cracking at strain concentrations. The parameter λ can also be suggested as a relevant measure of creep ductility. The dependences of parameter λ on the time to fracture are in Fig. 4. The parameter λ is not constant and exhibits two regions inherent to type of respective Co superalloys. The lower values of parameter λ for Co Stelit alloy indicate lower ductility of this alloy in comparison with alloy Co Ursa Stelit, even if the Co Stelit specimens exhibit longer time to fracture and lower values of minimum creep rate. According to Ashby and Dyson [3] the lower the value of parameter λ corresponds to more brittle mode of fracture of material. The values of parameter λ for alloy Co Stelit are about 2, which indicate that damage is due to the cavity growth resulting from the combined effect of power–law and diffusion creep (Fig. 5). The higher values of λ for Co Ursa Stelit show that the material can withstand strain concentration without local cracking; the fracture is more ductile.

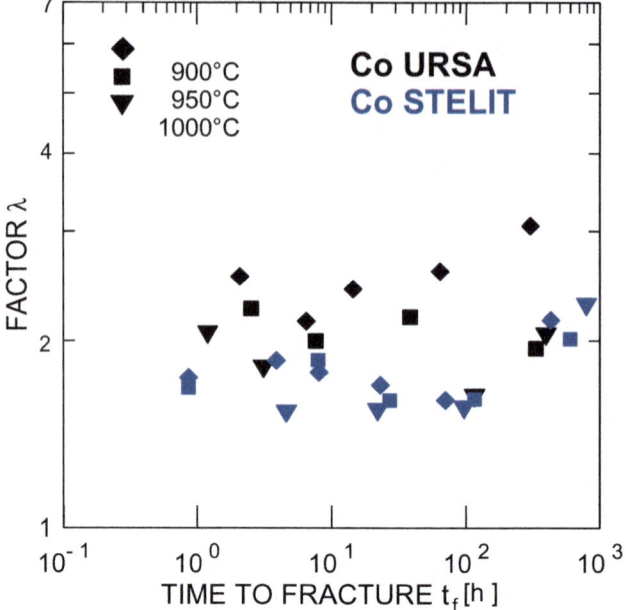

Fig. 4. The dependences of parameter λ on the time to fracture

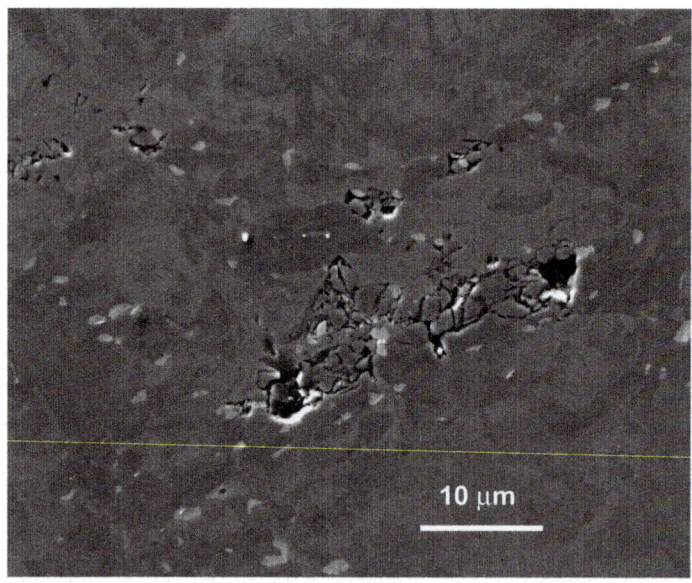

Fig. 5. Microstructure of Co Ursa Stelit after creep at 900 °C and 80 MPa.

A question naturally arises what is the reason for such reduction of creep ductility of Co Stelit superalloy. Besides different kinetics of creep cavitation and microcracks formation and growth in both superalloys we cannot excluded that the additional reason for decreasing of creep ductility of Co Stelit superalloy can be connected with presence of morphologically complicated shapes of plate-like particles and an eutectic inter-dendritic skeletons of primary carbides based on C, Cr, W and Co [4, 5]. The systematic quantitative metallographic investigation of the link between creep ductility and occurrence of secondary phases are in progress.

4   Conclusions

Creep damage processes in two Co based superalloys were studied at three temperatures 900 °C, 950 °C and 1000 °C within the interval of applied stress from 40 to 300 MPa. It was shown that the Co Stelit superalloy exhibits considerably lower ductility than the Co Ursa Stelit superalloy even if Co Stelit alloy has longer times to fracture. It should be explained the different mode of fracture behaviour and/or occurrence of different type and morphology of secondary strengthening phases.

Acknowledgements. We acknowledge financial support for this work provided by the Ministry of Industry and Trade of the Czech Republic within the framework programme MPO CR Trio under the Project FV 10699. We are grateful to Ing. Bozena Podhorna and Dr. Jiri Zyka of UJP PRAHA, a.s., Czech Republic, for stimulating discussions.

References

1. Monkman, F.C., Grant, N.J.: An empirical relationship between rupture life and minimum creep rate in creep-rupture tests. Proc. ASTM **56**, 593–620 (1956)
2. Dobeš, F., Milička, K.: The relation between minimum creep. Rate and time to fracture. Met. Sci. **10**, 382–384 (1976)
3. Ashby, M.F., Dyson, B.F.: Advances if Fracture Research. In: Valuri, S.R., Taplin, D.M.R., Rao, P.R., Knott, J.F., Dubey, R. (eds.) Pergamon Press, Oxford (1984)
4. Podhorna, B., Andrsova, I., Dobrovska, J., Vodarek, V., Hrbacek, K.: Structure stability of Ni/base and Co/base alloys. Mat. Sci. For. **782**, 431 (2014)
5. Dvorak, J., Kvapilova, M., Kucharova, M., Hrbacek, K., Kral, P., Sklenicka, V.: Creep properties of cast superalloys for application in glass industry. In: Kern, T.-U., Di Gianfrancesco, A. (eds.) Proceedings of the 14th Conference on Creep & Fracture (ECCC 2017), Paper # ID 54. The European Creep Collaborative Committee, Düsseldorf (2017). ISBN 978-3-514-00832-8

Fracture of Composite Material at Compression Along Two Parallel Cracks

Mykhailo Dovzhyk[1](✉), Vyacheslav Bogdanov[2],
and Vladimir Nazarenko[1,2]

[1] S.P. Timoshenko Institute of Mechanics, National Academy of Sciences
of Ukraine, Nesterov Str. 3, Kyiv 01057, Ukraine
Dovzhyk.M.V@ukr.net
[2] National Academy of Sciences of Ukraine,
Volodymyrska 54, Kyiv 01030, Ukraine

Abstract. Nonclassical problem of fracture mechanics for two parallel cracks under the action of compressive loads, directed along cracks were investigated. The axisymmetrical problem for penny-shaped crack is considered. There are two approaches that are used to investigate such problems "beam approximation" and three-dimensional linearized theory of stability of deformable bodies for finite and small subcritical strains. Within the limits of the offered in second approach the problem is reduced to the solution of system of integral equations Fredholm with a side condition. Using the Bubnov-Galerkin method and numerically analytic technique, the problem was reduced to system of linear equations. As an example numerical research for a composite material was conducted. Critical loads are obtained for small and large distance between cracks. Results for the composite materials behavior are also present and discussed.

Keywords: Composite materials · Compression along two parallel cracks
Stress intensity factors

1 Compression Along Cracks

Fracture of material at compression along cracks is one of the nonclassical problems of fracture mechanics. In this case, the classical approaches of fracture mechanics such as Griffiths-Irwin don't work. Currently, there are two approaches that are used to investigate such problems [1]. The first of them is the use of approximate calculation schemes and approximate theories [2]. Within the framework of this approach, has the greatest application the "beam approximation", when the part between the crack and the free surface (between the cracks) is replaced by a thin-walled element: beam, plate or shell, which are investigated in the framework of the applied theory of stability of thin systems. However, this method has significant drawbacks: it is necessary to carry out separate investigations to determine the possibilities of its application depending on the distance between the cracks, but even having determined this distance, there remains the question of choosing the conditions for fixing of thin-walled element. The second approach is based on the basic relationships and methods of the

© Springer International Publishing AG, part of Springer Nature 2019
E. E. Gdoutos (Ed.): ICTAEM 2018, SI 5, pp. 190–193, 2019.
https://doi.org/10.1007/978-3-319-91989-8_42

three-dimensional linearized theory of stability of deformable bodies for finite and small subcritical strains [3]. At the same time, the destruction process is identified as the moment of local stability loss within the framework of a rigorous linearized theory of elasticity.

1.1   Problem Formulation

We considered a composite material containing two parallel penny-shaped crack of radius a which is situated in the plane $x_3 = 0$ and $x_3 = -2h$ with center on Ox_3. The initial stresses that operated along a crack correspond to biaxial uniform compression and defined from [4].

Within the limits of the offered A.N. Guz's approach the problem is reduced to the solution of system of integral equations Fredholm with a side condition

$$f(\xi) + \frac{1}{\pi k} \int_0^1 M_1(\xi, \eta) f(\eta) d\eta + \frac{1}{\pi k} \int_0^1 N_1(\xi, \eta) g(\eta) d\eta = 0,$$

$$g(\xi) + \frac{1}{\pi k} \int_0^1 M_2(\xi, \eta) g(\eta) d\eta + \frac{1}{\pi k} \int_0^1 N_2(\xi, \eta) f(\eta) d\eta + \tilde{C}_1 = 0, \tag{1}$$

$$\int_0^1 g(\xi) d\xi = 0 \qquad (0 \le \xi \le 1, 0 \le \eta \le 1),$$

$$f(\xi) \equiv \varphi(a\xi), \quad g(\xi) \equiv \psi(a\xi).$$

1.2   Exploratory Procedure

We used the procedure on the basis of a method Bubnov-Galorkin for solve integral equations (1) and search of critical shortening and stress. As system of coordinate functions power functions were used.

$$f(x) = \sum_{i=0}^{N} F_i x^i, \ g(x) = \sum_{i=0}^{N} G_i x^i. \tag{2}$$

Unlike the previous works [5, 6] where after substitution of coordinate functions (2) to system the numerical integration was executed. Here the procedure which allows analytically to calculate integrals for the chosen system of coordinate functions using a package of symbolic computations is used. It has allowed to achieve at the further numerical calculations higher exactitude of evaluations at the expense of a numerical integration lapse exclusion. For acceleration of integrals solutions the recurrence relations were used.

Using a method offered in [7] Fredholm integral equations (1) have been transformed to system of the equations with corresponding factors F_{1ji}, G_{1ji} and variables F_i, G_i, $\tilde{C}_1$, $i,j \in [0, N]$.

$$\sum_{i=0}^{N} F_i F_{1ji} + \sum_{i=0}^{N} G_i G_{1ji} = 0;$$

$$\sum_{i=0}^{N} F_i F_{2ji} + \sum_{i=0}^{N} G_i G_{2ji} + \tilde{C}_1 = 0; \qquad (3)$$

$$\sum_{i=0}^{N} \frac{1}{i+1} G_i = 0, \quad 0 \le j \le N.$$

1.3   Results

As an example the task for two parallel penny-shaped crack in laminate composite with isotropic layers is conducted.

In macrovolumes such composite may be considered transversely-isotropic medium. In the case considered, cracks are located in plane $x_3 = const$, parallel to interface boundary of layers. Dependence of critical dimensionless compressive stress σ_{11}^0/E on ratio of the dimensionless distance between a cracks $\beta = h/a$ are given in Figs. 1 and 2 and Table 1 (for $v = v_1 = v_2$, fiber concentration c_1 is 0.7 and fiber aspect ratio is 10).

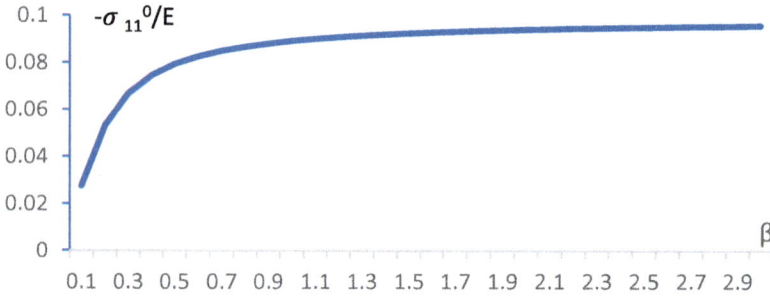

Fig. 1. Dependence of critical dimensionless compressive stress on ratio of the dimensionless distance between a cracks for big distance.

Critical values of σ_{11}^0/E at $\beta \to \infty$ go asymptotically to values 0.0965, which are equal to respective critical values at surface instability of half-space. For a thin plate critical stress can be found as $\sigma_{cr} = A_{cr}\beta^2$ and for small dimensionless distance between a cracks coefficient A found in Table 1.

Critical compressive stress were obtained for composite materials for large and small distances between the cracks. Analysis of the results allowed to determine the conditions of applicability of the "beam approximation". Beam approximation good work for small distance between the crack and the free surface (when $\beta < 0.01$

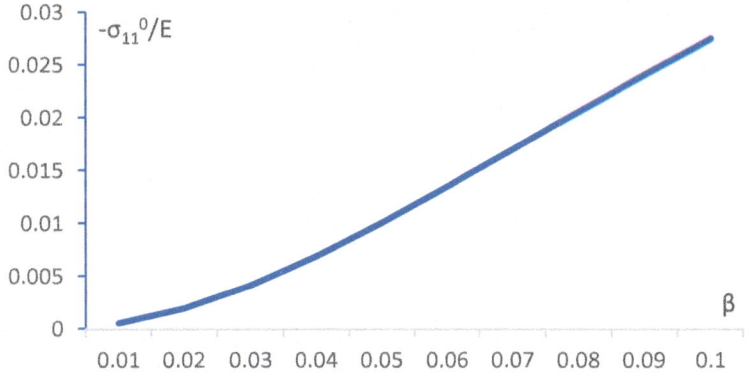

Fig. 2. Dependence of critical dimensionless compressive stress on ratio of the dimensionless distance between a cracks for small distance.

Table 1. Critical compressive stress for small dimensionless distance between a cracks.

β	Critical compressive stress (σ^0_{11}/E)	A
$1 \cdot 10^{-2}$	$-5.01367 \cdot 10^{-4}$	-5.01367
$1 \cdot 10^{-3}$	$-5.16309 \cdot 10^{-6}$	-5.16309
$1 \cdot 10^{-4}$	$-5.15527 \cdot 10^{-8}$	-5.15527
$1 \cdot 10^{-5}$	$-5.14160 \cdot 10^{-10}$	-5.14160
$1 \cdot 10^{-6}$	$-5.13770 \cdot 10^{-12}$	-5.13770
$1 \cdot 10^{-9}$	$-5.13672 \cdot 10^{-18}$	-5.13672

computing error less than 1%) and bad work in else case (when $\beta > 0.1$ computing error more than 5%).

References

1. Guz, A.N.: Establishing the foundations of the mechanics of fracture of materials compressed along cracks (review). Int. Appl. Mech. **50**, 1–57 (2014)
2. Obreimoff, I.W.: The splitting strength of mica. Proc. Roy. Soc. Lond. **127**, 290–297 (1930)
3. Guz, A.N.: On one criterion of fracture of solids in compression along the cracks. Plane problem. Docl. Akad. Nauk SSSR. **259**, 1315–1318 (1980). (in Russian)
4. Guz, A.N., Nazarenko, V.M.: Mechanics of fracture of materials in compression along the cracks (review). Highly elastic materials. Sov. Appl. Mech. **25**(9), 3–32 (1989)
5. Guz, A.N., Knukh, V.I., Nazarenko, V.M.: Compressive failure of materials with two parallel cracks: small and large deformation. Theor. Appl. Fract. Mech. **11**, 213–223 (1989)
6. Bohdanov, V.L.: Mutual influence of two parallel coaxial cracks in a composite material with initial stresses. Mater. Sci. **44**, 530–540 (2008)
7. Guz, A.N., Dovzhik, M.V., Nazarenko, V.M.: Fracture of a material compressed along a crack located at a short distance from the free surface. Int. Appl. Mech. **47**(6), 627–635 (2011)

Rate-Dependent Crack Propagation in Polyethylene

Martin Kroon[(⊠)]

Linnaeus University, 22000 Växjö, Sweden
martin.kroon@lnu.se

Abstract. Crack growth in semi-crystalline polymers, represented by poly-ethylene, is considered. The material considered comes in plates that had been created through an injection-molding process. Hence, the material was taken to be orthotropic. Material directions were identified as MD: molding direction, CD: transverse direction, TD: thickness direction. Uniaxial tensile testing was performed in order to establish the direction-specific elastic-plastic behaviour of the polymer. In addition, the fracture mechanics properties of the material was determined by performing fracture mechanics testing on plates with side cracks of different lengths. The fracture mechanics tests were filmed using a video camera. Based on this information, the force vs. load-line displacement could be established for the fracture mechanics tests, in which also the current length of the crack was indicated, since crack growth took place. Crack growth was modelled using a rate-dependent cohesive zone. The problem was analyzed using Abaqus, and the crack growth experiments were simulated. The experiments could be well reproduced. Furthermore, the direction-specific work of fracture had been established from the experiments and these energies could be compared to the values of the J-integral from the simulations for the different crack lengths.

Keywords: Fracture · Cohesive · Polyethylene · Rate-dependent

1 Introduction

The present paper concerns fracture mechanics of soft polymers. More specifically, the work of fracture of a low density polyethylene (LDPE) is investigated. Polyethylene is a polymer that may exhibit different degrees of crystallinity, mainly depending on the degree and type of branching of the polymer chains, which governs the ability of the chains to crystallize.

The energy required to propagate a crack in such a material may be divided into (at least) two parts. The first part, sometimes denoted the essential work of fracture, is associated with the fracture process at the very crack tip. For polymers, such as polyethylene, this process includes void formation and coalescence, and fibril formation and failure in a craze zone ahead of the crack tip. The second part of the work of fracture is associated with viscous dissipation in the region surrounding the crack tip. This second process includes plastic work required to cause crack tip blunting but also other dissipative plastic, viscous, and damage processes that may take place

© Springer International Publishing AG, part of Springer Nature 2019
E. E. Gdoutos (Ed.): ICTAEM 2018, SI 5, pp. 194–198, 2019.
https://doi.org/10.1007/978-3-319-91989-8_43

around the deforming crack tip. Both of these contributions to the total work of fracture are expected to be rate-dependent.

In the present work, we adopt a computational framework for assessing the work of fracture in soft polymers undergoing finite deformations. The material is taken to be anisotropic, and crack propagation is enabled by the inclusion of a cohesive zone in the model. The present study also includes experimental work, where the total work of fracture of an LDPE material is examined.

2   Experiments

We study crack growth in polyethylene both experimentally and numerically, and the problem to be analyzed is illustrated in Fig. 1. A polyethylene sheet of thickness $B = 0.6$ mm is clamped between two identical and rigid grips.

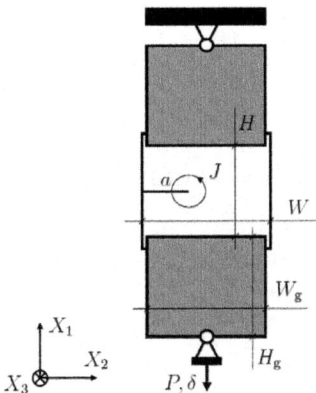

Fig. 1. Experimental setup.

The width and height of the grips are $W_g = H_g = 65$ mm respectively, and the corresponding dimensions of the polyethylene sheet are $W = 79$ mm and $H = 65$ mm. The polyethylene sheet is mounted symmetrically between the grips. The applied load is denoted by P, and the associated load-line displacement is denoted by δ. The J-integral is evaluated along contours around the crack tip. The sheet contains a side crack of length a.

The polymer sheet is loaded in displacement control, and the force vs. load-line displacement and crack length vs. load-line displacement was registered during the testing.

3 Numerical Analysis

The experiments were simulated using a 3D finite element model. The polyethylene sheet was modelled as an orthotropic material. The elastic behaviour was governed by the Young's modulus $E_{PE} = 210$ MPa, the Poisson's ratio $\nu_{PE} = 0.4$, and the shear modulus $G_{PE} = 46$ MPa. The plastic response was modelled by use of Hill's plasticity model and yield criterion. The elastic and plastic properties were determined in a previous study by the present group (Kroon et al. 2018).

The constitutive response of the cohesive zone in the polyethylene sheet was modelled by use of a rate-dependent traction-separation law, see Fig. 2.

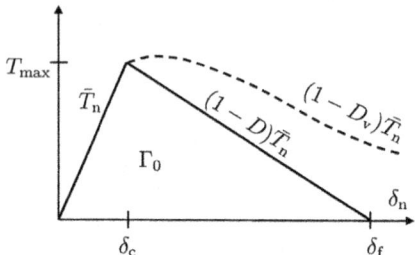

Fig. 2. Rate-dependent cohesive law.

The main parameters of the cohesive law are the surface energy, Γ_0, the maximum traction, T_{max}, the normal traction, T_n, the rate-independent damage, D, and the rate-dependent damage, D_v. The rate-dependent damage variable evolves according to

$$\dot{D}_v = \frac{1}{\tau}(D - D_v), \qquad (1)$$

where τ is a time constant describing the relaxation in the cohesive zone.

4 Results

One example of a comparison between the simulations and the experiments is shown in Fig. 3. The case studied is a specimen where the initial crack length was $a = 20$ mm, and where the current crack length is about a = 30 mm. The completely damaged cohesive elements can be seen in the simulation image, indicating the extent of crack growth. The overall kinematics of the simulation agrees well with the experiment.

In Fig. 4, model predictions for a number of Γ_0 values are shown for $T_{max} = 15$ MPa and $\tau = 0.05$ s, i.e. τ has been set to a relatively low value, in order to decrease the influence of viscosity in the cohesive law. As can be seen in Fig. 4(a), the simulation for the lowest value, $\Gamma_0 = 2$ kJ/m^2, severely underestimates the force response compared to the experimental results, and crack propagation takes place too early in the loading process, see Fig. 4(b). Increasing Γ_0 enables a better prediction, and for

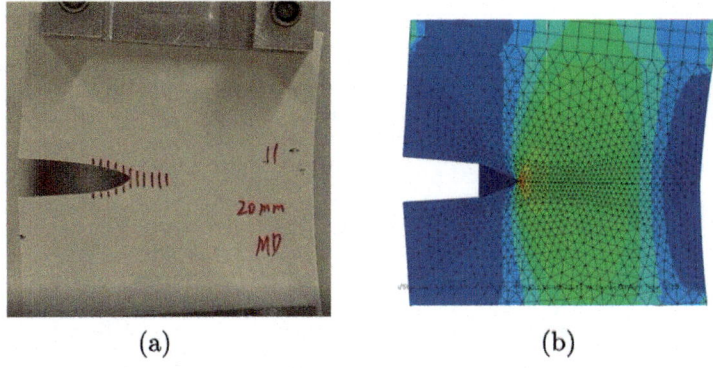

(a)   (b)

Fig. 3. Comparison between experiments and simulations.

$\Gamma_0 = 10$ kJ/m^2, both the force and crack propagation responses agree fairly well with the experimental results. However, if Γ_0 is increased further ($\Gamma_0 = 12$ kJ/m^2), crack tip blunting takes place, and the crack is arrested, which implies that the force in Fig. 4(a) increases monotonically and only a minimal crack growth (which should rather be seen as crack tip blunting) is predicted in Fig. 4(b).

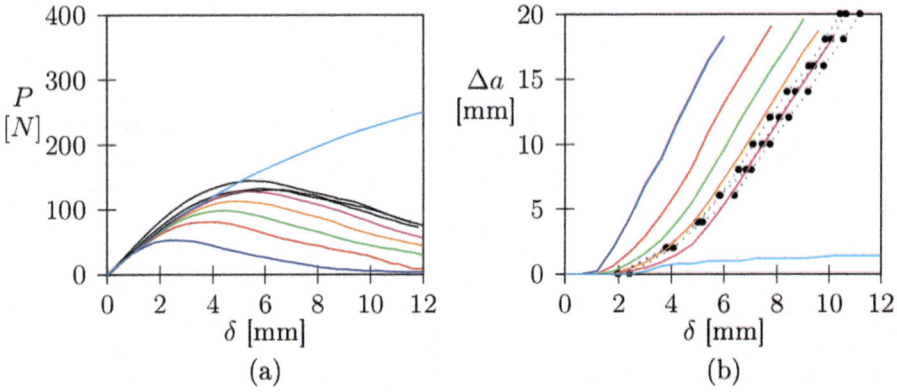

(a)   (b)

Fig. 4. Force vs. load-line displacement from experiments (black lines) and numerical simulations: $\Gamma_0 = 2$ kJ/m^2 (blue lines), $\Gamma_0 = 4$ kJ/m^2 (red lines), $\Gamma_0 = 6$ kJ/m^2 (green lines), $\Gamma_0 = 8$ kJ/m^2 (orange lines), $\Gamma_0 = 10$ kJ/m^2 (magenta lines), and $\Gamma_0 = 12$ kJ/m^2 (cyan lines).

Next, we consider the influence of the time constant, τ. In Fig. 5, predictions for a few values of τ are shown, where $\Gamma_0 = 2$ kJ/m^2 was applied. As can be seen in Fig. 5 (a), increasing the value of τ causes the dissipation in the cohesive zone to increase, which in turn increases the force response for a given load-line displacement. The choices $\tau = 0.15$ s and 0.20 s enable reasonable predictions of both force and crack propagation response, as can be seen in Fig. 5(a) and (b).

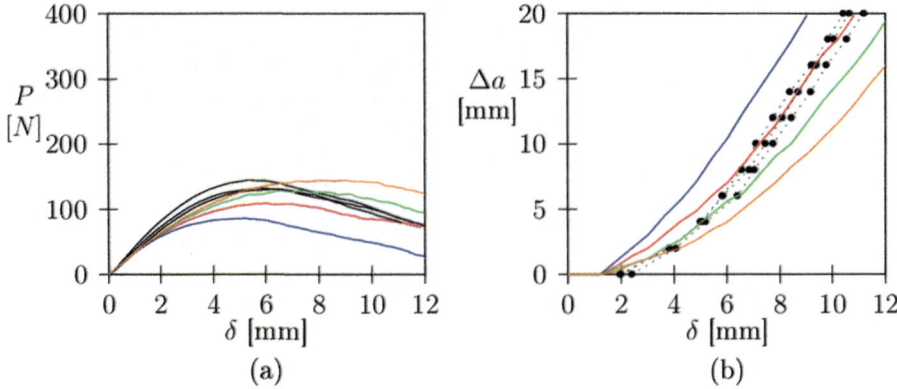

Fig. 5. Force vs. load-line displacement from experiments (black lines) and numerical simulations: $\tau = 0.10$ s (blue lines), $\tau = 0.15$ s (red lines), $\tau = 0.20$ s (green lines), and $\tau = 0.25$ s (orange lines).

5   Discussion and Concluding Remarks

The fracture mechanics properties of PE are of great practical interest. Crack initiation and growth in polyethylene seem to take place through formation and subsequent breakdown of a craze zone ahead of the crack tip. This process is driven by localized plastic strains, initiation and coalescence of voids, and formation of fibrils that eventually fail.

The cohesive zone approach, adopted in the present study, is a powerful computational tool for modelling crack initiation and growth. This approach requires (among other things) knowledge of the local fracture energy of the material. One of the outcomes of the present study is an estimate of this local fracture energy for LDPE.

An experimental and numerical study of the fracture mechanics behaviour of injection-moulded low-density polyethylene has been performed in the present work. The total work of fracture was determined experimentally, where the testing was performed on a sheet with a side crack. The local work of fracture was estimated through numerical analyses, where the initiation of crack growth was simulated. In the numerical formulation, initiation of crack growth was modelled by use of a cohesive zone and an associated traction-separation law. The estimated local work of fracture is consistent with previous experimental measurements of the essential work of fracture of the material in question, and the total work of fracture, retrieved from the present experiments, agreed well with the calculated far field values of the J-integral from the numerical analyses.

Reference

Kroon, M., Andreasson, E., Persson Jutemar, E., Petersson, V., Persson, L., Dorn, M., Olsson, P.A.T.: Exp. Mech. **58**, 75–86 (2018)

Improving the Reliability of Design of a Hinge Kit System Subjected to Repetitive Stresses

Seong-woo Woo[(⊠)]

Reliability Association of Korea, 146, Sunyoo-ro,
Yeongdeungpo-gu, Seoul 150-103, Korea
twinwoo@yahoo.com

Abstract. Based on field data and a tailored set of accelerated life tests, the hinge kit system (HKS) of a closing door in a refrigerator was redesigned to improve its lifetime. Using a force and moment balance analysis, the mechanical impact loads of HKS were calculated in closing the refrigerator door. At first ALT the kit housing in the HKS fractured. When breaking up HKS, oil damper was leaked. The failures of 1st ALT were similar to those of the failed samples obtained in the field. The HKS housing and oil damper was modified. At 2nd ALT, the cover housing fractured. The cover housing in HKS was changed from plastics to Aluminum. After two rounds of parametric ALT with corrective action plans, reliability of the new HKS is guaranteed to be over B1 life 10 years with a yearly failure rate of 0.1%.

Keywords: Reliability design · Hinge kit system · Fracture
Parametric accelerated life testing · Missing design parameter

1 Introduction

When a consumer uses the refrigerator door, one want to close it gently. For this (intended) function, the hinge kit system that consists of a kit cover, shaft, spring, and oil damper, etc. is designed (Fig. 1).

In field the malfunction of the HKS had often been reported by endusers of the refrigerator. The data of the failed products were important for figuring out the usage conditions of consumers and helping to pinpoint design changes that needed to be modified in the product. Thus, based on field data, failure analysis and its reproduction was required to know what parameters in the defective HKS needed to be redesigned.

A typical pattern of repeated load or overloading due to daily consumer usage may cause structural failure in product lifetime. Many engineers think such possibility can be assessed: (1) mathematical modeling like Newtonian method, (2) the time response of system simulation for dynamic loads, (3) the rain-flow counting method, and (4) miner's rule that the system damage can be estimated. However, because there are a lot of assumptions, this analytic methodology is exact but complex to reproduce the product failures due to the design flaws.

The purpose of this study is to present the reliability methodology to the mechanical system like HKS subjected to repetitive loading. The reliability methodology includes (1) load analysis, (2) a tailored of parametric ALTs, and (3) modifying

© Springer International Publishing AG, part of Springer Nature 2019
E. E. Gdoutos (Ed.): ICTAEM 2018, SI 5, pp. 199–204, 2019.
https://doi.org/10.1007/978-3-319-91989-8_44

(a)   **(b)**

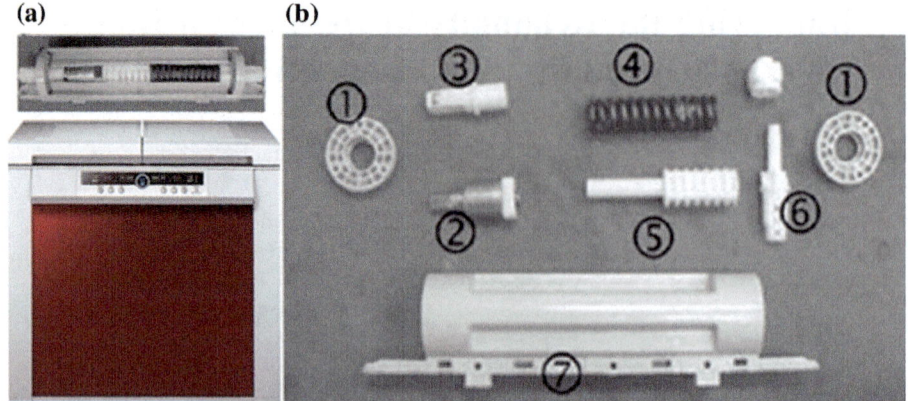

Fig. 1. (a) Refrigerator and HKS (b) Mechanical parts of HKS: kit cover (1), oil damper (2), spring (4), shaft (6), and HKS housing (7)

the designs of the HKS. If modified mechanical system achieves the required reliability target, parametric ALT will terminate.

2   Materials and Methods

2.1   Load Analysis

In the field, HKS parts in a refrigerator were cracking and fracturing due to design failure. Based on the consumer usage conditions, HKS were subjected to different loads during the opening and closing of the refrigerator door (Fig. 2).

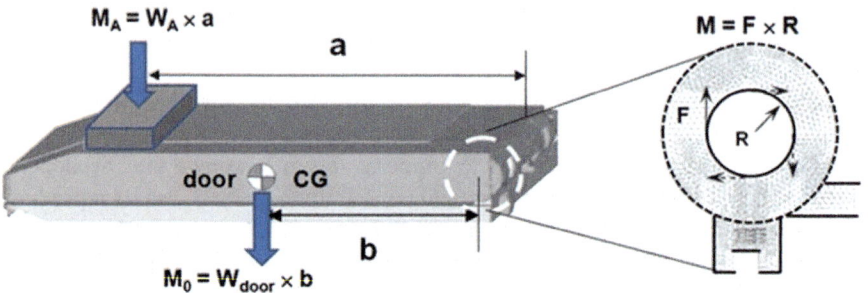

Fig. 2. Design concept of HKS

The moment balance around the HKS can be expressed as

$$M_0 = W_{door} \times b = T_0 = F_0 \times R \tag{1}$$

If accelerated weight on the refrigerator door is added, the moment balance around the HKS can be modified as

$$\sum M = M_1 = M_0 + M_A = W_{door} \times b + W_A \times a = T_1 = F_1 \times R \qquad (2)$$

Under the same environmental conditions, the life-stress model (LS model) can be modified as:

$$TF = A(S)^{-n} = AT^{-\lambda} = A(F \times R)^{-\lambda} = B(F)^{-\lambda} \qquad (3)$$

We know that product lifetime depends on applied impact force. Therefore, the acceleration factor (AF) can be derived as

$$AF = \left(\frac{S_1}{S_0}\right)^n = \left(\frac{T_1}{T_0}\right)^{\lambda} = \left(\frac{F_1 \times R}{F_0 \times R}\right)^{\lambda} = \left(\frac{F_1}{F_0}\right)^{\lambda} \qquad (4)$$

2.2 Parametric Accelerated Life Testing in HKS

To carry out parametric ALT, the sample size equation with the acceleration factors in Eq. (5) might be expressed as [1]:

$$n \geq (r+1) \cdot \frac{1}{x} \cdot \left(\frac{L_{BX}^*}{AF \cdot h_a}\right)^{\beta} + r \qquad (5)$$

If the reliability of the new HKS was targeted to be B1 life 10 years, the number of required test cycles can be obtained for given sample pieces. In parametric ALTs the missing parameters of HKS in the design phase can be identified to achieve the reliability target.

3 Results and Discussion

Generally, the operating conditions for the HKS in a refrigerator were approximately 0–43 °C with a relative humidity ranging from 0% to 95%, and 0.2–0.24 g's of acceleration. The opening and closing of the door occurred in an estimated average of 3 to 10 times per day. With a life cycle design point for 10 years, the life of HKS incurs about 36,500 usage cycles for worst case.

For the worst case, the impact force around the HKS was 1.10 kN which was the maximum force applied by the typical consumer. For the ALT with accelerated weight the impact force of HKS was 2.76 kN. Using a stress dependence of 2.0, the acceleration factor was found to be approximately 6.3 in Eq. (4).

For the B1 life 10 years, the required test cycles for sample six pieces calculated in Eq. (5) were 23,000 cycles if shape parameter is supposed to be 2.0. This parametric ALT was designed to ensure a B1 life 10 years with about a 60% level of confidence that it would fail less than once during 23,000 cycles.

Figure 3 shows a photograph comparing the failed product from the field and from 1st accelerated life testing, respectively. In 1st ALT the housing of HKS fractured at

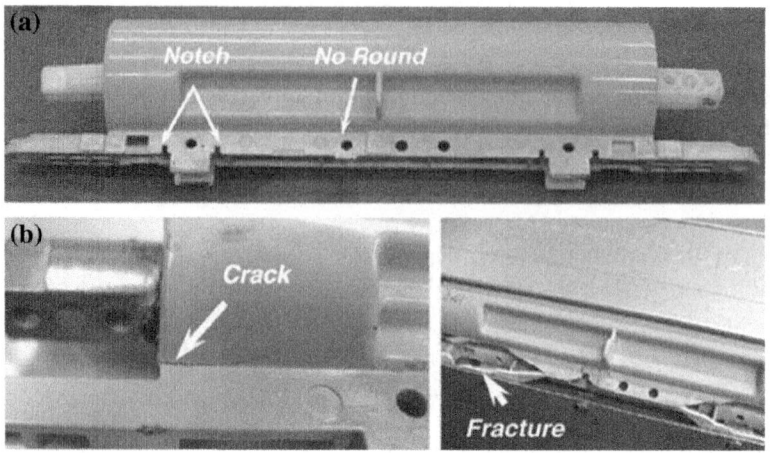

Fig. 3. (a) Failed products in field (b) crack after 1st ALT

3,000 cycles and 15,000 cycles. When breaking down HKS, the oil damper in the hinge kit assembly was spilled at 15,000 cycles (Fig. 4).

Fig. 4. Spilled oil damper in 1st ALT

In the second ALTs the fracture of hinge kit cover occurred in the cover housing of HKS at 8,000, 9,000, and 14,000 cycles (Fig. 5).

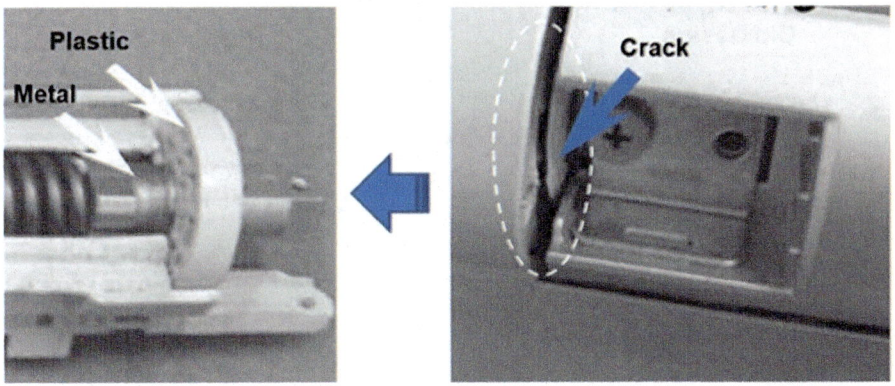

Fig. 5. Structure of problematic products at 2nd ALT

To withstand the HKS design problems due to the repetitive impact loads, the HKS system were redesigned as follows: (1) reinforcing the housing design of HKS, C1 (Fig. 6); (2) changing the sealing structure in the oil damper, C2 (Fig. 7); (3) changing cover housing material, C3, from plastics to the Al die-casting. With these design changes, the refrigerator could also be opened and closed more comfortably in product lifetime because there is no problems till 23,000 mission cycles.

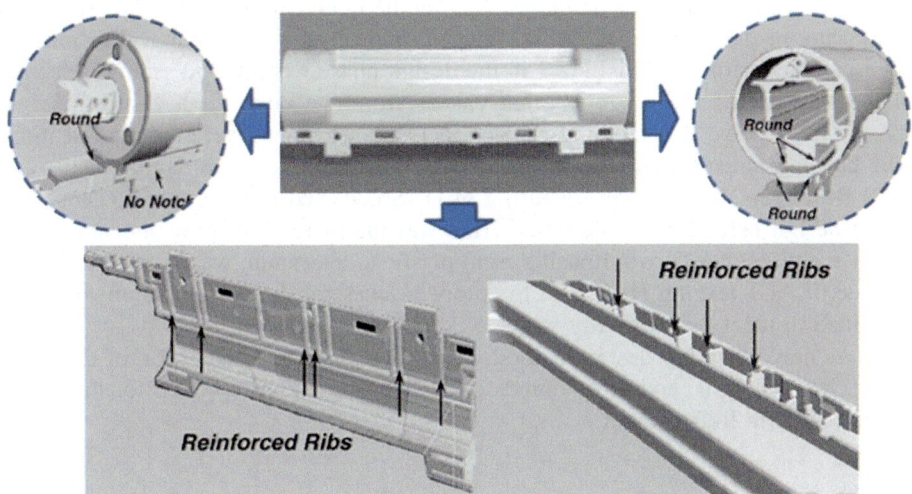

Fig. 6. Redesigned HKS housing structure

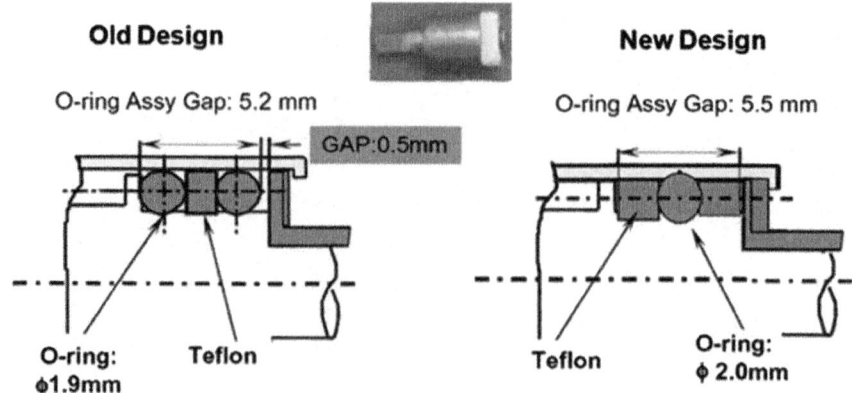

Old Design

O-ring Assy Gap: 5.2 mm

GAP:0.5mm

O-ring: Teflon
φ1.9mm

New Design

O-ring Assy Gap: 5.5 mm

Teflon   O-ring:
φ 2.0mm

Fig. 7. Redesigned oil damper

4 Conclusion

To improve the reliability of a newly designed HKS in refrigerators, we have utilized reliability methodologies – identifying the failure modes and mechanism investigation of fractured HKS in field, conducting a series of accelerated life testing, and redesigning the HKS with action plans. Based on the products that failed both in the field and in the ALTs, the primary failure of the HKS occurred as following:

(1) Based on the products that failed both in the field and in 1st ALT, the failure of HKS occurs in the fracture of the kit housing and oil damper leaking. The missing design parameters of the failed HKS in the design phase were the oil sealing structure and the housing of the hinge kit that causes from the concentrated stress due to improper fillets, ribs, and notching. The corrective action plans were the modifications of the housing hinge kit and the redesigned sealing structure of oil damper.

(2) Based on 2nd ALT, the fracturing of HKS occurred in the cover housing. The missing design parameter of the failed HKS was the material of cover housing. As a corrective action plans, cover housing from plastic to aluminum was modified. After a sequence of ALT testing, HKS with the proper values for the design parameters were determined to meet the reliability target – B1 life 10 years.

(3) As new reliability design methodologies, we knew that inspection of the failed product, load analysis, and three rounds of ALT was greatly improved for the newly designed HKS in Refrigerator.

Reference

1. Woo, S., O'Neal, D.: Reliability design of mechanical systems subject to repetitive stresses. Recent Pat. Mech. Eng. **8**(4), 222–234 (2015)

Investigation into the Breakdown of Continuum Fracture Mechanics at the Nanoscale: Synthesis of Recent Results on Silicon

Pasquale Gallo$^{(\boxtimes)}$, Takashi Sumigawa, Takahiro Shimada,
Yabin Yan, and Takayuki Kitamura

Department of Mechanical Engineering and Science, Kyoto University,
Kyoto-daigaku-Katsura, Nishikyo-Ku, Kyoto-Shi 615-8540, Japan
pasquale.gallo@aalto.fi

Abstract. The present contribution reviews some recent results on the experimental characterisation of the nanoscale fracture toughness of silicon by using pre-cracked specimens and alternatively the theory of critical distances (TCD). Later, the results are discussed to provide the ultimate dimensional limit of the continuum fracture mechanics at the nanoscale in the light of sophisticated discrete atomic simulations at the onset of brittle fracture. The results show that the fracture toughness of Si is independent of the scale, crystal orientation and the singular stress field length. This confirms the atomistic nature of the brittle fracture. Moreover, the continuum fracture mechanics fails below a singular stress field approaching 2 nm.

Keywords: Fracture nanomechanics · Nanoscale · Silicon
Quantum mechanics

1 Introduction

Conventional fracture mechanics [1] has been primarily investigated starting from the II World War, leading to the well-known linear elastic fracture mechanics (LEFM). In the last decade, the miniaturisation of electronic devices driven by the increasing demand for high-density integrations have brought problems of material behaviour at very small scales into the domain of fracture mechanics. At micro and nanoscale, indeed, the material still fails by nucleation and propagation of cracks, but the continuum assumption basing the LEFM has to face the discrete nature of atoms [2]. The so-called *fracture nanomechanics* poses entirely new challenges, both from the experimental and theoretical point of view. However, the nanoscale investigation of fracture has sparked the interest of the scientific community [3–5].

In the above background, the experimental characterisation of fracture behaviour at the nanoscale and the investigation of the low limit of the continuum theory are at the present relevant and challenging aspects. Therefore, the present work presents recent experimental tests on the evaluation of single crystal Si fracture toughness at the nanoscale [6, 7]. Results obtained from pre-cracked samples and by using the theory of

© Springer International Publishing AG, part of Springer Nature 2019
E. E. Gdoutos (Ed.): ICTAEM 2018, SI 5, pp. 205–210, 2019.
https://doi.org/10.1007/978-3-319-91989-8_45

critical distances (TCD) [8] are compared to highlight pros and cons of each procedure. Later, in Sect. 3, the experimental results are discussed in the light of complex discrete atomic simulations at the onset of brittle fracture [9] to provide the ultimate dimensional limit of the continuum fracture mechanics.

2 Experimental Characterization of Si Fracture Toughness at the Nanoscale

2.1 Procedure Based on Pre-cracked Samples

Recently, the K_{IC} of Si has been investigated accurately by Sumigawa et al. [6]. Those authors employed pre-cracked specimens accurately realised to avoid large curvature at the crack tip. First, a V-notch was realised by focused ion beam (FIB) processing system along the cleavage plane Si(011), and later a pre-crack was generated by gradually opening the notch. The opening displacement was controlled and applied through a wedge-shaped diamond indenter. Image of a sample is reported in Fig. 1. Four samples with several pre-crack lengths were generated by cutting the upper part of the notch. Details are reported in [6].

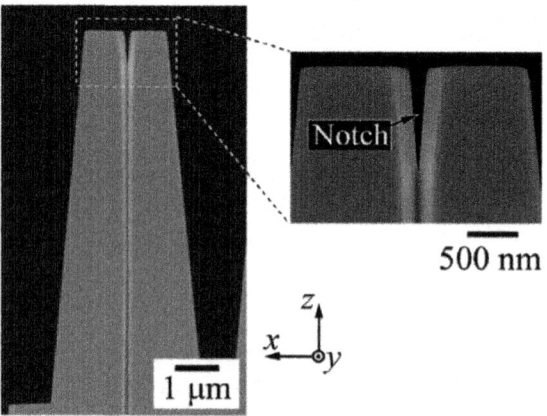

Fig. 1. Example of pre-cracked specimen

The four specimens were tested in a transmission electron microscope (TEM) provided by *in situ* observation camera. Experimental results were later re-analysed by finite element method (FEM) to determine the stress state accurately at the crack tip and the final average K_{IC} of about 1 MPa$^{0.5}$. This value is in agreement with the bulk silicon K_{IC} that is 0.75–1.08 MPa$^{0.5}$. The FEM simulations showed, besides, that all the geometries presented a singular field of $r^{0.5}$, while its length varied from 23 nm to 60 nm. The K_{IC} seems therefore independent of the scale and the singular field length, while the local fracture mechanisms take place at least at those or smaller sizes.

2.2   Procedure Based on Theory of Critical Distances

The fabrication of pre-cracks in nanoscale specimens, even if feasible it is clearly difficult. To avoid strict control on the fabrication process, an alternative strategy using the TCD and notched specimens has been employed by Gallo et al. [7]. TCD in the form of the point method states indeed that failure occurs when the linear elastic maximum principal stress σ_1 at a given distance $L/2$ from the notch root equals the inherent material strength σ_0 [8]. L is called material characteristic length, it is a material property and is defined as follows:

$$L = \frac{1}{\pi} \left(\frac{K_{IC}}{\sigma_0} \right)^2 \qquad (1)$$

As stated by Susmel and Taylor [10], the intersection of at least two linear elastic stress field in incipient failure of specimens with *different sharpnesses* gives a correct estimation of both σ_0 and L. Once these two parameters are defined, (1) can be used to the determine $K_{IC} = \sigma_0 \sqrt{\pi L}$. This strategy simplifies the experimental procedure since it is based on notches instead of cracks, and a strict control on the final notch geometry is not needed [10].

According to the procedure briefly explained above, notches of different sharpness have been introduced in two nano cantilever beams by FIB processing system. The sharp notch had a radius ρ of about 6.3 nm and an opening angle 2α of 68°; the blunt notch had $\rho = 20.2$ nm and $2\alpha = 59°$. The sharp specimen had a notch depth a of 144 nm while the blunt sample had a notch depth a of 179 nm. The geometry is depicted in Fig. 2(a) while additional details can be found in [7].

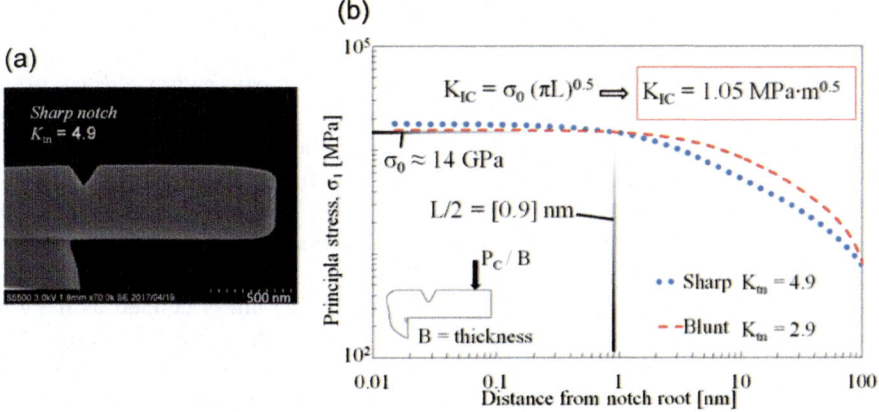

Fig. 2. Example of nanoscale notched specimen (a), and stress fields under the critical load P_C at failure (b)

The loading experiments are conducted in a TEM, and the load is applied through an indenter after accurate alignment. Displacements and loads are recorded. The stress fields at incipient failure are later re-analysed by FEM and overlapped in a single graph in Fig. 2(b) together with stress concentration factors (net section). The intersection point revealed a σ_0 of about 14 MPa and $L/2$ of about 0.9 nm. By employing (1) the final K_{IC} was 1.05 MPa$^{0.5}$. The results are therefore in agreement with [6] and macro counterpart values (0.75–1.08 MPa$^{0.5}$). Moreover, being the material characteristic length L a representative scale parameter of the mechanisms involved in the fracture process, the procedure provides a more precise information compared to Sect. 2.1 and suggests that those mechanisms take place at least at 2 nm or even smaller size [6].

3   Discussion: Breakdown of Continuum Theory

The results of Sect. 2.1 suggests K_{IC} is independent of the singular stress field length and therefore that the fracture mechanisms take place below 23–60 nm [6]. However, a precise estimation of the size was not provided. The TCD material characteristic length L evaluated in Sect. 2.2 and extended recently in [7] is, instead, a representative length scale parameter that assumes different values as the fracture mechanisms involved change scale [11]. The results briefly reported here suggest that the fracture develops at a scale of about 2 nm. At this small scale, one must deal with atomic structure, and further considerations can be done only with atomic simulations. These have been conducted recently by Shimada et al. [9]. Fracture tests were performed *in situ* using center cracked nanoscale plates of Si. The width of the specimens $2W$ varied from 2.1 to 276 nm while the crack length $2a$ was kept equal to one-third of $2W$. The experiments were first re-analysed from a continuum theory point of view, and it was shown that the singular stress field of $r^{0.5}$ shrinks as the plate width does. However, when the width is very small and the stress intensity factor (SIF) K region approaches 1 nm, the fracture is no longer governed by the merely SIF.

To investigate this statement and to highlight the atomistic nature of fracture, the experiments have been re-analysed by instability mode analyses [12] and density-functional theory (DFT). With these approaches, the deformation mode of atoms at the onset of brittle fracture is accurately determined. Both DFT and instability mode analyses showed that the crack advances along the cleavage plane (110) by atomic *bond breaking*. This phenomenon is observed for all of the specimens, regardless of the size, as Fig. 3 shows.

Moreover, a fracture-process zone Λ_f of about 0.4–0.6 nm is defined as the zone where the discrete motion of atoms is concentrated. When the K-dominant zone Λ_K is geometrically too much small and close to the fracture-process zone (i.e. very small specimens), continuum theory breaks down. Concluding, the ultimate dimensional limit for the single crystal silicon is quantified as the range 3–$6\Lambda_f$ roughly 2 nm. It should be pointed out that the TCD employed in Sect. 2.2 estimated a material characteristic length of 1.8 nm, and therefore very close to ≈ 2 nm. This confirms that the material characteristic length is representative of the mechanisms scale involved in the fracture process.

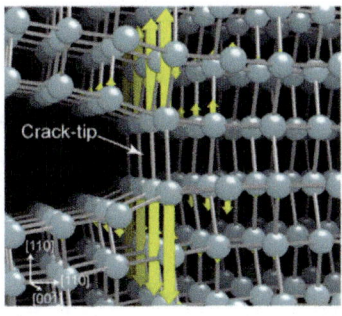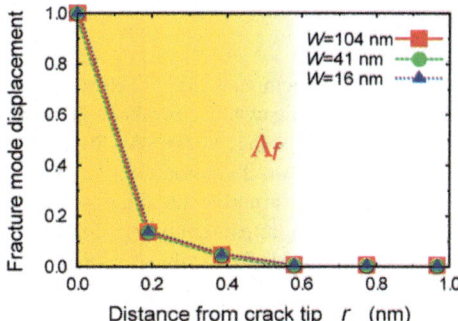

Fig. 3. Fracture mode at the onset of brittle fracture (left), and fracture mode displacement normalized by the maximum displacement at the crack-tip for different size of specimens (right); reproduced from [9] under CC BY 4.0 license permission (https://creativecommons.org/licenses/by/4.0/)

4 Conclusion

The results showed that the fracture toughness of Si is independent of the size and the crystal orientation. Regardless of the methods employed, i.e. using accurate pre-cracked samples or notched components combined with TCD, the K_{IC} has a value of approximately 1 $MPa^{0.5}$. Since the value is in good agreement with the macro counterpart, it can be concluded that from the macroscale to nanoscale the mechanisms involved in the fracture process are the same.

If pre-cracked samples are employed, even a small curvature at the crack tip can generate large discrepancy on the final fracture toughness.

On the other hand, the TCD has shown several advantages: (i) it simplifies the experimental procedure since the realisation of an effective crack tip is not necessary; (ii) it yields an approximated magnitude of fracture process provided by the material characteristic length $L = 1.8$ nm. This value has been verified by discrete atomic simulations and instability mode analysis that captured the deformation mode of atoms at the onset of a brittle fracture. The analyses showed, for the silicon, a fracture process zone of 0.4–0.6 nm and the low limit of the continuum fracture mechanics of about 2 nm, in agreement with L. For singular stress fields size approaching the lower limit, continuum theory fails and discrete nature of atoms should be considered.

Acknowledgements. This work was supported by the Japan Society for the Promotion of Science (JSPS) International Research Fellow program (Grant No. 16F16366); JSPS KAKENHI (Grant No. JP15H02210, JP26630009, and JP25000012); JSPS Grant-in-Aid for Specially Promoted Research (Grant No. 25000012).

References

1. Liebowitz, H.: Fracture. Academic Press, New York (1968)
2. Kitamura, T., Sumigawa, T., Hirakata, H., Shimada, T.: Fracture Nanomechanics, 2nd edn. Pan Stanford Publishing, Singapore (2016)
3. Gallo, P., Sumigawa, T., Kitamura, T., Berto, F.: Evaluation of the strain energy density control volume for a nanoscale singular stress field. Fatigue Fract. Eng. Mater. Struct. **39**, 1557–1564 (2016). https://doi.org/10.1111/ffe.12468
4. Huang, K., Shimada, T., Ozaki, N., Hagiwara, Y., Sumigawa, T., Guo, L., Kitamura, T.: A unified and universal Griffith-based criterion for brittle fracture. Int. J. Solids Struct. **128**, 67–72 (2017). https://doi.org/10.1016/j.ijsolstr.2017.08.018
5. Sumigawa, T., Shimada, T., Tanaka, S., Unno, H., Ozaki, N., Ashida, S., Kitamura, T.: Griffith criterion for nanoscale stress singularity in brittle silicon. ACS Nano **11**, 6271–6276 (2017). https://doi.org/10.1021/acsnano.7b02493
6. Sumigawa, T., Ashida, S., Tanaka, S., Sanada, K., Kitamura, T.: Fracture toughness of silicon in nanometer-scale singular stress field. Eng. Fract. Mech. **150**, 161–167 (2015). https://doi.org/10.1016/j.engfracmech.2015.05.054
7. Gallo, P., Yan, Y., Sumigawa, T., Kitamura, T.: Fracture behavior of nanoscale notched silicon beams investigated by the theory of critical distances. Adv. Theory Simul., 1700006 (2017). https://doi.org/10.1002/adts.201700006
8. Taylor, D.: The Theory of Critical Distances: A New Perspective in Fracture Mechanics. Elsevier, Oxford (2007)
9. Shimada, T., Ouchi, K., Chihara, Y., Kitamura, T.: Breakdown of continuum fracture mechanics at the nanoscale. Sci. Rep. **5**, 8596 (2015). https://doi.org/10.1038/srep08596
10. Susmel, L., Taylor, D.: The theory of critical distances as an alternative experimental strategy for the determination of K_{Ic} and ΔK_{th}. Eng. Fract. Mech. **77**, 1492–1501 (2010). https://doi.org/10.1016/j.engfracmech.2010.04.016
11. Taylor, D.: The theory of critical distances: a link to micromechanisms. Theor. Appl. Fract. Mech. **90**, 228–233 (2017). https://doi.org/10.1016/j.tafmec.2017.05.018
12. Kitamura, T., Umeno, Y., Fushino, R.: Instability criterion of inhomogeneous atomic system. Mater. Sci. Eng. A **379**, 229233 (2004). https://doi.org/10.1016/j.msea.2004.02.061

Evolution of Size Distribution of Pores in Metal Melts at Tension with High Strain Rates

Polina Mayer[(⊠)] and Alexander Mayer

Chelyabinsk State University, Bratiev Kashirinykh Street 129, 454001
Chelyabinsk, Russia
polina.nik@mail.ru

Abstract. Interest to mechanical behavior of metal melts is associated with both the development of experimental technique and the possible technological applications. One of the essential properties is the dynamic tensile strength of melts, that is, the level of negative pressure, which leads to cavitations and further fracture of the melt. The tensile state of melt is metastable, like for solid. Molecular dynamic (MD) simulations show that the complete fragmentation of melt occurs when the pore volume fraction reaches 80% or more. The work done by the negative pressure maintained in the melt at the stage of bubbly liquid can exceed the work to reach the cavitations limit due to the longer exposure time. In this paper we investigate with the help of MD the late stages of melt fracture with special attention to the study of the character of the size distribution of cavities and its evolution during tension.

Keywords: Metal melts · Tensile fracture · Pores · Size distribution

1 Introduction

The interest in the behavior of metals under extreme loading conditions, including high temperatures and high strain rates (up to 1/ns) is associated with both the appearance of experimental techniques for creating such loading conditions and the prospects for developing of new types of facilities for energy production. To simulate the operation of such facilities, data on the properties and mechanical behavior are needed for not only solid metals, but also molten ones. One of the essential properties is the dynamic tensile strength of the melts, that is, the level of negative pressure, which at a given strain rate and temperature leads to cavitations and further fracture of the melt. The tensile state of melt is metastable, like for solid. This metastable state decays through cavitations. Only the restricted number of experimental data is available for the dynamic tensile strength of melts [1–4], which raises importance of theoretical investigations [5, 6].

Molecular dynamic (MD) simulations of late stages of tensile fracture of aluminum [6] and iron melts show that the state of a metal melt with cavities occurs during a sufficiently long evolutionary stage, beginning with the formation of cavities and ending with complete fragmentation, which occurs when the pore volume fraction reaches 80% or more. Despite the fact that the negative pressure maintained in the melt

© Springer International Publishing AG, part of Springer Nature 2019
E. E. Gdoutos (Ed.): ICTAEM 2018, SI 5, pp. 211–214, 2019.
https://doi.org/10.1007/978-3-319-91989-8_46

at this stage is much less than the tensile strength limit, the work done by it before fragmentation, on the contrary, can in many cases exceed the work to reach the ultimate strength due to the longer exposure time. This work, in particular, contributes to the additional absorption of the tension pulse. This stage requires a more detailed study. In particular, for metal melts, it was shown that in addition to the average cavity size, their size distribution is essential from the viewpoint of calculating the effective negative pressure.

Here we perform the MD study of the character of the size distribution of cavities in metal melts and its evolution at the late stages of tension. Using the obtained data we develop the continuum model for the late stage of tension, which takes into account the cavity size distribution and the presence of resistance to tension even at large volume fractions of pores.

2   MD Simulations and Models of Cavities Evolution

Classical MD simulations are performed using the program package LAMMPS [7]. Program OVITO [8] is used for visualization and analysis of atom configurations. The "Construct surface mesh" algorithm [9] is applied for searching the surfaces of pores.

The considered problem consists in uniform tension of a sample of molten metal by scaling of atom coordinates with the help of "deform" command of LAMMPS. This mode of tension corresponds to the tension by inertial, which is typical for expansion of a metal layer heated by laser or electron beam, as well as for reflection of a shock wave from a free surface. The true strain rate is constant in simulations; the system temperature is maintained on a constant level by Nose-Hover thermostat. The strain rate (above 1/ns) and the size of the MD system (several million atoms) are chosen to provide formation of multiple cavities within the considered volume of melts. The sample preparation consists of creation of an ideal crystal lattice of corresponding metal followed by heating to the temperature considerably above the melting temperature with further cooling down to the test temperature.

An example of the obtained atomic configuration is presented in Fig. 1. One can see formation and development of the bubble structure in iron melt at high-rate tension. During the tension, both the growth and the collapse of cavities are observed.

We compare the obtained MD results with two types of model. The direct stochastic model includes the Rayleigh-Plesset equation for description of cavities growth or collapse applied for each separate cavity, which nucleation is random according to the nucleation rate, which follows from thermal fluctuation nature of the process. The continuum model uses a similar technique, but applies it to separate generations of cavities [5].

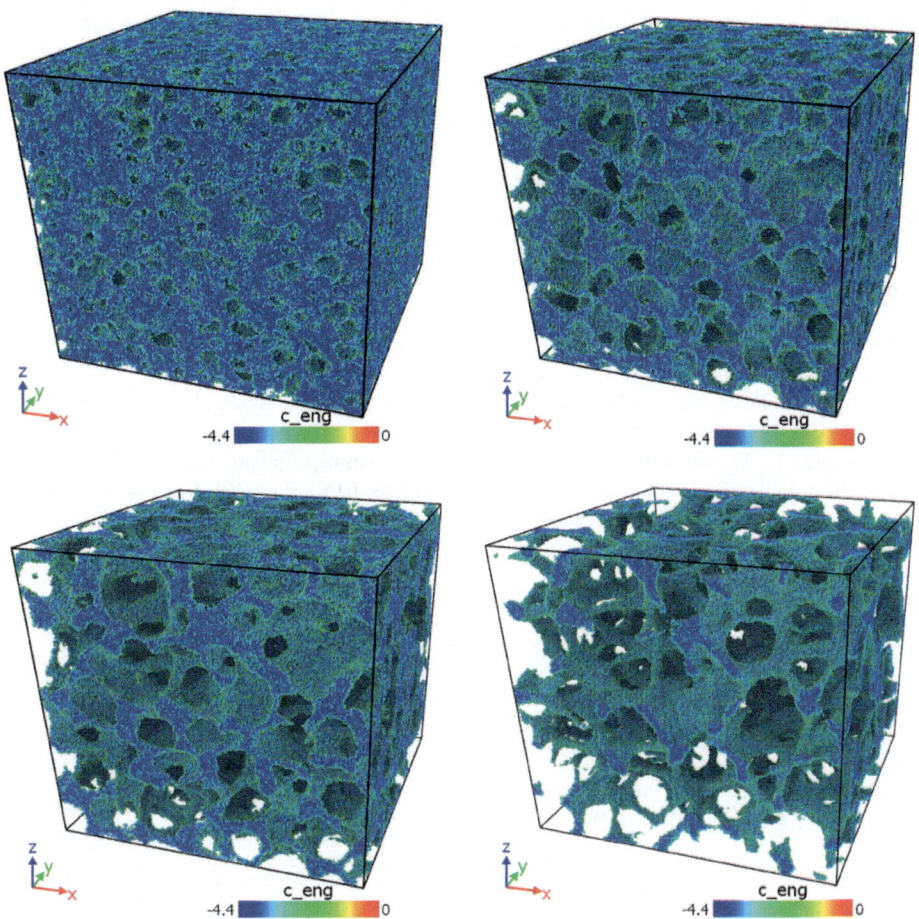

Fig. 1. Evolution of porous structure in molten iron at tension with the strain rate of 100/ns at the constant temperature of 2000 K: atomic configurations obtained in MD simulations for time moments of 4, 8, 12 and 20 ps.

3   Conclusions

We perform MD investigation of size distribution of cavities in metal melts under the high-rate tension and compare the results with two types of models of the process.

This work is supported by the Russian Science Foundation (Project 17-71-10205).

References

1. Agranat, M.B., Anisimov, S.I., Ashitkov, S.I., Zhakhovskii, V.V., Inogamov, N.A., Komarov, P.S., Ovchinnikov, A.V., Fortov, V.E., Khokhlov, V.A., Shepelev, V.V.: Strength properties of an aluminum melt at extremely high tension rates under the action of femtosecond laser pulses. JETP Lett. **91**(9), 471–477 (2010)

2. Ashitkov, S.I., Komarov, P.S., Ovchinnikov, A.V., Struleva, E.V., Agranat, M.B.: Strength of liquid tin at extremely high strain rates under a femtosecond laser action. JETP Lett. **103**(8), 544–548 (2016)
3. Struleva, E.V., Ashitkov, S.I., Komarov, P.S., Khishchenko, K.V., Agranat, M.B.: Strength of iron melt at high extension rate during femtosecond laser ablation. J. Phys: Conf. Ser. **774**, 012098 (2016)
4. Kanel, G.I., Savinykh, A.S., Garkushin, G.V., Razorenov, S.V.: Dynamic strength of tin and lead melts. JETP Lett. **102**(8), 548–551 (2015)
5. Mayer, A.E., Mayer, P.N.: Continuum model of tensile fracture of metal melts and its application to a problem of high-current electron irradiation of metals. J. Appl. Phys. **118**(3), 035903 (2015)
6. Mayer, P.N., Mayer, A.E.: Late stages of high rate tension of aluminum melt: molecular dynamic simulation. J. Appl. Phys. **120**(7), 075901 (2016)
7. Plimpton, S.: Fast parallel algorithms for short-range molecular dynamics. J. Comput. Phys. **117**, 1–19 (1995). http://lammps.sandia.gov
8. Stukowski, A.: Visualization and analysis of atomistic simulation data with OVITO–the Open Visualization Tool. Model. Simul. Mater. Sci. Eng. **18**, 015012 (2010). http://www.ovito.org
9. Stukowski, A.: Computational analysis methods in atomistic modeling of crystals. JOM **66**(3), 399–407 (2014)

Particle-Based DEM Model for Simulating Brittle Cracks Evolution in Rock-like Materials During the Tensional Fracturing Process

Piotr Klejment$^{(\boxtimes)}$, Wojciech Dębski, and Alicja Kosmala

Institute of Geophysics Polish Academy of Sciences,
Ksiecia Janusza Street 64, 01-452 Warsaw, Poland
pklejment@igf.edu.pl

Abstract. The aim of this study was to recreate cracks network in a rock material during hydraulic fracturing process using the Discrete Element Method implemented in the ESyS-Particle software. We have investigated the behavior of brittle material under tensional stress induced by fracking fluid pumped either with constant velocity or with constant pressure. With DEM approach it was possible to check how the fluid expands inside the rock formation under different conditions. Consequently, we have calculated the injection velocities for different values of pressure as well as a spatial extension of cracks in the material. It gave an insight into hydraulic fracturing process which is possible only using computer simulations.

Keywords: Particle-based numerical modeling · Discrete Element Method
Brittle failure · Cracks extension · ESyS-Particle · Rock fractures
Hydraulic fracturing

1 Research Methodology

1.1 Numerical Modeling Approaches

Numerous numerical modeling approaches have been applied for simulating hydraulic fracturing phenomena. Some of them are based on the Displacement Discontinuity Method to take advantage of the efficiency provided by a Boundary Element Method. The Galerkin's Finite-Element approach is often used for fluid-flow modeling and the rock mass deformation [7]. The less popular Elastic Displacement Discontinuity Method (DDM) has also been applied to hydrofracturing processes. However, various Finite-Element Method-based approaches, the most popular techniques in engineering applications, when applied to fragmentation/fracturing processes exhibits many undesired features highly limiting the efficiency and accuracy of the method [6]. The totally different approaches based on discrete particle-like models of a medium in hand is a base of an another class of numerical algorithms generally referred to as the Discrete Element Methods. In this presentation we discuss an application of such a method to geophysically-orientated problem of hydraulic fracturing of the britle materials [12].

© Springer International Publishing AG, part of Springer Nature 2019
E. E. Gdoutos (Ed.): ICTAEM 2018, SI 5, pp. 215–220, 2019.
https://doi.org/10.1007/978-3-319-91989-8_47

1.2   The Discrete Element Method and Its Advantages

The Discrete Element Method (DEM) is one of numerical methods for simulating materials consisting of a large number of particles [4]. The basic idea behind DEM is to represent the material as an assemblage of discrete interacting particles mimicking the "atomistic" structure of materials [8, 10]. With advances in computer power and numerical algorithms for nearest neighbor sorting, it has become possible to numerically simulate behavior of material samples built of millions of particles on a single processor. At each time step, the calculations performed in DEM alternate between integrating equations of motion for each particle, and applying the force – displacement law at each particle contacts (Fig. 1).

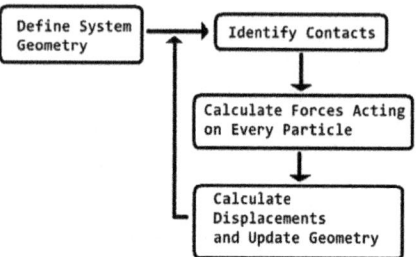

Fig. 1. The DEM simulation scheme.

From the perspective of the discussed application the main advantages of DEM are: (a) depicting the fracture geometry in details and (b) considering the progressive degradation of material integrity during this process [13]. Due to these two features of the DEM technique we are using this technique in our research.

1.3   ESyS-Particle – Open Source Implementation of DEM

The ESyS-Particle is open source implementation of DEM developed by the Earth Systems Science Computational Centre (ESSCC) at the University of Queensland [1]. The engine of this software is written in C++ programming language and it has got implemented the Message Passing Interface (MPI). The MPI parallelization technique allows to split performed simulations between multiple computers or multiple processor or cores within the same computer. This provides the most efficient way of performing the simulation [2].

The three main advantages of the ESyS-Particle software are: MPI parallel engine and so called BrittleBeamPrms particle-pair interaction model (Fig. 2) and finally the Open Source software distribution model. The most important numerical aspects of this software, the BrittleBeamPrms interaction model has been specifically designed for rock fracture and fragmentation simulations. It correctly simulates Griffith crack propagation as well as wing cracks formation. This type of interaction incorporates

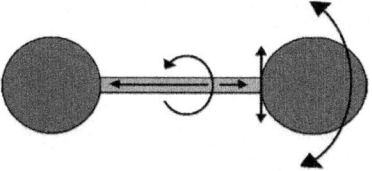

Fig. 2. The BrittleBeamPrms particle-pair interaction.

both translational and rotational degrees of freedom. As illustrated in Fig. 2, two bonded particles may undergo normal and shear forces, as well as bending and twisting moments.

2   DEM Model

2.1   Numerical Sample Preparation and Our Expectations

Preparation of rock-like samples with DEM is very demanding and time-consuming. In order to obtain sample which behavior correspond to the real material, iterative method is usually used. [14] Properties of particles' parameters and bonds between particles should recreate behavior of material with specific macro-parameters, like Poisson's Ratio or Young's Modulus and Unconfined Compressive Strength. This usually requires many trials and testing simulations what in practice can only be achieved with HPC computational resources [5, 11].

We have used DEM model for simulating cracks and fracking fluid propagation in rocks, pressure distribution in material during fracking, etc. [15] This data is very desirable since it cannot be obtained during real observations – fracking is being performed in very deep formations, even a few kilometers under the Earth surface, and it is impossible to observe that phenomena in 'in situ' conditions (a camera can be used, but only visible results can be recorded, not physical) [9].

2.2   The Basic Assumptions

The model for our simulation was created and it represents fluid injected into a rock through a single hole in the main tube. It consists of two blocks of particles – upper block represented by smaller grains stands for fracturing fluid, meanwhile block of bigger particles represents brittle rock material. Densities of rock particles and density of fracturing fluid were chosen to correspond to real data. Free fluid particles undergo frictional interactions and they can easily penetrate fractures inside particles of the rock. Consequently, friction between small particles stands for fluid viscosity. Injection of these small particles into the block of bigger particles represents, in a simplified way, injection of fracking fluid during real hydraulic fracturing.

3   Cracks Extension in Rocks Samples

3.1   The Study of Pressure and Stress Distribution in the Material

We checked numerous velocities of injection in order to find out how pressure changes during the process. We distinguished three parts of the simulation. In the first one the fluid is pumped into the rock formation what requires a high pressure to break bonds between particles of the rock formation. Shortly after that, pressure falls down significantly – the fluid start to percolate through the created pores/fractures. At this stage the fluid can flow easily into the newly opened space and, in this part of the simulation, nearly no force/pressure is needed. In the last, third part of the simulation all empty spaces in the rock sample are filled up with the fluid. There is no left space for new ones so force is needed to pump another portion of the fluid. We have studied behavior of the rock under fluid injected with constant pumping rate (velocity) and under a constant pressure.

3.2   How Far Can Cracks Propagate?

In our simulations we checked how far fluid particles can penetrate the rock. That gives the information how big cracks fracking phenomena are generated. For this purpose, the farthest positions of the fluid-like particles were found, and based on that information, the area of created cracks was calculated. The obtained results suggest that the area in which fluid's particles occurred is not directly related to the applied pressure. The dependence between pressure and volume of cracked rock is not linear. This opens the space for further studies directed towards an optimizing the relationship between the applied pressure and the extent of the generated fractures.

4   Summary

In this contribution we report our effort of a numerical simulation of processes occurring during hydraulic fracturing. We have used the Discrete Element Method implemented in ESyS-Particle Software as well as we took advantage of the Okeanos Supercomputer (Cray XC30) of the ICM UW supercomputing center [3]. Our model consists of millions of particles in size of tenths of millimeter. We have represented the fracturing fluid as an assembly of particles which undergo frictional interaction only. On the other hand, the rock-matrix was built of larger particles interacting each other by BrittleBeamPrms particle-pair interaction. Numerical model of the rock was previously carefully calibrated by comparison with macroscopic parameters of real materials (Young's Modulus, Poisson Ratio, Unconfined Compressive Strength).

We have investigated the behavior of brittle material into which fracking fluid was pumped with either constant velocity or constant pressure. Fluid, after breaking bonds between particles creating the rock, enters almost freely in created voids/pores. It was shown that the biggest pressure is needed to move the fluid into rock masses in the first phase of fracturing while in the second phase (fluid penetration) the pressure drops significantly.

Finally, it can be concluded that DEM approach is a useful tool for studying hydraulic fracturing process. Hydraulic fracturing is extremely expensive technology. By using DEM simulations, it was possible to check how fracking fluid behaves inside rock formation under different constant pressures or velocities. Consequently, injection velocities were calculated for different values of pressure. It gives an insight into the process which is possible only using computer simulations.

Acknowledgments. This research was supported by Young Scientist Grant Institute of Geophysics PAS no 6d/IGF PAN/2017 mł.

This research was carried out with the support of the Interdisciplinary Centre for Mathematical and Computational Modelling (ICM) Warsaw University, Computational Research Grant on The Okeanos Supercomputer, Project no GC70-6.

This paper was partially supported by research grant No. 2015/17/B/ST10/01946 of NCN, Poland.

References

1. Abe, S., Boros, V., Hancock, W., Weatherley, D.: ESyS – Particle Tutorial and User's Guide Version 2.3.1 (2014)
2. Abe, S., et al.: Simulation of the influence of rate- and state-dependent friction on the macroscopic behavior of complex fault zones with the lattice solid model. Pure. Appl. Geophys. **159**, 1967–1983 (2002)
3. Computational Research Grant on The Okeanos Supercomputer ICM UW, Project GC70-6, Simulations of Brittle Fracturing with The Discrete Element Method (2017)
4. Cundall, P.A., Strack, O.D.L.: A discrete numerical model for granular assemblies. Geotechnique **29**(1), 47–65 (1979). https://doi.org/10.1680/geot.1979.29.1.47
5. Damjan, B., Cundall, P.: Application of distinct element method to simulation of hydraulic fracturing in naturally fractured reservoirs. Comput. Geotech. **71**, 283–294 (2016)
6. Fakhimi, A., Alavi Gharahbagh, E.: Discrete element analysis of the effect of pore size and pore distribution on the mechanical behavior of rock. Int. J. Rock Mech. Min. Sci. **48**(1), 77–85 (2011). https://doi.org/10.1016/j.ijrmms.2010.08.007
7. Ghassemi, A.: Application of rock failure simulation in design optimization of the hydraulic fracturing. In: Shojaei, A.K., Shao, J. (eds.) Porous Rock Fracture Mechanics with Application to Hydraulic Fracturing, Drilling and Structural Engineering. Woodhead Publishing Series in Civil and Structural Engineering, United Kingdom, pp. 3–23 (2017)
8. Klejment, P., Dębski, W.: Brazilian test – a microscopic point of view on the tensile fracture generation. In: Conference Proceedings, V International Conference on Particle-based Methods – Fundamentals and Application (2017)
9. Klejment, P., Kosmala, A., Foltyn, N., Dębski, W.: Discrete element method as the numerical tool for better understanding the hydraulic fracturing process. In: GeoPlanet: Earth & Planetary Sciences (accepted). https://doi.org/10.1007/978-3-319-71788-3
10. Klejment, P., Kosmala, A., Foltyn, N., Dębski, W.: Symulacje komputerowe parametrów mikroskopowych i makroskopowych materiałów przy pomocy Metody Elementów Dyskretnych. Innowacje w polskiej nauce w obszarze matematyki i informatyki, pp. 5–19
11. Lisjak, A., Grasselli, G.: A review of discrete modeling techniques for fracturing processes in discontinuous rock masses. J. Rock Mech. Geotech. Eng. **6**(4), 301–314 (2014)

12. Onate, E., Rojek, J.: Combination of discrete element and finite element methods for dynamic analysis of geomechanics problems. Comput. Methods Appl. Mech. Eng. **193**, 3087–3128 (2004)
13. O'Sullivan, C.: Particulate Discrete Element Modeling: A Geomechanics Perspective. Spon Press/Taylor & Francis, New York (2011)
14. Potyondy, D., Cundall, P.: A bonded-particle model for rock. Int. J. Rock Mech. and Min. Sci. **41**(8), 1329–1364 (2004). https://doi.org/10.1016/j.ijrmms.2004.09.011
15. Tan, P., et al.: Analysis of hydraulic fracturing initiation and vertical propagation behavior in laminated shale formation. Fuel **206**, 482–493 (2017)

Controlling Dynamic Fracture and Structural Disintegration

Shintaro Sakaguchi[1]([⊠]), Koji Uenishi[1], Hiroshi Yamachi[2],
and Junichiro Nakamori[2]

[1] Department of Aeronautics and Astronautics,
University of Tokyo, Tokyo 113-8656, Japan
sakaguchi@dyn.t.u-tokyo.ac.jp
[2] Sumitomo Mitsui Construction Co. Ltd., Chiba 270-0132, Japan

Abstract. In recent years, dynamic techniques to partially or fully disintegrate structures utilizing electrical energy have been proposed, which may be more controllable than other conventional dynamic demolition methods, for instance, blasting by explosives. So far, by applying electric discharge impulses in the field, we have performed fracture experiments of relatively small brittle concrete specimens, and together with our finite difference numerical investigations, we have pointed out that the development of fracture network depends very sensitively on the geometrical settings in the specimens (e.g. positions of blast holes (energy sources), empty dummy holes, free surfaces and interfaces to reflect and/or diffract waves and control crack propagation directions). Here, we continue our study and observe that even for dynamic disintegration in more realistic, large concrete slabs, the geometrical settings do play a crucial role and in this case, pre-set slits have strong effect on wave propagation and eventual dynamic fracture pattern. Our three-dimensional numerical code for a PC may well explain the experimental results, suggesting that further research to find the optimal geometrical settings for more predictable and controlled dynamic structural disintegration should be conducted from the wave dynamics point of view.

Keywords: Controlled dynamic disintegration · Wave dynamics
Fracture mechanism

1 Introduction

Understanding the fundamental mechanics of dynamic fracture is significantly important not only to avoid (often unexpected) devastating failures of structures but also to develop faster, more precise and quantitatively controllable techniques for partial or full disintegration of old structures. However, the mechanism of dynamic fracture, especially that due to impact loading, has not been fully clarified yet. When impact load is given to structures made of solid materials, waves may be induced and stress field inside the structures may change quickly and drastically with time. The field may become very complicated, for example, because of wave reflection and interaction near interfaces. This complexity makes it difficult to analyze wave/crack propagation and development of ultimate fracture network.

© Springer International Publishing AG, part of Springer Nature 2019
E. E. Gdoutos (Ed.): ICTAEM 2018, SI 5, pp. 221–226, 2019.
https://doi.org/10.1007/978-3-319-91989-8_48

Structural disintegration by blasting (detonation of explosives) has been one of the most widely used techniques for dynamic demolition, but when applied to construction/destruction sites in densely populated urban regions, it still has problems in light of safety (controllability) including noise. In order to enhance the safety of dynamic disintegration work, recently, fracture technology utilizing electrical energy has been proposed. In the electric discharge impulse crushing system (EDICS) developed by Nichizo Tech, Inc., for instance, the electric energy stored in a capacitor (typically 3000 V) is released in a self-reactive liquid (deflagration agent like nitromethane) in several hundreds of microseconds through an electronic switch and high pressure is generated by the rapid evaporation of the liquid. The produced electric discharge impulses (EDI) are considered to be suitable for precisely controlling structural disintegration, because their power can be regulated without danger simply by adjusting the volume of the self-reactive liquid contained in cartridges that are placed in blast holes (energy sources).

In our previous study, based on wave dynamics, some controllable fracture techniques with EDI have been developed for relatively small brittle concrete specimens with/without reinforcing steel bars. It has been found that cracks can really connect multiple blast holes as desired but the dynamic fracture mechanism is rather complicated: Although the final main crack plane looks as if it were developed uni-directionally, our high-speed digital video camera has revealed that the crack actually has extended multi-directionally from blast holes, free surfaces of the specimen, and if pre-set, from empty dummy holes that are introduced to guide the crack propagation direction [1]. Dummy holes, placed either on or outside the expected main crack path, can be used not only to direct crack propagation directions but also to effectively break only targeted areas [2]. Slight shifts in the positions of dummy holes can considerably change the final fracture pattern in the specimens, but in order to find the optimal geometrical settings for structural disintegration with EDI, further research is required. Therefore, here, we try to efficiently destruct only some portions of more realistic model structures, i.e. large concrete slabs with relatively smaller free surface areas compared with our earlier specimens. We examine whether our wave-based disintegration-control techniques are valid also for larger models or not.

2 Experimental Observations

As stated above, during the course of our previous experimental and numerical investigations [1–3], the effect of geometrical and loading conditions on dynamic destruction of given structures has been investigated. In this study, by setting in advance vertical slits inside large, unreinforced concrete slabs, we try to control wave interaction and dynamic structural disintegration. Figure 1 shows the slabs and a typical section (1,050 mm × 300 mm), prepared and fractured by EDI. The experimental area is composed of two types of concrete slabs (3,500 mm × 2,700 mm × 120 mm and 3,500 mm × 2,700 mm × 150 mm), and they are split into sections with slits (depth 70 mm) as archetypally shown in Fig. 1 (right). Every section has six vertical blast holes (diameter 12 mm, depth 110 mm) along its center line at intervals of 150 mm, and in each blast hole a cartridge (diameter 10 mm, length 80 mm)

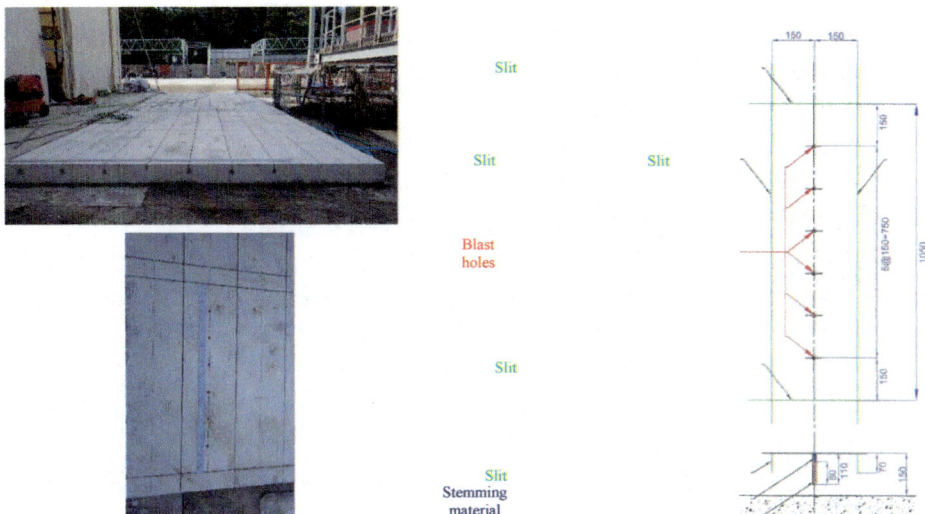

Fig. 1. The concrete slabs prepared for the field fracture experiments utilizing electric discharge impulses (EDI) (top left) and typical rectangular section (1,050 mm × 300 mm) without reinforcing steel bars (bottom left). The cartridges indicated in red contain the self-reactive liquid and they are set in the blast holes and covered by stemming material (CCR tamper) that is marked in blue. The positions of pre-set slits to control fracture extension are shown in green. Each section has six blast holes along the center line and they are simultaneously pressurized with EDI [unit: mm] (right).

containing the self-reactive liquid (2 cm³) is placed, covered by stemming material (CCR tamper) and connected to the control unit of EDICS. The previous series of investigation suggests that inside this section, a main crack connecting the blast holes, bifurcated cracks running from the blast holes at both ends to the slits will emerge owing to the dynamic tension induced by waves reflected at the top horizontal free surfaces and the vertical slits. At the same time, it is expected that a horizontal fracture plane (approximately) connecting the bottoms of the vertical slits may be developed due to wave diffraction around them. As a result, only the rectangular zone surrounded by the slits, with a depth comparable to that of the slits (70 mm), will be fractured. Figure 2 shows the typical final fracture network generated by the simultaneous application of EDI to the slabs. As expected, a main crack with bifurcated ends can be observed in Fig. 2 (top).

The surprisingly flat horizontal fracture plane near the bottoms of the slits is also recognized, and it is located at a depth slightly larger than the slit depth (Fig. 2, bottom). Thus, only a rectangular part of concrete slabs can be effectively and quickly removed. Although in addition to blast holes, slits must be prepared before the dynamic EDI application, we believe that the abovementioned method, namely, setting energy sources inside slits with some depths to remove only the block section surrounded by the slits smoothly, is of practical importance in complex sites in urban regions where extremely careful treatment concerning structural disintegration is required.

Fig. 2. Typical final fracture network (top) generated by the simultaneous application of EDI to the slabs illustrated in Fig. 1. The fractured section after removal of broken fragments (bottom left) and the main fragment (bottom right) are shown. Crack connecting the six blast holes and the fracture plane constituting the bottom face of the main fragment are observable.

3   Numerical Speculations and Discussion

In order to numerically investigate wave propagation and interaction induced by EDI and comprehend the effect of relative positions of the blast holes to the top free surface and the slits on fracture evolution, a three-dimensional finite difference simulator for a PC has been developed with the spatiotemporally second order accuracy. In the preliminary simulations, homogeneous, isotopic and linear elastic concrete (mass density 2,270 kg/m^3, Young's modulus 33.3 GPa and Poisson's ratio 0.28) is assumed. This combination of material properties gives the longitudinal (P) and shear (S) wave speeds as $V_P \approx 4,300$ m/s, $V_S \approx 2,400$ m/s, respectively. Orthogonal 211 × 211 × 81 grid points with constant grid spacing $\Delta x = 10$ mm are set for the concrete slab model (2,100 mm × 2,100 mm × 800 mm). For graphical (presentation) clarity, the stemming material covering the cartridges in the blast holes is presumed to have the same physical characteristics as concrete material. The (cross-sectional) width of all blast holes is the same (square cross-section with sides of length $2\Delta x$) and waves may propagate cylindrically from the middle parts and (hemi) spherically from the top and bottom ends of the cartridges placed in the blast holes. The time history of the EDI-related pressure $P(t)$ is assumed to have the form $P(t) = A \sin^2(\pi t/T_0)$ for $(0 \leq t \leq T_0)$ and 0 (otherwise) and is applied to the walls of the blast holes (cartridges), with $A = 1.0$ GPa and the duration $T_0 = 260$ μs based on the previous study [1, 2]. No specific fracture criterion is incorporated at this moment because propagation and interaction of waves generated by the new technique, EDI application, must be understood first. Usually, body waves propagate earlier than fractures and therefore, before fracture is initiated, our simulations to find the most endangered sections in the model are valid even without introducing any fracture criterion. In Fig. 3, snapshots of contours of volumetric strain at time 40, 80, 120 and 160 μs after the simultaneous application of EDI are shown. It is noticeable that, upon wave interaction with the top horizontal free surfaces and pre-set slits (vertical free surfaces), dilatational and compressive zones develop and move considerably inside the model. In Fig. 3, the waves generated around the cartridges are initially compressive (a), and then tensile zones due to reflection at the top free surface and the slits emerge (a, b, c, d), which may induce cracks there as observed in Fig. 2. Plotting only the zones of larger tension (dilatation between 5.0×10^{-5} and 5.0×10^{-4}, not shown here due to the page limitation but roughly corresponding to the zones surrounded by yellow regions in Fig. 3) clearly indicates that tensile zones which are likely to be more easily fractured can be identified only inside the rectangular section bordered by the top horizontal free surface, vertical slits and the rather flat horizontal plane near the bottoms of the slits, as observed in Fig. 2, and thus fracture generation mechanism may be qualitatively explained.

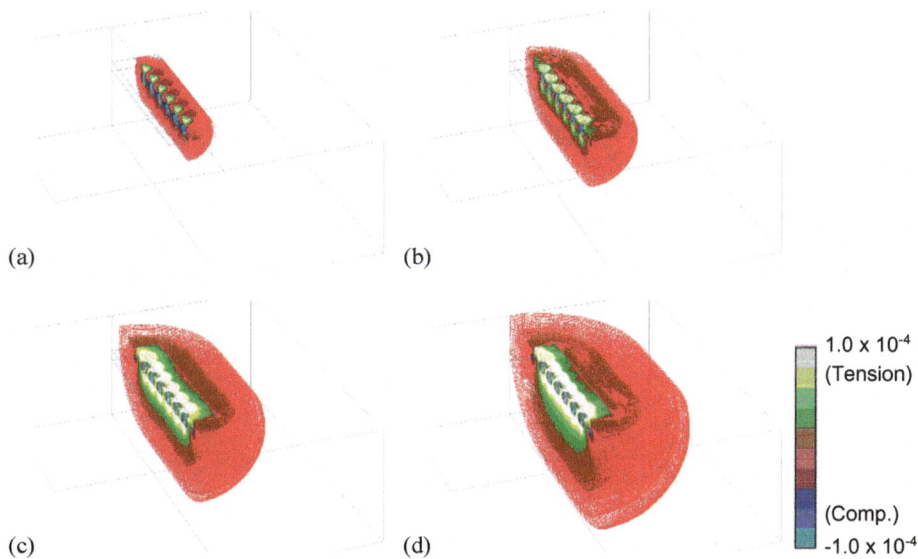

(a)   (b)

(c)   (d)

1.0 x 10^{-4}
(Tension)

(Comp.)
-1.0 x 10^{-4}

Fig. 3. Snapshots of dynamic disturbances moving in the rectangular section in concrete slabs with six EDI sources, generated by our fully three-dimensional finite difference numerical calculations. For the initial speculation of wave propagation and interaction in the structure, no fracture criterion is introduced. Contours of volumetric strain at time (a) 40, (b) 80, (c) 120 and (d) 160 μs after the simultaneous application of EDI are illustrated for the right half part of the rectangular section.

4 Conclusions

The field experiments of dynamic disintegration of sections in large concrete slabs without reinforcing steel bars have shown that a main crack running through the blast holes can be generated as desired and, together with the horizontal fracture plane connecting the bottoms of the slits, only a rectangular part can be efficiently removed from concrete slabs. The relative positions of the blast holes to the top free surface and pre-set slits (compared with the wavelengths of EDI-induced waves) may play an important role in generating final fracture pattern. Our fully three-dimensional finite difference numerical code for a PC may well explain the real field observations qualitatively. The experimentally and numerically obtained results will be immediately applicable to practical construction/destruction work. However, in order to achieve more precisely controlled dynamic structural disintegration, further research is required, and especially a careful study on an appropriate physical criterion for three-dimensional fracture of brittle solids will be needed.

Acknowledgements. The authors are grateful to the technical support provided by Nichizo Tech, Inc.

References

1. Uenishi, K., Yamachi, H., Yamagami, K., Sakamoto, R.: Dynamic fragmentation of concrete using electric discharge impulses. Constr. Build. Mater. **67**(B), 170–179 (2014)
2. Uenishi, K., Shigeno, N., Sakaguchi, S., Yamachi, H., Nakamori, J.: Controlled disintegration of reinforced concrete blocks based on wave and fracture dynamics. Procedia Struct. Integrity **2**, 350–357 (2016)
3. Uenishi, K., Takahashi, H., Yamachi, H., Sakurai, S.: PC-based simulations of blasting demolition of RC structures. Constr. Build. Mater. **24**(12), 2401–2410 (2010)

Degradation of the Strength of a Grain Boundary in Ni-Base Superalloy Under Creep-Fatigue Loading

Wataru Suzuki[1], Akari Sawase[2], Ken Suzuki[2], and Hideo Miura[2(✉)]

[1] Department of Finemechanics, Tohoku University,
Sendai, Miyagi 9808579, Japan
[2] Fracture and Reliability Research Institute, Tohoku University, Sendai, Japan
hmiura@rift.mech.tohoku.ac.jp

Abstract. Electron back-scatter diffraction method was applied to analyze the degradation process of Ni-base superalloy Alloy 617 under various loading conditions at elevated temperatures from the view point of the change of the order of atom arrangement in grains and grain boundaries. The local damage evolution around grain boundaries was evaluated by applying the intermittent creep and creep-fatigue tests. The creep-fatigue test was operated at 800 °C under the stress of 150 MPa with strain rate of 0.4%/s and hold time of 10 min in inert gas (99.9999% Ar). Each test was stopped at a certain strain and the micro texture was evaluated step by step at the same area in each test sample continuously. The crystallinity of grains and grain boundaries was evaluated by applying image quality (IQ) and confidence index (CI) values obtained from EBSD analyses. It was found that degradation of the crystallinity was accelerated at certain grain boundaries drastically and the first crack always initiated at a grain boundary when the IQ value decreased to a critical value. It was also confirmed that this degradation decreased the strength of the grain boundary. Finally, it was concluded that the strength of a grain boundary was dominated by its crystallinity and there was a critical IQ value at which a crack initiated at the grain boundary.

Keywords: Crystallinity · Strength · Grain boundary · Creep-fatigue
Ni-base superalloy

1 Introduction

In order to further improve the thermal efficiency of thermal power plants and reduce CO_2 emissions, various R&D projects have been conducted to develop A-USC (advanced ultra-supercritical) power plants of the 700 °C-class. Ni-base superalloy is a candidate alloy for boiler tubes and pipes of A-USC power plants [1, 2]. Since it has higher coefficient of thermal expansion than conventional ferritic steels, however, the increase in the thermal stress is concerned in the power plant components. In addition, it is important to consider the effect of creep-fatigue loading on the lifetime of the components in the design and maintenance of the components. This is because frequent output change is concerned in the power plants from the viewpoint of the stable total

© Springer International Publishing AG, part of Springer Nature 2019
E. E. Gdoutos (Ed.): ICTAEM 2018, SI 5, pp. 227–232, 2019.
https://doi.org/10.1007/978-3-319-91989-8_49

power supply with the consideration of unstable change of the output of various renewable energies. In order to assure the safe and reliable coexistence of thermal power plants and renewable energy systems, it is indispensable to operate the thermal power plants under the random frequent change of output at elevated temperatures. The increase in the creep-fatigue damage should degrade the strength and reliability of heat-resistant alloys used in the plants. It is well-known that the fracture lifetime of heat-resistant alloys decreases drastically under creep-fatigue loadings and the main crack propagation mode changes from trasgranular to intergranular [3]. Therefore, it is very important to clarify the degradation mechanism quantitatively and establish its evaluation method for estimating the residual life of the alloys. In this study, electron back-scatter diffraction method was applied to analyze the degradation process of Ni-base superalloy Alloy 617 under various loading conditions at elevated temperatures from the view point of the change of the order of atom arrangement in grains and grain boundaries.

2 Evaluation Method of Crystallinity

The crystallinity of grains and grain boundaries was evaluated by applying image quality (IQ) and confidence index (CI) values obtained from EBSD analyses [4–6]. The IQ value is the average sharpness of the obtained diffraction (Kikuchi) pattern as shown in Fig. 1, and it was calculated by averaging the Hough-transform of each Kikuchi line. Since the sharpness of Kikuchi line is dominated by the order of atom arrangement in the observed area, it was used for estimating the order of atom arrangement [7]. Since the order of atom arrangement is degraded by various defects such as vacancies, dislocations, impurities, local distribution of strain, and so on, the change of IQ value was attributed to the change of crystallinity of the observed grains and grain boundaries in the analysis. The density of dislocation was evaluated by using Kernel average misorientation (KAM). The EBSD analysis was performed by using an OIM (Orientation Imaging Microscopy) system commercialized by TSL. In this study, a grain boundary was defined as the area within 6 μm from the center position of the conventionally defined grain boundary, which was determined by CI value. This was because all the defects were found to mainly accumulate in this area.

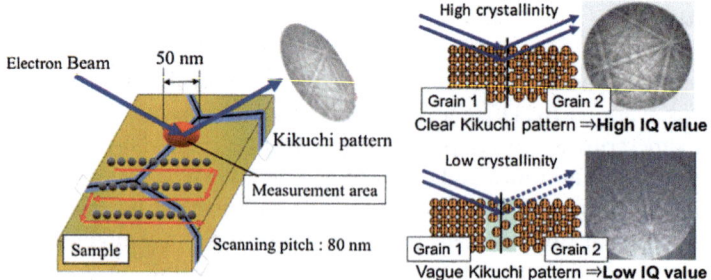

Fig. 1. Crystallinity evaluation by applying EBSD analysis

The local damage evolution around grain boundaries was evaluated by applying the intermittent creep and creep-fatigue tests as shown in Fig. 2. The creep test was performed at 800 °C under the stress of 150 MPa in inert gas (99.9999% Ar). The creep-fatigue test was operated at 800 °C under the stress of 150 MPa with strain rate of 0.4%/s and hold time of 10 min in inert gas (99.9999% Ar). Each test was stopped at a certain strain and the micro texture was evaluated step by step at the same area in each test sample continuously. This procedure was effective for analyzing the change of micro texture in the same area in a sample during the test. The surface of the test specimen was carefully polished before the analysis by using 3-μm diamond paste for flattening the sample surface. The final polish was performed using 50-nm colloidal silica to eliminate the surface damage layer and the thin oxide layer. Table 1 shows the chemical composition of the Alloy 617. Solution annealing at 1150 °C for 1.12 h was performed before each test.

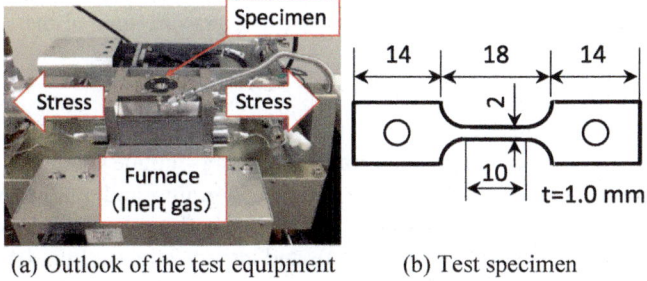

(a) Outlook of the test equipment   (b) Test specimen

Fig. 2. Intermittent creep-fatigue test with continuous monitor of micro texture

Table 1. Composition of alloy 617 (mass%)

C	S		Cr	Ni	Mn	Si	Mo
0.06	<0.002		21.98	54.71	0.03	0.06	8.8
Ti	Cu	Fe	P		Al	Co	B
0.4	0.02	0.8	<0.002		1.14	11.69	0.001

3   Change of the Crystallinity During the Creep-Fatigue Tests

Figure 3 shows the observed example changes of the micro texture (inverse pole figure map) of the alloy during the intermittent creep-fatigue tests. In these maps, each color indicates different crystallographic orientation. By using these maps, it is easy to find out the position of each grain boundary. All the cracks initiated at grain boundaries and the crack propagated along grain boundaries even though heavy plastic deformation was observed in each grain. The upper row shows a change of micro texture where a crack initiated at $t/t_f = 0.5$, where t is test time and t_f is the final fracture time of this specimen. The lower row shows another area where the first crack initiated at $t/t_f = 0.8$.

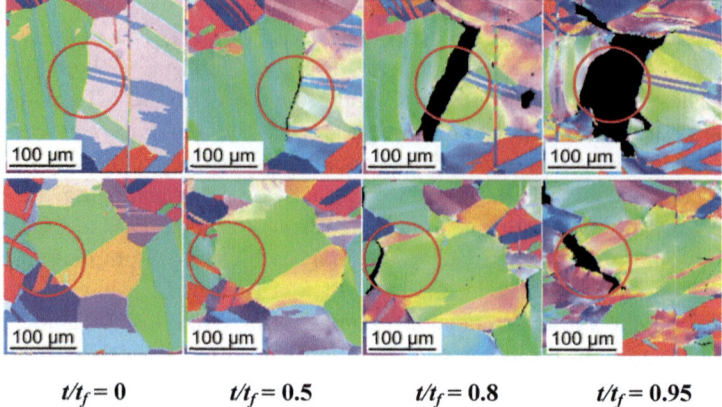

$t/t_f = 0$ $t/t_f = 0.5$ $t/t_f = 0.8$ $t/t_f = 0.95$

Fig. 3. Two examples of the continuous observation of the initiation and propagation of a crack at grain boundaries under the creep-fatigue loading. t_f is fracture time of the sample and t is loading time.

The initial cracks were observed at $t/t_f = 0.98$ under the creep loading. Therefore, the crack initiation was clearly accelerated under the creep-fatigue loading.

The changes of the crystallinity (IQ value) of grains and grain boundaries are summarized in Fig. 4. The crystallinity of grains and grain boundaries decreased monotonically during the creep (Fig. 4(a)) and creep-fatigue (Fig. 4(b)) tests. The decrease rate under the creep-fatigue loading was slightly faster than that under the creep test. It was found that degradation of the crystallinity was accelerated at certain grain boundaries drastically and the first crack always initiated at a grain boundary when the IQ value decreased to about 85% of the initial value in this study. This result indicates that the there is a critical IQ value at which a crack starts to initiate at the grain boundary. It was also confirmed that this degradation decreased the strength of the

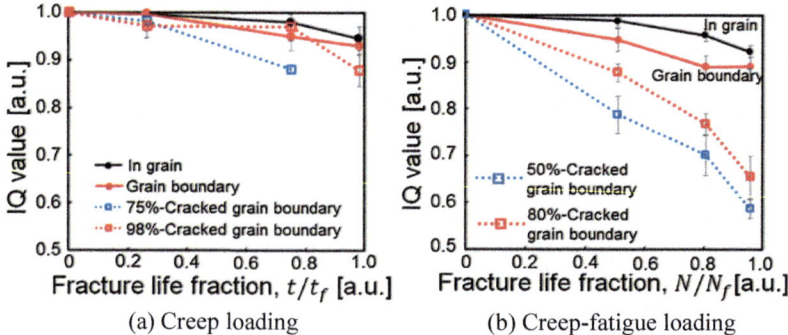

(a) Creep loading   (b) Creep-fatigue loading

Fig. 4. Degradation of the crystallinity of grains and grain boundaries during the creep and creep-fatigue tests. The crystallinity of a cracked grain boundary decreased drastically.

grain boundary. Even though the initial quality of grain boundaries was almost the same, there were grain boundaries where the damage was accumulated faster than other grain boundaries. The reason for the decrease of IQ value is attributed to the increase of total defects such as dislocations and vacancies.

Figure 5 shows the changes of KAM value in grains and grain boundaries under the creep loading (Fig. 5(a) and creep-fatigue loading (Fig. 5(b)), respectively. During the both tests, it was confirmed KAM value, in other words, plastic deformation increased monotonically in grains and grain boundaries. The increase ratio in grains under the creep loading was almost the same as that under the creep-fatigue loading. The increase rate in grain boundaries under the creep-fatigue loading, however, was much larger than that under creep loading. The accumulation of dislocations was clearly accelerated under the creep-fatigue loading. This was the reason for the decrease in the fracture time under the creep-fatigue loading comparing with that under the creep loading.

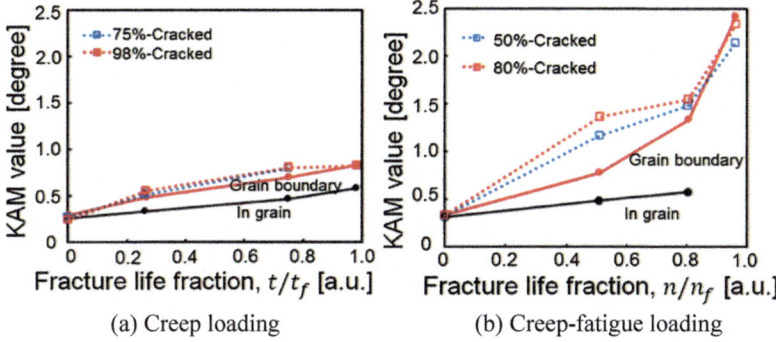

(a) Creep loading   (b) Creep-fatigue loading

Fig. 5. Changes of KAM value of grains (black line), uncracked grain boundaries (red line), and cracked grain boundaries (blue and red dashed lines)

By comparing the observed results shown in Fig. 5 and those in Fig. 4, it was confirmed that the main reason for the drastic decrease of IQ value in cracked grain boundaries was the increase of dislocation density in those grain boundaries. Accumulation of dislocations around grain boundaries was the main reason for the initiation of a crack at a certain grain boundary. The strength of a grain boundary was degraded by this accumulation of dislocations. Since there was no clear difference in the accumulation rate of dislocations in grains as shown in Figs. 5(a) and (b), the average diffusion constant of dislocations under the creep loading was almost the same as that under the creep-fatigue loading. Therefore, it was assumed that the nucleation density of dislocations under the creep-fatigue loading was much larger than that under the creep loading. It is necessary to clarify the acceleration mechanism of the accumulation of dislocations around grain boundaries under the creep-fatigue loading.

4 Summary

The local damage evolution around grain boundaries in Ni-base alloy, Alloy 617 under creep-fatigue loadings was evaluated by applying the intermittent creep and creep-fatigue tests with EBSD analyses. It was found that degradation of the crystallinity was accelerated at certain grain boundaries drastically and the first crack always initiated at a grain boundary when the IQ value decreased to a critical value. The decrease rate of IQ value in cracked grain boundaries under the creep-fatigue loading was much faster that that under the creep loading. The main reason for the rapid decrease was attributed to the drastic accumulation of dislocations around the cracked grain boundaries. The strength of a grain boundary was degraded by this accumulation of dislocations.

Since there was no clear difference in the accumulation rate of dislocations in grains regardless of the loading mode, it was assumed that the nucleation density of dislocations under the creep-fatigue loading was much larger than that under the creep loading.

Finally, it was concluded that the strength of a grain boundary was dominated by its crystallinity and there was a critical IQ value at which a crack initiated at the grain boundary.

Acknowledgments. This research was supported partly by JSPS KAKENHI Grant Number JP16H06357.

This research activity has been also supported partially by Japanese special coordination funds for promoting science and technology, Japanese Grants-in-aid for Scientific Research, and Tohoku University.

References

1. Masuyama, F.: Current status and future trends of Mo-based silicide alloys for ultrahigh temperature applications. JSPS Report on the 123rd Committee on Heat Resisting Materials and Alloys (2007)
2. American Society of Mechanical Engineers: Boiler and Pressure Vessel Code, Sec. III NH (2010)
3. Murakoshi, T., Suzuki, K., Nonaka, I., Miura, H.: Microscopic analysis of the initiation of high-temperature damage of Ni-based heat-resistant alloy. In: Proceedings of ASME 2016 International Mechanical Engineering Congress and Exposition, no. IMECE2016-67599, pp. 1–6 (2016)
4. Dunne, F.P.E., Wilkinson, A.J., Allen, R.: Int. J. Plast. **23**, 273–295 (2007)
5. Othon, M.A., Morra, M.M.: 182 weld joints. Microsc. Microanal. **11**, 522–523 (2005)
6. Kamaya, M., Wilkinson, A.J., Titchmarsh, J.M.: Nucl. Eng. Des. **235**, 713–725 (2005)
7. Murata, N., Suzuki, K., Miura, H.: Quantitative evaluation of the crystallinity of grain boundaries in polycrystalline materials. In: Proceedings of ASME IMECE 2012, no. 87426, pp. 1–6 (2012)

Analysis of Fracture of Composites Loaded Along Cracks

Viacheslav Bogdanov[(✉)]

S.P. Timoshenko Institute of Mechanics of National
Academy of Sciences of Ukraine, Kiev 02053, Ukraine
bogdanov@nas.gov.ua

Abstract. The results of studying the problems on the fracture of composites containing interacting cracks under the action of forces directed along the cracks are reported. Given are the descriptions of two non-classical failure mechanisms – the fracture of compressed bodies under the action of forces directed in parallel to the planes containing cracks and the fracture of composites with residual stress acting along cracks. We use a combined approach to investigate the abovementioned fracture mechanisms within the framework of linearized solid mechanics. Spatial problems on composites that contain interacting circular cracks are considered. Problems on an infinite body containing two parallel co-axial cracks and on a space with the periodical set of co-axial parallel cracks as well as those on a half-space with subsurface crack are solved. Several patterns of loading on the crack faces (normal loading, radial shear and torsion) are considered. The effects of residual stresses on stress intensity factors are analyzed for some types of composite materials. Critical fracture parameters for composites with interacting cracks compressed along the cracks are calculated using the approach mentioned. The effect of geometrical parameters of the problems as well as physical and mechanical properties of materials on these critical parameters is analyzed.

Keywords: Residual stress · Compression along cracks · Composite materials

1   Introduction

Processes of manufacturing composite materials often induce residual stress-strain fields which can considerably affect the values of the fracture loads for composites containing crack-like defects. In the case when the residual stresses act along cracks, the approaches of classical fracture mechanics turn out to be invalid. Another group of non-classical problems is the fracture of composites compressed along crack planes which occurs because of the loss of material stability in the vicinity of cracks. For study of abovementioned problems of fracture in works of A.N. Guz an approach based on the three-dimensional linearized mechanics of deformable bodies was elaborated.

In the paper a new method of calculation of critical parameters of fracture of composites with cracks under compression along the cracks planes is proposed. These parameters are calculated as the values of residual compressive stresses which cause sharp "resonance-like" changes in the stress-strain state parameters, in particular, in

© Springer International Publishing AG, part of Springer Nature 2019
E. E. Gdoutos (Ed.): ICTAEM 2018, SI 5, pp. 233–234, 2019.
https://doi.org/10.1007/978-3-319-91989-8_50

stress intensity factors. We demonstrate the employment of this method by the investigation of 3d problems on the fracture of composites containing interacting penny-shaped cracks (two parallel co-axial cracks, periodical set of co-axial parallel cracks as well as near-surface crack). Herewith it is assumed that crack sizes are significantly larger than those of the structural elements of composites (i.e. macrocracks are considered); besides, we only analyze fracture processes in which the composite does not demonstrate properties of the piecewise-uniform medium (like fracture on the media interface etc.). Under such assumptions, we use the continuum composite model with the abovementioned characteristics of a transversally isotropic body.

2   Results

The mathematical statements of problems for a pre-stressed unbounded composite containing two parallel co-axial cracks or a periodical set of co-axial parallel cracks as well as for a semi-infinite composite with a subsurface crack are given. The analysis involves the representation of stresses and displacements of the linearized theory through harmonic potential functions. Through the use of the integral Fourier–Hankel transformations we reduce the problems first to systems of dual integral equations and then to Fredholm integral equations of the second kind, which can be effectively analyzed numerically. From the analysis of the stress distribution in the vicinity of the cracks we obtain values of stress intensity factors and investigate their dependence on the initial stresses, mechanical properties of composite components and geometrical parameters of the problems. By using the method proposed we calculate the critical parameters of fracture of composites containing interacting cracks under compression along the cracks. The influence of geometrical parameters of the problems as well as physical and mechanical properties of materials on these critical parameters is analyzed.

Several patterns of loading on the crack faces (normal loading, radial shear and torsion) are considered. Numerical results are obtained for two types of composites, namely, laminated two-component composites with isotropic layers and a composite with stochastic reinforcement by short ellipsoidal fibres.

The results obtained allow the following conclusions:

– for all the problems considered the stress intensity factors are substantially dependent on the residual stresses;
– the values of the stress intensity factors change abruptly (the "resonance-like" effect) when the residual stresses tend to the values with which there is a local loss of stability in the crack's vicinity. According to the method proposed here this effect enables one to determine critical parameters of loading in problems on composites compressed along the cracks;
– the geometrical parameters of the problems and mechanical properties of composite materials had great effect both on the values of the stress intensity factors and on the critical compressive stresses.

Studying of Fracture of the Orthotropic Elastic and Visco-Elastic Plates with Periodic System of Collinear Cracks

Olga Bogdanova[✉]

S.P. Timoshenko Institute of Mechanics of National Academy
of Sciences of Ukraine, Nesterov Str. 3, 02053 Kyiv, Ukraine
o.bogdanova@i.ua

Abstract. On the basis of a modified δ_c-model of crack, the limiting state of an orthotropic plate made of a material satisfying the general strength condition and weakened by a system of collinear cracks is studed. The relations for the determination of major parameters of the model of cracks (the size of process zones, stresses in these zones, and the crack-tip opening displacements) are deduced. The mechanism of fracture of the plate containing a periodic system of collinear cracks is investigated. The influence of the degree of anisotropy and geometric parameters of the problem on the formation of the process zones and limiting state of the plate is revealed. The region of safe loading of an orthotropic viscoelastic plate with cracks is determined. The influence of the rheological parameters of the material on the region of safe loading is analyzed.

Keywords: Periodic system of collinear cracks · Critical loading
Safe loading · Orthotropic materials

On the basis of a modified δ_c-model of crack a thin orthotropic plate with periodic system of collinear cracks of length $2l$ located along the axis of Ox-axis is considered. The plate is stretched by a homogeneous load applied at infinity $\sigma_y = p > 0$, $\sigma_x = 0$, $\tau_{xy} = 0$ by $z \to \infty$ ($z = x + iy$).

The field of ultimate loads p_* can be defined by δ_c-criterion as

$$
\sigma_y^0(p_*) \frac{\tilde{l}}{\arcsin(\tilde{l})} \int_1^{\tilde{L}/\tilde{l}} F(t, \rho_*) \frac{dt}{\sqrt{1 - t^2 (\tilde{l})^2}} = 2\sigma_y^0\left(p_*^{(1)}\right) \ln \sec \frac{\pi p_*^{(1)}}{2\sigma_y^0\left(p_*^{(1)}\right)},
$$

$$
F(x, r) = \ln \frac{\left(1 + x\cos^2 r + \sqrt{1 - x^2 \cos^2 r} \sin r\right)(x - 1)}{\left(1 - x\cos^2 r + \sqrt{1 - x^2 \cos^2 r} \sin r\right)(x + 1)}, \qquad \rho_* = \frac{\pi p_*}{2\sigma_y^0(p_*)}.
$$

(1)

where $p_*^{(1)}$ - the limiting load in the case of a plate with a single crack.

E. E. Gdoutos (Ed.): ICTAEM 2018, SI 5, pp. 235–236, 2019.
https://doi.org/10.1007/978-3-319-91989-8_51

The load $p_*^{(d)}$ at which the process zones of neighboring cracks occur is defined by

$$\frac{D}{l} = \frac{\pi}{2 \arcsin(\cos \rho^{(d)})}, \quad \rho^{(d)} = \frac{\pi p_*^{(d)}}{2\sigma_y^0\left(p_*^{(d)}\right)} \tag{2}$$

Thus, the boundary condition of the plate with the periodic system of collinear cracks is determined by the set of conditions (1) and (2). The dependence of the critical load on the dimensionless distance between the cracks D/l obtained on the basis of the relations (1) and (2) was constructed. It is obvious that for arbitrary crack lengths and the distance between them, there is such a "critical" level of load when there is a partial destruction of the entire intersection along the line of location of the cracks. This "critical" load is higher, the smaller the length of the cracks.

A thin orthotropic plate of viscoelastic material weakened by a periodic system of collinear cracks located along one of the axes of orthotropy axis is considered.

For bodies with limited creep the "safe" loads p_S for which the value $\delta(l_0, t)$ can't reach the critical value δ_c and the development of the crack does not happen is determined

$$\sigma_y^0(p_S) \frac{\tilde{l}}{\arcsin(\tilde{l})} \int_1^{\tilde{L}/\tilde{l}} F(\tau, \rho_S) \frac{d\tau}{\sqrt{1 - \tau^2(\tilde{l})^2}} = 2 \frac{T_0}{T_\infty} \sigma_y^0\left(p_*^{(1)}\right) \ln \sec \frac{\pi p_*^{(1)}}{2\sigma_y^0\left(p_*^{(1)}\right)}. \tag{3}$$

The dependence of the safe load p_S on the dimensionless distance between the cracks and the rheological parameters of the material is obtained. It can be noted that the change in the rheological parameters of the material λ and γ significantly affects the area of safe load: the decrease λ or increase $|\gamma|$ in other permanent characteristics leads to an increase in the level of safe loading, the parameter α does not affect the level of safe load. Thus, by changing the rheological parameters of the material, it is possible to achieve a decrease in the area of loading, in which the development of cracks occurs.

1 Conclusions

The proposed modification of the Leonov-Panasyuk-Dagdale crack model to the case of orthotropic materials allows to effectively solve problems of the destruction of orthotropic bodies with cracks, the material of which satisfies the condition of the strength of the general form. With this model, the boundary state of the orthotropic plate, relaxed by the periodic system of collinear cracks, the material of which is elastic or visco-elastic, is investigated. The mechanism of fracture of the plate with cracks and the influence on the boundary and safe loading of the properties of the material have been investigated.

Generalized Stress Intensity Factors Determination by Overdeterministic Method in Case of Bi-material Junction

Ondřej Krepl$^{(\boxtimes)}$ and Jan Klusák

CEITEC IPM, Institute of Physics of Materials AS CR,
Zizkova 22, 616 62 Brno, Czech Republic
krepl@ipm.cz

Abstract. Generalized stress intensity factors H_k (GSIFs) are the necessary parameters to describe stress state near bi-material junction tip by means of asymptotic series. GSIFs can be determined as a least square method solution of overdetermined system of linear equations. The equations consist of analytical eigenfunctions, which are determined as a solution of eigenvalue problem and results of finite elements analysis (FEA), in form of displacement or stress components. The study presents the application of the overdeterministic method on bi-material junction problem. The effect of number of input values from FEA and the radius of their extraction on resulting H_k is studied. The study also shows how number of calculated singular and non-singular terms affects values of individual factors. The results are compared with results of the Psi-integral method. Once the GSIFs are determined, stress solution on particular diameter by asymptotic series is compared with pure FEA results.

Keywords: Bi-material junction · Overdeterministic method
Non-singular terms

1 Introduction

As in the case of V-notch or bi-material notch, sharp material inclusion tip is place where singular stress concentration and thus potential crack initiation occur. The bi-material junction is a model for such geometric and material discontinuity. Stress distribution near bi-material junction is given by asymptotic stress series:

$$\sigma_{ij}(r, \theta) = \sum_{k=1}^{n} \left\{ H_k f_{ijk}(\theta) r^{\lambda_k - 1} + \overline{H}_k \overline{f}_{ijk}(\theta) r^{\overline{\lambda}_k - 1} \right\} \tag{1}$$

The H_k are generalized stress intensity factors, $f_{ijk}(\theta)$ are angular eigenfunctions, λ_k are eigenvalues and the overline denotes the complex conjugate. Terms of the stress series can be either singular or non-singular. Stress description is commented in [1–3]. The overdeterministic method (ODM) is based on the least-squares solution of overdetermined system of linear equations. The method was proposed by Seweryn in [4] as the method of analytical constraints. In [5] Ayatollahi and Nejati applied this method to calculate SIFs of a crack and in [6] to a sharp notch. Then Ayatollahi et al.

© Springer International Publishing AG, part of Springer Nature 2019
E. E. Gdoutos (Ed.): ICTAEM 2018, SI 5, pp. 237–239, 2019.
https://doi.org/10.1007/978-3-319-91989-8_52

studied ODM in bi-material notch application [7]. The ODM takes large number of results, FEA displacements or stresses to compute chosen number of GSIFs. The displacements are often preferred since the majority of FE codes is displacement based. The system of equations is written:

$$\mathbf{F}_{[2m \times n]} \mathbf{H}_{[n]} = \mathbf{u}^{FE}_{[2m]} \tag{2}$$

where the matrix on the left hand side is formed of the known analytical eigenfunctions. There is also the vector of unknown n GSIFs. The right hand side consists of FEA of radial and tangential displacement components. Since $2m > n$ no exact solution exists and the vector $\mathbf{H}_{[n]}$ is calculated by least square method. The FEA displacements are extracted on a circle surrounding the singular point. The effect of the circle radius on the values of H_k is studied. From mathematical point of view, the values of H_k depend on the number n of terms of vector $\mathbf{H}_{[n]}$. This effect is studied and minimal number of n to reach convergence is established. The results of ODM are compared with results of Psi-integral method (Fig. 1).

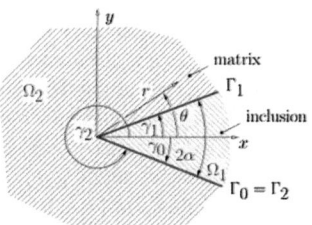

Fig. 1. Bi-material junction as the sharp material inclusion model.

Acknowledgements. The authors would like to thank Czech Science Foundation: Grant 16/18702S and Ministry of Education, Youth and Sports of the Czech Rep.: CEITEC 2020 [LQ1601].

References

1. Paggi, M., Carpinteri, A.: On the stress singularities at multimaterial interfaces and related analogies with fluid dynamics and diffusion. Appl. Mech. Rev. **61**, 1–22 (2008)
2. Krepl, O., Klusák, J.: Reconstruction of a 2D stress field around the tip of a sharp material inclusion. Procedia Struct. Integrity **2**, 1920–1927 (2016)
3. Krepl, O., Klusák, J.: The influence of non-singular terms on the precision of stress description near a sharp material inclusion tip. Theor. Appl. Fract. Mech. **90**, 85–99 (2017)
4. Seweryn, A.: Modeling of singular stress fields using finite element method. Int. J. Solids Struct. **39**, 4787–4804 (2002)
5. Ayatollahi, M.R., Nejati, M.: An over-deterministic method for calculation of coefficients of crack tip asymptotic field from finite element analysis. Fatigue Fract. Eng. Mater. Struct. **34**, 159–176 (2010)

6. Ayatollahi, M.R., Nejati, M.: Determination of NSIFs and coefficients of higher order terms for sharp notches using finite element method. Int. J. Mech. Sci. **53**, 164–177 (2011)
7. Ayatollahi, M.R., Mirsayar, M.M., Nejati, M.: Evaluation of first non-singular stress term in bi-material notches. Comput. Mater. Sci. **50**, 752–760 (2010)

Fatigue of Single-Crystal Gold Micro-specimen by Resonant Vibration

Takashi Sumigawa$^{(\boxtimes)}$ and Takayuki Kitamura

Department of Mechanical Engineering and Science, Kyoto University,
Kyoto-daigaku-katsura, Nishikyo-ku, Kyoto 615-8246, Japan
sumigawa@cyber.kues.kyoto-u.ac.jp

Abstract. A fatigue testing method for micro-metals using resonant vibration was developed. For the control of the fatigue cycle, we designed a gold micro-cantilever specimen with a weight at the tip, which reduced the resonant frequency. The tension-compression fatigue cycle was applied to the specimen using a piezoelectric actuator. The characteristic slip bands along the primary slip system, which possesses the highest Schmid factor, were generated on the specimen surface. Although the slip bands had similar morphologies to those of persistent slip bands (PSBs) in bulk, they were much narrower (width: approximately 50 nm) and needed a higher formation stress.

Keywords: Fatigue · Micro-specimen · Single crystal · Gold
Resonant vibration

1 Introduction

In cyclically deformed metals, dislocation self-organization develops characteristic fatigue structure. Many studies have investigated persistent slip bands (PSBs) [1], because they act as preferential crack initiation sites in fatigue. The PSB has a thickness in the range of some microns [2], which is independent of material size. However, micrometer- or nanometer-scale metals cannot contain the PSB. Thus, they should possess a characteristic fatigue behavior different from that of the bulk counterpart. Although cyclic loading experiments for small dimensions were widely performed for thin films [3], they are small in only one-dimension. The research of fatigue properties of three-dimensionally micrometer- or nanometer-scale metallic components is not enough.

This work develops a fatigue testing method for micro-specimen and investigates the fatigue behavior of single-crystal gold (Au) micro-specimen.

2 Experimental Procedure

The specimen is fabricated by focused ion beam (FIB) processing from a polycrystalline gold (Au) plate after annealing it in vacuum at 973 K for 24 h (Fig. 1(a)). The crystal orientation and shape of each grain on the surface are analyzed by electron backscatter diffraction pattern. Figure 1(b) shows a stereographic projection of the specimen. The slip planes and directions are respectively labeled A–D and 1–6.

© Springer International Publishing AG, part of Springer Nature 2019
E. E. Gdoutos (Ed.): ICTAEM 2018, SI 5, pp. 240–241, 2019.
https://doi.org/10.1007/978-3-319-91989-8_53

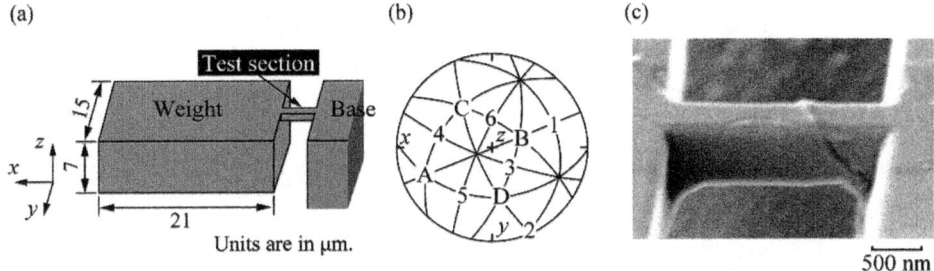

Fig. 1. (a) Schematic illustration of specimen for fatigue experiment, (b) stereographic projection, and (c) SEM observation image after fatigue.

The specimen is mounted on a piezoelectric actuator by cyanoacrylate adhesive. The oscillation displacement is measured by *in-situ* SEM observations. A high-cycle fatigue experiment is conducted at the resonant frequency for an input voltage amplitude, $\Delta V_{in}/2$, of 0.1 V. The oscillations are terminated after 10^7 cycles and the resonant frequency is measured. The surface morphology of the test section is observed by scanning electron microscopy (SEM). If there is no change in the surface, oscillation is applied with an amplitude that is approximately 200 nm larger than the previous one. Above steps are repeated until a remarkable change is observed on the surface.

3   Results and Discussion

Figure 1(c) shows a field-emission SEM (FE-SEM) image of the upper surface of the specimen after fatigue ($\Delta V_{in}/2 = 4.0$ V). The straight traces are crystallographic slip bands that are approximately 220 nm wide and consist of narrow steps. The slip bands is generated by the activation of the B4 slip system, which possesses the maximum resolved shear stress amplitude, $\Delta\tau_{mrss}/2$. The critical $\Delta\tau_{mrss}/2$ on the B4 is approximately 150 MPa, which is about six times larger than that of the PSB in Au bulk ($\Delta\tau_{mrss}/2 = 23.4$ MPa [4]). These results indicate the existence of the characteristic fatigue behavior in micro-scale metals.

References

1. Winter, A.T., Pedersen, O.R., Rasmussen, K.V.: Dislocation microstructures in fatigued copper polycrystals. Acta Metall. **29**(5), 735–748 (1981)
2. Mughrabi, H.: The Strength of Metals and Alloys. Pergamon Press, Oxford (1980). Haasen, P., Gerold, V., Kostorz, G. (eds.)
3. Schwaiger, R., Dehm, G., Kraft, O.: Cyclic deformation of polycrystalline Cu films. Phil. Mag. **83**(6), 693–710 (2003)
4. Li, P., Li, S.X., Wang, Z.G., Zhang, Z.F.: Mater. Sci. Eng. A **527**, 6244–6247 (2010)

Dynamic Fracture and Fragmentation of Ice Materials

Koji Uenishi[1(✉)], Toshio Hasegawa[1], Tomoya Yoshida[1],
Shintaro Sakauguchi[1], and Kojiro Suzuki[2]

[1] Department of Aeronautics and Astronautics, University of Tokyo,
Tokyo 113-8656, Japan
uenishi@dyn.t.u-tokyo.ac.jp
[2] Department of Advanced Energy, University of Tokyo, Chiba 277-8561, Japan

Abstract. By simultaneously operating high-speed digital video cameras, we have been experimentally investigating the mechanical characteristics of ice spheres that impinge upon a fixed elastic plate consisting of ice/polycarbonate. We have found that, when ruptured under dynamic impact, ice spheres show two specific fracture patterns: "top" and "orange segments." Our three-dimensional finite difference calculations simulating impact on linear elastic spheres indicate that "top" ("orange segments") fracture pattern is generated due to a shorter (longer) contact time during the impact process, respectively. Here, using pressure sensors, we try to clarify more quantitatively the generation mechanism of two dynamic fracture patterns. The new experimental observations suggest that the rise time (the time needed to reach a certain pressure owing to impact) is basically shorter when the generated fracture is the "orange segments"-type. This shorter rise time renders a longer effective contact time and hence waves of longer lengths, consistent with our earlier speculations.

Keywords: Ice · Impact · Dynamic fragmentation · Wave propagation

1 The Two Specific Fracture Patterns of Ice Spheres Under Dynamic Impact

Ice may be one of the most familiar brittle solid materials to us and its related physical phenomena range vastly, even in light of fracture only, from cracks in ice cubes dropped in water, avalanches and fracture of icebergs on the earth to the generation of the rings of the planet Saturn. However, not only fracture but also dynamic characteristics of ice materials are not well understood yet. Therefore, in our earlier study [1–3], we have experimentally traced the free fall and collision of ice spheres (diameter: 25, 50 or 60 mm) against a fixed plate made of ice or polycarbonate using two high-speed digital video cameras at a frame rate of up to 150,000 frames per second. In addition to the dependence of the coefficient of inelastic restitution on the relative speed of ice spheres before collision (speed of impact), we have found two specific fracture patterns induced by the collision (impact): (1) "Top"-type fracture, usually with a relatively smaller speed of impact, in which only the bottom, near-surface areas of an ice sphere are broken into small fragments upon impact at the bottom and a relatively

E. E. Gdoutos (Ed.): ICTAEM 2018, SI 5, pp. 242–243, 2019.
https://doi.org/10.1007/978-3-319-91989-8_54

large top-shaped part remains unfractured; and (2) "Orange segments"-type where fracture planes develop roughly along the central axis of the sphere and the sphere is split into three or four larger segments of comparable size.

2 Pressure Induced by Impact: Discussion and Conclusions

The two fracture patterns can be repeatedly recognized, and comparison of the experimental observations with the linear elastic transient wave fields numerically generated with our fully three-dimensional finite difference simulator for a PC has indicated that the "top" fracture pattern is due to the propagation of surface waves with relatively shorter lengths from the bottom along the free surface of the sphere, which will generate fracture only at the bottom, near-surface areas. On the other hand, in the case of the "orange segments"-type fracture, stressed regions expand more widely (and more quasi-statically-like) along the central axis of the sphere owing to a longer contact time, giving larger fracture planes along the central axis [1, 2]. Our new series of laboratory experiments using pressure sensors together with laser displacement sensors suggest that the rise time, here, the time needed to reach a certain pressure upon impact, is generally shorter when fracture is the "orange segments"-type. This shorter rise time renders a longer effective contact time with higher impact pressure and hence waves of longer lengths while the rise time for the "top"-type fracture is relatively longer, causing a shorter effective contact time with higher pressure and waves with shorter wavelengths. These experimental observations are consistent with our earlier numerical speculations mentioned above. We are currently conducting numerical simulations to reproduce the experimentally observed patterns by incorporating these experimental new findings and theoretical criteria that will be effective for real three-dimensional fracture.

Acknowledgements. KU is grateful to the financial support by the Ministry of Education, Culture, Sports, Science and Technology of Japan (MEXT) through the "KAKENHI: Grant-in-Aid for Scientific Research (C)" Program (No. 16K06487).

References

1. Uenishi, K., Yano, R., Yoshida, T., Yamagami, K., Suzuki, K.: Inelastic collisions between icy bodies: dependence on impact velocity and its fluctuations. In: Proceedings of the Japan Geoscience Union Meeting 2013, PPS21-P20. Japan Geoscience Union, Tokyo (2013)
2. Uenishi, K., Shigeno, N., Sakaguchi, S., Yano, R., Suzuki, K.: Dynamic impact-induced fracture development in ice spheres. In: Floryan, J.M. (ed.) Contributions to the Foundations of Multidisciplinary Research in Mechanics, pp. 2170–2171. IUTAM 2016 International Program Committee, Montreal (2016)
3. Uenishi, K., Yoshida, T., Sakaguchi, S., Suzuki, K.: Dynamic fracture patterns formed in transparent ice spheres by impact loading. In: Proceedings of the M&M 2017 Materials and Mechanics Conference, pp. 1828–1829. Materials and Mechanics Division, Japan Society of Mechanical Engineers, Tokyo (2017)

Collective Mechanical Behavior and Stability of a Group of Cracks in Brittle Solids

Koji Uenishi[1(✉)], Yuki Fukuda[1], Tomoya Yoshida[1],
Shintaro Sakauguchi[1], and Ioan R. Ionescu[2]

[1] Department of Aeronautics and Astronautics,
University of Tokyo, Tokyo 113-8656, Japan
uenishi@dyn.t.u-tokyo.ac.jp
[2] Galilee Institute, University Paris 13, 93430 Villetaneuse, France

Abstract. Structural failures are often related to mechanical destabilization of fracture in brittle solids that are subjected to some loading conditions. Either dynamically or quasi-statically, fracture areas (cracks) may be initiated and developed in or between solid structural elements, which may serve as sources of catastrophic failures. In reality, multiple cracks, not only a single one, may expand simultaneously and interact with each other mechanically. The collective behavior of such multiple cracks, however, does not seem to have been intensively studied so far. This contribution summarizes our recent experimental investigation into the nonlinear deformation of a solid material that is initially linear elastic but containing sets of cracks. Together with the earlier mathematical prediction [1], it is shown that the prescribed constant rate of strain, externally applied to a solid with multiple cracks, plays a crucial role in determining the overall stress-strain relation and eventually the tensile strength of the solid material.

Keywords: Collective behavior · Stability of cracks
Mechanical destabilization · Strain rate

1 Theoretical Background

Preventing catastrophic failures of solid structures is one of the crucial issues not only in the field of engineering (aerospace, civil, mechanical, etc.) but also in other research areas such as geoscience. For instance, an earthquake occurring in the solid earth is usually regarded as a physical phenomenon of mechanical destabilization of fracture in brittle solid materials that are under tectonic loading. Fracture related to an earthquake may occur fast (corresponding to a normal earthquake recorded by seismometers) or slowly (associated with the so-called slow slip event), but in each case, some fracture areas (cracks), nucleated in solids, expand and function as sources of ensuing failures. Cracks may enlarge concurrently in the failure source region and nucleated multiple cracks may mechanically interact with one another, either quasi-statically or dynamically. Although the overall collective mechanical behavior of these multiple cracks in brittle solid materials may be of certain importance in comprehending the physics of structural failures, it seems to have been hardly understood theoretically as well as

© Springer International Publishing AG, part of Springer Nature 2019
E. E. Gdoutos (Ed.): ICTAEM 2018, SI 5, pp. 244–245, 2019.
https://doi.org/10.1007/978-3-319-91989-8_55

experimentally. In this situation, recent mathematical study on the deformation of an initially linear elastic solid with sets of cracks [1] has indicated the clear dependence of the overall stress-strain relation of the solid on the prescribed, externally applied constant strain rate. In contradiction to our initial guess, a higher strain rate is associated with a "normal" smooth stress-strain curve while a lower one induces an abrupt stress drop in the stress-strain relation. Also, the overall tensile strength increases with the strain rate externally applied to the cracked solid.

2 Experiments Scrutinizing the Validity of the Theoretical Prediction and Conclusions

In order to validate the abovementioned mathematical prediction of the dependence of the collective characteristics of a solid with multiple cracks on the applied strain rate, photoelastic experiments have been performed using originally elastic specimens, here, transparent polycarbonate plates with sets of parallel cracks prepared by a digitally controlled laser cutter [2]. The specimens have been elongated by a tensile testing machine at different values of constant strain rates (to be more specific, displacement rates), and together with a high-speed digital video camera, the local and global stress evolution in the specimens has been simultaneously recorded. Our preliminary observations have shown distinct abrupt stress drops with relatively smaller applied displacement rates. Increase of the overall tensile strength with the rate has been also comfirmed. Although more systematic experiments with specimens containing optimally distributed cracks and more sophisticated techniques for increasing the applied displacement rate to much higher values may be needed to give more conclusive remarks, the newly found rate dependence in the overall stress-strain diagram might give a clue in grasping the relation between, for example, quasi-static diastrophism and dynamic faulting in geoscience.

Acknowledgements. We would like to kindly acknowledge the financial support by the Ministry of Education, Culture, Sports, Science and Technology of Japan (MEXT) through the "KAKENHI: Grant-in-Aid for Scientific Research (C)" Program (No. 16K06487) as well as that by the "Invitation Fellowship Program for Research in Japan" of the Japan Society for the Promotion of Science (JSPS).

References

1. Gomez, Q.: Discontinuous Galerkin modeling of wave propagation in damaged materials. Ph.D. thesis, University Paris 13, Villetaneuse (2017)
2. Uenishi, K., Yoshida, T., Sakaguchi, S., Fukuda, Y., Ionescu, I.R., Gomez, Q.: On the collective behavior of a group of cracks in brittle solids and its implications in earthquake source physics. In: Proceedings of the 2017 Seismological Society of Japan Fall Meeting, S08-P25, Seismological Society of Japan, Tokyo (2017)

Miscellaneous (Biomechanics, Compu-tational mechanics, Dynamics, Nano-mechanics, Plasticity, Structures, Wave propagation)

Analysis of Materials Systems Represented with Graphs

Andrey P. Jivkov[(⊠)]

Mechanics and Physics Research Group, School of Mechanical,
Aerospace and Civil Engineering, The University of Manchester,
Oxford Road, Manchester M13 9PL, UK
andrey.jivkov@manchester.ac.uk

Abstract. Presented is a rigorous mathematical formulation of boundary value problems defined on discrete systems described by mathematical graphs. The formulation is applicable to mechanical and physical problems and includes an effective algebraic framework and efficient computational implementation. Mechanical problems involving damage initiation and evolution are soled to illustrate the proposed method. It is concluded that the graph-theoretical approach to discrete systems offers substantial benefits in terms of conceptual clarity and computational efficiency.

Keywords: Discrete analysis · Boundary value problems · Damage and failure

1 Introduction

Representing materials sub-continuum structures with mathematical graphs, or 1-complexes as they are known in algebraic topology, takes inspiration from the classical mechanics of atom interactions. A specific structure, e.g. a set of grains forming a polycrystalline material, which can be generally represented with a 3-complex with cells, faces, edges and vertices, is simplified or reduced to a 1-complex (a graph) by a problem-specific selection of vertices or sites and edges or bonds. There is a considerable body of research where beams [1, 2] and springs [3, 4] have been investigated as bonds in order to capture the mechanical response of materials at different scales.

The attraction of graph-based, also known as lattice, models is their discrete nature, which allows for modifying the behavior of individual bonds or sets of bonds to capture heterogeneity at microstructural level, e.g. by incorporating measurable features of variable sizes, such as pores or particles, and measurable properties of variable magnitudes, such as elastic constants or grain boundary excess energies. Further, specifically for deformation and fracture analysis, the discrete representation allows for capturing explicitly micro-crack initiation, growth and coalescence through bond deletion. It is worth noting that lattice models are not limited to mechanics but have also been used for analysis of heat and mass transfer in solids [5, 6], and transport through porous media [7, 8]. Finally, the graph-based approach offers substantial computational efficiency compared to representations with 3-complexes, which can also incorporate spatially distributed features and properties, but at this stage require substantial new developments for analysis of damage evolution.

© Springer International Publishing AG, part of Springer Nature 2019
E. E. Gdoutos (Ed.): ICTAEM 2018, SI 5, pp. 249–254, 2019.
https://doi.org/10.1007/978-3-319-91989-8_56

The aim of this work is to present a rigorous mathematical description of graph-based models of materials structures and their analysis, applicable to any field of study. Proposed is a new formulation of boundary value problems on graphs and associated efficient numerical approach for construction and analysis of such problems. This is implemented in a C++ library capable of tracking topological and geometric modifications arising from various physical and mechanical problems. Numerical examples illustrating the benefits of the proposed framework are given at the end.

2   Analysis on Graphs

2.1   Mathematical Basis – Graph Topology

An abstract graph $G = (V, E)$ is a tuple of finite sets V and E, where elements $e_i \in E$, called edges, are pairs of elements $v_j \in V$, called vertices. For the purposes of this work we consider only graphs, for which elements of E are unordered and distinct pairs (v_j, v_k) with $j \neq k$, i.e. undirected graphs without multiple edges and loops. The number of vertices is given by $|V| = n$ and edges by $|E| = m$.

A graph from the class considered here can be equipped with orientation by specifying an arbitrary but fixed order of the vertices of each edge $e_i = (v_j, v_k) \in E$, e.g. by selecting v_j to be the first vertex (origin, denoted by $v_j = o(e_i)$) and v_k to be the second vertex (terminus, denoted by $v_k = t(e_i)$), the oriented edge is $e_i = [v_j, v_k]$.

The topology of an oriented graph can be encoded into a $m \times n$ matrix $\mathbf{A}$, referred to as the incidence matrix, with the following components

$$a_{ij} = \begin{cases} +1, & \text{if } v_j = t(e_i) \\ -1, & \text{if } v_j = o(e_i) \,. \\ 0, & \text{otherwise} \end{cases} \qquad (1)$$

The significance of the incidence matrix cannot be overemphasized, as together with the graph topology it provides the boundary operator on nodes, while its transpose provides the boundary operator on edges. Specifically, for a discrete function over nodes $\mathbf{u}$, the boundary operator gives a discrete function over edges $\mathbf{A}\mathbf{u}$, computing the differences between nodal values at edges, hence a discrete gradient. Inversely, for a discrete function over edges f, the transposed boundary operator gives a discrete function over nodes $\mathbf{A}^T f$, computing the algebraic sums of edge values at nodes, hence a discrete divergence. Thus the incidence matrix establishes the basis for analysis on graphs.

2.2   Geometry and Physics – Graph Metric

Formulation of mechanical or physical problems on graphs requires a geometric realisation of G, i.e. embedding the graph into a metric space, for example $\mathbb{R}^d$, by associating a point (coordinates) $\mathbf{x}_j \in \mathbb{R}^d$ with every vertex $v_j \in V$. If the edges are considered as straight segments described by $\mathbf{y}_i = \lambda \mathbf{x}_j - (1 - \lambda)\mathbf{x}_k, \lambda \in [0, 1]$, the embedding induces a standard graph metric, given by edge lengths $l_i = \mathbf{x}_j - \mathbf{x}_k$ for

$e_i = [v_j, v_k]$. It is required that the embedding does not lead to vertices in the interior of edges, i.e. $\mathbf{x}_l \neq \lambda \mathbf{x}_j - (1 - \lambda)\mathbf{x}_k$, for $\lambda \in (0, 1)$ and $\forall l$, and to edges of zero size, i.e. $l_i > 0 \forall i$. In practice the embedding will be dictated by a specific selection of vertices and edges that represents given material system, so the selection should obey these restrictions. In what follows consideration will be given to embedding in $\mathbb{R}^3$.

Each edge e_i of an embedded graph can be assigned a triad of unit vectors – tangent, $\mathbf{t}_i$, normal, $\mathbf{n}_i$, and bi-normal, $\mathbf{b}_i$ – which can be expressed in local and global coordinate systems by the matrices $\widehat{\mathbf{B}}_i$ and $\mathbf{B}_i$, respectively, related by $\widehat{\mathbf{B}}_i = \mathbf{Q}_i^{\mathrm{T}} \mathbf{B}_i$ and $\mathbf{B}_i = \mathbf{Q}_i \widehat{\mathbf{B}}_i$, where $\mathbf{Q}_i$ is the matrix of direction cosines $\mathbf{Q}_i \mathbf{Q}_i^{\mathrm{T}} = \mathbf{Q}_i^{\mathrm{T}} \mathbf{Q}_i = \mathbf{I}_{3x3}$. Further, the incidence matrix is expanded into a $3m \times 3n$ matrix $\mathbf{A}$ with components

$$a_{ij} = \begin{cases} +\mathbf{I}_{3x3}, & \text{if } v_j = t(e_i) \\ -\mathbf{I}_{3x3}, & \text{if } v_j = o(e_i) \, . \\ \mathbf{O}_{3x3}, & \text{otherwise} \end{cases} \tag{2}$$

In addition to the geometric embedding, mechanical and physical problems require assignment of edge weights, which represent resistances to the particular phenomenon studied. Clearly they depend on problem-specific intrinsic physical properties, as well as on local geometric characteristics, and are typically specified with respect to the local coordinate system. This requires a weight function $E \rightarrow \mathbb{R}^3$ assigning to each edge e_i resistances $(\alpha_i, \beta_i, \gamma_i)$ in the directions of its unit vectors. Given weight function introduces a metric on the graph, described in the local and global coordinate systems, respectively, by the matrices

$$\widehat{\mathbf{M}}_i = \begin{bmatrix} \alpha_i \mathbf{t}_i \\ \beta_i \mathbf{n}_i \\ \gamma_i \mathbf{b}_i \end{bmatrix} \text{ and } \mathbf{M}_i = \mathbf{Q}_i \widehat{\mathbf{M}}_i \mathbf{Q}_i^{\mathrm{T}}. \tag{3}$$

It should be noted that for physical problems, such as heat and mass transport, as well as mechanical problems involving spring networks $\alpha_i \neq 0, \beta_i = \gamma_i = 0$, while for mechanical problems with more complex interactions $\alpha_i \neq 0, \beta_i = \gamma_i \neq 0$. The full graph metric is given by the block-diagonal matrix $\mathbf{M} = \mathrm{diag}(\mathbf{M}_1, \mathbf{M}_2, \dots, \mathbf{M}_m)$.

2.3   Boundary Value Problems – Graph Analysis

For a given vector-valued function on vertices $\boldsymbol{u} : V \rightarrow \mathbb{R}^3$, the product $\mathbf{A}\boldsymbol{u}$ provides the discrete gradient – a vector-valued function on edges $\mathbf{A}\boldsymbol{u} : E \rightarrow \mathbb{R}^3$. The graph metric transforms the discrete gradient into flux vector $\boldsymbol{f} = \mathbf{M}\mathbf{A}\boldsymbol{u}$, the discrete divergence of which must satisfy equilibrium with external sources, $\boldsymbol{b}$, at nodes, i.e. $\mathbf{A}^{\mathrm{T}}\boldsymbol{f} = \boldsymbol{b}$. This leads to the following expression for the equilibrium of the system

$$\mathbf{A}^{\mathrm{T}}\mathbf{M}\mathbf{A}\boldsymbol{u} = \mathbf{L}\boldsymbol{u} = \boldsymbol{b}, \tag{4}$$

where $\mathbf{L}$ denotes the Laplacian of the weighted graph, which can be shown to be symmetric, positive semi-definite, and singular.

In order to apply boundary conditions, the degrees of freedom of graph vertices $k = 3n = 3|V|$ are sorted into two sets, k_1 with essential, and k_2 with natural conditions. This allows for expressing Eq. (4) in the form

$$\begin{bmatrix} \mathbf{L}_{11} & \mathbf{L}_{12} \\ \mathbf{L}_{21} & \mathbf{L}_{22} \end{bmatrix} \begin{pmatrix} \boldsymbol{u}_1 \\ \boldsymbol{u}_2 \end{pmatrix} = \begin{pmatrix} \boldsymbol{b}_1 \\ \boldsymbol{b}_2 \end{pmatrix}, \tag{5}$$

where $\boldsymbol{u}_1$ and $\boldsymbol{b}_2$ collect all components with prescribed values, and $\boldsymbol{u}_2$ and $\boldsymbol{b}_1$ collect all components with unknown values. Notably $\mathbf{L}_{12} = \mathbf{L}_{21}$. Unknowns are found by solving the following two steps

$$\mathbf{L}_{22}\boldsymbol{u}_2 = \boldsymbol{b}_2 - \mathbf{L}_{21}\boldsymbol{u}_1, \tag{6}$$

$$\boldsymbol{b}_1 = \mathbf{L}_{11}\boldsymbol{u}_1 + \mathbf{L}_{12}\boldsymbol{u}_2 \tag{7}$$

Efficient construction of the sub-matrices $\mathbf{L}_{ij}$ can be achieved by splitting the incidence matrix $\mathbf{A}$ into a matrix $\mathbf{U}$ containing only k_1 columns corresponding to the essential boundary conditions, and a matrix $\mathbf{V}$ containing the remaining k_2 columns; both matrices have the same rows as $\mathbf{A}$. It can be shown that $\mathbf{L}_{11} = \mathbf{U}^\mathrm{T}\mathbf{M}\mathbf{U}$, $\mathbf{L}_{22} = \mathbf{V}^\mathrm{T}\mathbf{M}\mathbf{V}$, and $\mathbf{L}_{12} = \mathbf{V}^\mathrm{T}\mathbf{M}\mathbf{U}$. Proof and procedure details can be found in [9].

2.4 Computational Implementation

The algebraic framework for graph analysis described above has been implemented in an in-house code. Of particular interest is to apply the analysis on problems of damage initiation and propagation, which requires handling evolution of the graph topology, e.g. by deleting edges to represent local failures. To this end, a Boost Graph Library (BGL) [10] data structure has been employed to maintain a record of the connectivity information, from which $\mathbf{A}$ and $\mathbf{M}$ can be easily extracted. Therefore, if an edge is to be removed during analysis, the matrices can be output from the data structure once again or modified accordingly. The definition of the physics problem is inherent to the matrix $\mathbf{M}$, therefore minimal user interaction is required to provide the function associated with each edge.

Given a sufficiently large number of vertices and edges, it can be seen that the oriented incidence matrix $\mathbf{A}$ and $\mathbf{M}$ are very sparse. Therefore, a variant of the compressed column storage scheme available in Eigen [11] has been used. Additionally, Eigen is equipped with a number of iterative solvers, of which the conjugate gradient method has been used for tackling the linear form of Eq. (4). For a non-linear Eq. (4), an implementation of Newton's method and Broyden's method [12] has been made.

For a number of problems it is desirable to construct the graph from a set of randomly distributed vertices, with connectivity and other geometric characteristics, potentially required for $\mathbf{M}$, determined by the Voronoi diagram around the vertex set. For such cases the Voro++ [13] has been used to obtain the necessary information.

3   Example Problems

Two problems are selected to illustrate the performance of the discrete approach in analysis of damage initiation and fracture: Figs. 1 and 2. For simplicity the graphs are made as regular as the geometry allowed and the edges are modeled as trusses, i.e. with graph metric defined by $\alpha_i = EA/l_i, \beta_i = \gamma_i = 0$, where E and A are the modulus and elasticity and a predefined cross-section (this can be linked to area of the dual face in the Voronoi tessellation if required). Damage initiation and propagation are represented by edge deletion when a critical value of the discrete strain is reached. More details on the procedures used, further examples on physical and mechanical problems and the code developed can be found at [14].

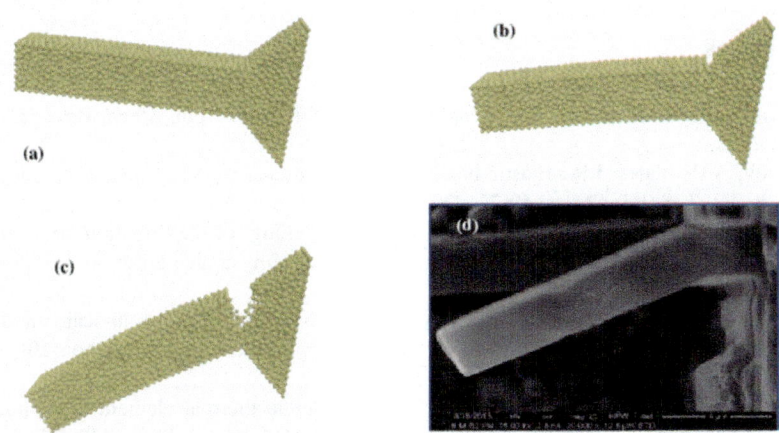

Fig. 1. Process of damage evolution and failure (a)–(c) of graph-modeled beam in comparison with experimentally observed behavior in nuclear graphite micro-cantilever beam (d).

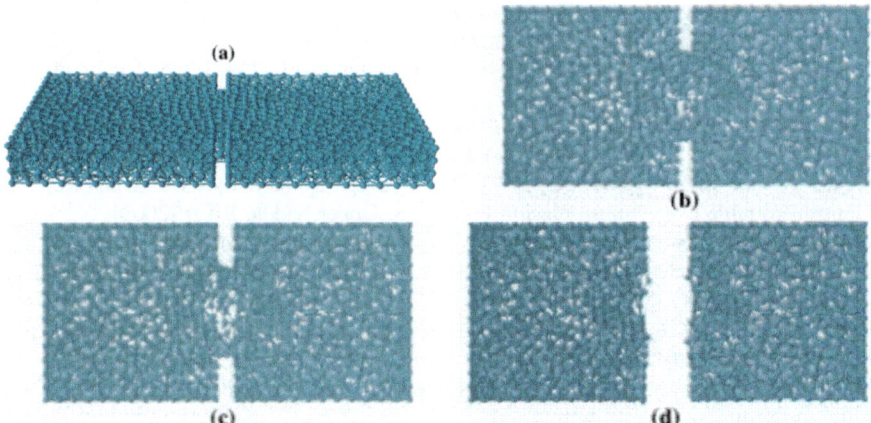

Fig. 2. Graph model of double edge notched plate (a), and evolution of damage and failure (b)–(d) under tension normal to notches.

4 Discussion and Conclusions

The results presented demonstrate that the computational model is performing well in terms of capturing the initiation and tracking the propagation of fracture, close to the observed or expected behavior. The graph-theoretical approach offers a clear and concise way for describing discrete systems and allows for computationally efficient algorithms to be employed. Further work is necessary to investigate the applicability of the approach to cases with complex interactions.

Acknowledgments. The author appreciates highly the financial support of EPSRC via grants EP/K016946/1 "Graphene-based Membranes" and EP/N026136/1 "Geometric Mechanics of Solids".

References

1. Ostoja-Starzewski, M.: Lattice models in micromechanics. Appl. Mech. Rev. **55**(1), 35–60 (2002)
2. Jivkov, A.P., Yates, J.R.: Elastic behaviour of a regular lattice for meso-scale modelling of solids. Int. J. Solids Struct. **49**(22), 3089–3099 (2012)
3. Zhang, M., Jivkov, A.P.: Micromechanical modelling of deformation and fracture of hydrating cement paste using X-ray computed tomography characterisation. Compos. B Eng. **88**, 64–72 (2016)
4. Morrison, C.N., Jivkov, A.P.: Vertyagina, Ye., Marrow, T.J.: Multi-scale modelling of nuclear graphite tensile strength using the Site-Bond lattice model. Carbon **100**, 273–282 (2016)
5. Feng, Y.T., Han, K., Li, C.F., Owen, D.R.J.: Discrete thermal element modelling of heat conduction in particle systems: basic formulations. J. Comput. Phys. **227**(10), 5072–5089 (2008)
6. Jivkov, A.P., Yates, J.R.: A discrete model for diffusion-induced grain boundary deterioration. Key Eng. Mater. **592–593**, 757–760 (2014)
7. Xiong, Q., Jivkov, A.P., Yates, J.R.: Discrete modelling of contaminant diffusion in porous media with sorption. Microporous Mesoporous Mater. **185**, 51–60 (2014)
8. Xiong, Q., Joseph, C., Schmeide, K., Jivkov, A.P.: Measurement and modelling of reactive transport in geological barriers for nuclear waste containment. Phys. Chem. Chem. Phys. **17**, 30577–30589 (2015)
9. Codsi C., Jivkov, A.P.: Mixed boundary value problem for Poisson's equation on weighted graphs with physical application. SIAM J. Discrete Math. (2018, under review)
10. Siek, J.G., Lee, L.Q., Lumsdaine, A.: The Boost Graph Library: User Guide and Reference Manual. Portable Documents. Pearson Education, Upper Saddle River (2001)
11. Guennebaud, G., Jacob, B.: Eigen (2010). http://eigen.tuxfamily.org
12. Broyden, C.G.: A class of methods for solving nonlinear simultaneous equations. Math. Comput. **19**(92), 577–593 (1965)
13. Rycroft, C.H.: VORO++: a three-dimensional Voronoi cell library in C++. Chaos Interdisc. J. Nonl. Sci. **19**(4), 041111 (2009)
14. MaPoS Homepage. https://mapos.manchester.ac.uk/. Last accessed 31 Jan 2018

Numerical Issues Affecting the Eigenproblem Solution of Transversely Vibrating Segmented Structures

Lubov Andrusiv$^{(\boxtimes)}$

United States Air Force Academy,
2354 Fairchild Dr., USAF Academy, El Paso County, CO 80840, USA
lubov.andrusiv@usafa.edu

Abstract. The subject of this research is multi-segmented, telescoping, beam-like structures, which represent an important class of engineering systems with, in general, nonuniform geometric and physical parameters. The distributed parameter Euler-Bernoulli sectioning methodology applied to such structures produces a transcendental eigenvalue equation that requires a numerical root-funding algorithm. This study provides evidence that for continuous system upper modes, root finding algorithm is subject to numerical instability due to the finite precision associated with software and hardware used for the computations. The results obtained after extensive numerical computation and analysis of three-segment cantilever telescoping beams with two lap joints, yield new insights into the prediction of the numerical instability and the role computational issues may play in the solution of distributed parameter eigenproblems.

Keywords: Continuous systems · Telescoping structures
Numerical computation

1 Introduction

Segmented, telescoping structures, which are quite compliant and prone to vibrations, are characterized by segments that can retract inside of, or relative to one another during operation. Examples include variable focal-length telescopes, pneumatic and hydraulic cylinders, cranes and ladders, extensible booms, antennas, robotic arms, and deployable space structures.

From a mechanics perspective, the mathematical modeling of such systems substantially differs from that of corresponding single-segment system with fixed configurations. Because geometric properties, inertia, and stiffness are varying parameters, they introduce analytical and computational challenges and their effect on performance and stability needs to be understood. In this work, the transverse motion of the three-segment telescoping beams based on an Euler-Bernoulli formulation and the numerical stability issues related to the eigenproblem solution are addressed.

In the open literature several papers on beam vibration mention that evaluation of the natural frequencies may encounter with numerical instability. Goncalves et al. [1] pointed out that the most common form of the expression for the mode shape function

© Springer International Publishing AG, part of Springer Nature 2019
E. E. Gdoutos (Ed.): ICTAEM 2018, SI 5, pp. 255–261, 2019.
https://doi.org/10.1007/978-3-319-91989-8_57

of a uniform Euler-Bernoulli beam may permit the evaluation of only the first twelve modes or so due to "numerical issues." Singh et al. in [2] admitted that the presence of transcendental functions introduce errors in determining the higher eigenvalues, which in turn lead to inaccuracies in physical parameter estimation.

However, Amos and Qu [3] encountered numerical instability during the solution of the eigenproblem. Authors predicted very closely spaced higher eigenvalues, but interpreted these results as a result of interaction between the cantilevered modes and the free-free modes of the sliding segment brought on by the speed of the deployment process.

Determination of modal characteristics of a two-segment telescopically deploying beam is the subject of another paper by Amos [4]. Again, closely spaced roots for the eigenvalue equation, Amos continues to interpret as a physical phenomenon occurring due to non-uniform inertia and stiffness distributions created during beam deployment maneuvers.

As discussed in the following sections, Amos' interpretation is not consistent with the findings of the presented investigation.

2   Theoretical Model of Telescoping Systems and Solution

The three-segment telescoping system approximated as transversely vibrating beam, meaning that longitudinal and torsional motions are neglected. In this model the term "segment" refers to a physically contiguous portion of the beam structure that cannot telescope upon itself. A segment consists of a single component of fixed length, L_i.

Upon application of the Euler-Bernoulli method, the modeling procedure must proceed to a discretization process [5, 6] that divides the beam segments domain into sections as shown in Fig. 1, where Section 2 and Section 4 are overlap sections between segments 1 and 2 and 2 and 3, respectively. The length of Sections 2 and 4 denoted as a and b, and $w(x, t)$ represents the transverse deflection of the beam from its equilibrium position.

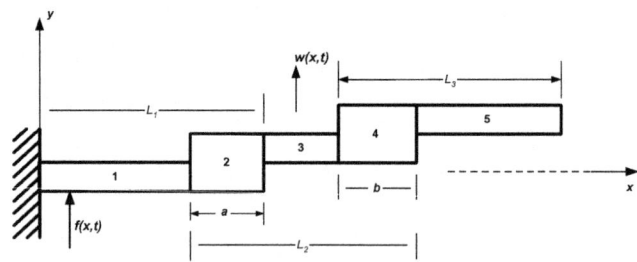

Fig. 1. Division of a three-segment cantilever beam into distributed parameter sections.

After segment discretization is complete, a distinct, unique, fourth-order partial differential equation of motion is defined for each section and for the five-section beam can be expressed as:

$$EI^{[i]}(x)\frac{\partial^4 w^{[i]}(x,t)}{\partial x^4} + m^{[i]}(x)\frac{\partial^2 w^{[i]}(x,t)}{\partial t^2} = f(x,t), \quad i \in [1,5] \tag{1}$$

The classical boundary conditions must be applied at the fixed end of Section 1 and at the free end of Section 5, and the continuity conditions must be formulated at the sections interfaces between Sections 1 and 2, 2 and 3, and 3 and 4 as described in [5].

2.1   Free Vibration Solution

Frequency Equation. In the absence of any excitations, the free vibration solution to the homogeneous part of (1) is found using a separation of variables approach as $w^{[i]}(x,t) = W^{[i]}(x)T(t)$. The general solution for eigenfunctions $W^{[i]}(x)$ within each section of the beam in the free vibration can be given by the following equation

$$W^{[i]}(x) = Z_1^{[i]}\cosh(\beta^{[i]}x) + Z_2^{[i]}\sinh(\beta^{[i]}x) + Z_3^{[i]}\cos(\beta^{[i]}x) + Z_4^{[i]}\sin(\beta^{[i]}x) \tag{2}$$

If the boundary conditions incorporate into (2), then a set of 20 linear homogeneous algebraic equations representing the eigenvalue problem can be obtained. In matrix form it can be written as $[A]\{Z\} = \{0\}$, where $[A]$ is a square matrix of order $4m$ (where m is a number of sections); the elements of $[A]$ are a function of the eigenvalue parameter $\beta_n^{[i]}l^{[i]}$ defined in [5] and $\{Z\}$ is a $4m$ column vector of eigenvalue coefficients. With the exception of trivial solution, this equation is satisfied only if

$$det([A]) = 0 \tag{3}$$

Equation (3) is recognized as the characteristic, or frequency equation of the telescoping system, and its solution consists of an infinite number of unique values, and consequently, natural frequencies, ω_n. Because (3) is transcendental, it must be solved numerically.

2.2   Implementation and Numerical Analysis

Root Finding Algorithm. The frequency Eq. (3) when expanded symbolically yields an extremely complicated algebraic equation composed of an amalgamation of hyperbolic and trigonometric terms. The computational algorithm for solving eigenvalue problem formulated for the segmented beam system based on the Matlab *fzero* function, is designed to find the real roots of a single equation. A simple representation of the function's syntax is *fzero*(function, x_0), which finds a zero of the function "function" near initial guess "x_0". The application of the proposed root finding algorithm to beams with the various segment overlap configurations provides evidence that it may be subject to the numerical instability regarding the solution of continuous eigenvalue problem.

Beam with Two Overlapped Sections. Consider a beam in Fig. 1. Assume the segments made of the same material with $E = 200$ GPa, $\rho = 7860$ kg/m^2, of equal length $L_1 = L_2 = L_3 = 1$ m, and rectangular cross sections of 0.05 m width and 0.00635 m height. Various configurations of the beam can be characterized by the overlap ratio parameter, γ, defined as $\gamma_1 = a/L_2$ for the Section 2 and as $\gamma_2 = b/L_3$ for the Section 4. The two representative beam configurations are shown in Fig. 2 (a, b) and the corresponding natural frequency results, calculated by the proposed algorithm in Tables 1 and 2. If a solution for a natural frequency value was not obtained due to instability, then in the table it is shown as "-*NI*-".

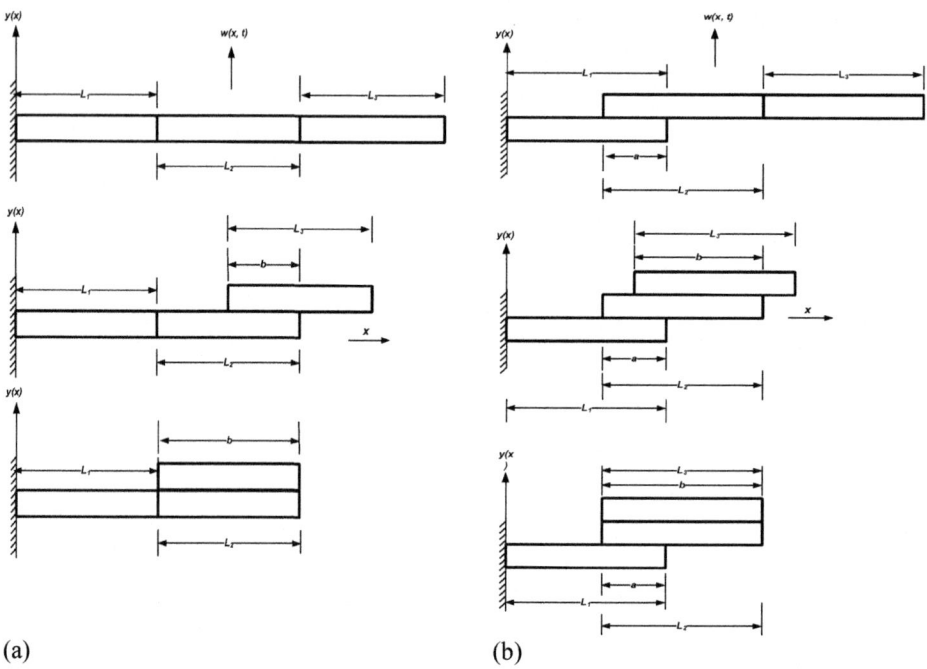

Fig. 2. Three-segment beam configurations: initial, intermediate, and final for (a) $\gamma_1 = 0$, (b) $\gamma_1 = 0.6$.

Example Case Study for $0 \le \gamma_2 \le 1$ and (a) $\gamma_1 = 0$ and (b) $\gamma_1 = 0.6$. As the telescoping beam changes length, the relative size of overlap sections and non-overlap change too, dramatically affecting the beam's physical characteristics. In Fig. 2 (a), the second segment is fully extended relative to segment 1 and the third segment varies from fully extended to fully retracted relative to segment 2. In Fig. 2 (b), the second segment is partially extended and the position of the third segment again varies relative to segment 2.

Table 1. Natural frequency values, f(Hz), for a telescoping beam with a fixed first interface ratio $\gamma_1 = 0$ and variable second interface ratio.

γ_2	f_1	f_2	f_3	f_4	f_5	f_6
0	0.575	3.603	10.088	19.769	32.680	48.819
0.1	0.603	3.898	10.906	21.185	35.070	52.755
0.2	0.634	4.244	11.996	22.843	37.908	58.389
0.3	0.665	4.643	13.421	24.901	41.082	*-NI-*
0.4	0.699	5.095	15.245	27.676	44.719	*-NI-*
0.5	0.735	5.595	17.483	31.761	49.226	*-NI-*
0.6	0.772	6.130	20.058	37.929	55.554	*-NI-*
0.7	0.812	6.726	22.864	45.853	65.910	*-NI-*
0.8	0.855	7.354	25.889	52.842	*-NI-*	*-NI-*
0.9	0.900	8.029	29.182	57.586	*-NI-*	*-NI-*
1.0	0.949	8.761	32.657	60.513	99.941	154.16

Table 2. Natural frequency values, f(Hz), for a telescoping beam with a fixed first interface ratio $\gamma_1 = 0.6$ and variable second interface ratio.

γ_2	f_1	f_2	f_3	f_4	f_5	f_6
0	1.076	5.506	17.932	34.142	54.703	89.716
0.1	1.152	6.243	19.767	36.669	61.288	100.487
0.2	1.233	7.335	21.904	40.114	70.031	113.159
0.3	1.318	9.073	24.420	45.458	83.843	126.531
0.4	1.408	12.307	27.501	59.437	100.536	*-NI-*
0.5	1.491	14.619	31.579	67.034	118.926	*-NI-*
0.6	1.579	17.236	38.176	77.613	136.053	*-NI-*
0.7	1.673	19.866	48.619	95.164	*-NI-*	*-NI-*
0.8	1.773	22.245	61.802	*-NI-*	*-NI-*	*-NI-*
0.9	1.880	24.191	73.763	*-NI-*	*-NI-*	*-NI-*
1.0	1.994	25.463	82.917	60.513	260.444	430.325

The results in Table 1 show that the solution algorithm had entered a numerically unstable regime first for the fifth mode with an "unbalanced" configuration defined by $\gamma_2 = 0.8$ and $\gamma_2 = 0.9$. The only configurations that converge for mode six are $\gamma_2 = 0$, 0.1, 0.2, and 1. The $\gamma_2 = 0.1$ and 0.2 represent the most balanced configurations.

The telescoping beam configurations in Fig. 2 (b) continue getting shorter and stiffer as γ_2 increases. However, numerical instability occurs earlier in this case than for case (a). As indicated in Table 2, numerical instability affected some specified configurations as low as the fourth mode.

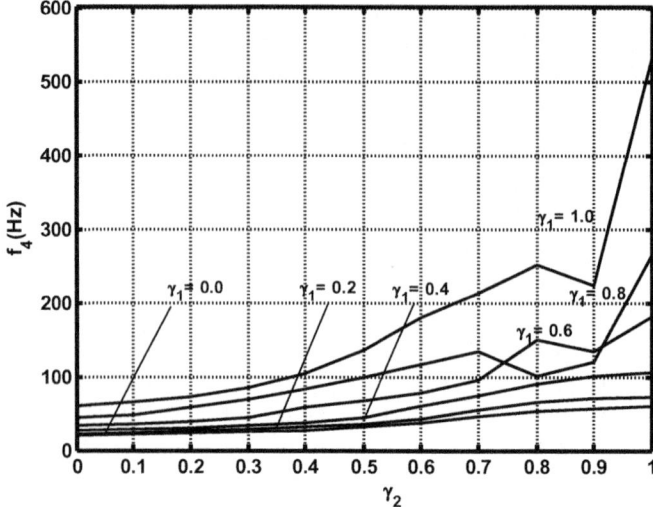

Fig. 3. Variation of the fourth natural frequency for a broad range of configurations defined by γ_2 as an independent variable and discrete γ_1 values.

3   Conclusion

This and other studies [5] demonstrate that as the overlap ratio vary during the course of segment deployment and retraction, the beam configuration and its flexibility change, which affects the inertia and stiffness properties of the Euler-Bernoulli sections and result in changes to the system's modal characteristics. However, formulation of the eigenproblem as a determinate equation, as occur with segmented beams, exacerbate the numerical issues associated with the solution and significantly reduce the number of eigenvalues that can be accurately determined. Such a result is inevitable with finite precision of software using for calculations and eigenvalues that may increase *ad infinitum*. It is possible to formally state this observation as the *Transcendental Eigenproblem Lemma (TEL)*:

For any physical structure modeled as a continuous vibratory system with an infinite number of eigenvalues, and having a frequency equation comprised of transcendental terms, there is a limit to the number of eigenvalues and eigenfunctions that can be determined using fixed-precision eigenproblem solution algorithm.

The effect of the varying overlap ratios on the instability of the eigensolution can be depicted graphically as in Fig. 3 to facilitate trend identification.

The graph for the fourth natural frequency exhibit discontinues associated with the extreme combinations of the overlap parameters γ_1 and γ_2. Plot shows irregularities for specific beam configurations. The natural frequencies for these configurations are incorrect, and should simply be ignored.

References

1. Goncalves, P., Brennan, M., Elliott, S.: Numerical evaluation of high-order modes of vibration in uniform Euler-Bernoulli beams. J. Sound Vib. **301**(3–5), 1035–1039 (2007)
2. Singh, K., Li, G., Pang, S.: Free vibration and physical parameter identification of non-uniform composite beams. J. Sound Vib. **74**(1), 37–50 (2006)
3. Amos, A., Qu, B.: The dynamic equations of motion for a deploying telescopic beam. In: The 34th AIAA/ASME/ASCH/AHS/ASC Structures, Structural Dynamics and Materials Conference, pp. 2028–2038 (1993)
4. Amos, A.: Variations in the modal characteristics of a telescopically deploying beam. Mech. Struct. Mach. **21**(3), 335–356 (1993)
5. Andrusiv, L., Richards, C.: Modeling and vibration response of segmented flexible structures. In: Kit, H. (ed.) International Conference on Mathematical Problems of Mechanics of In-Homogeneous Structures 2006, NASU, Ukraine, Lviv, vol. 2, pp. 157–160 (2006)
6. Andrusiv, L., Prater, G., Richards, C.: Dynamic behavior of segmented telescoping structures in automotive systems. SAE Int. J. Passeng. Cars **1**(1), 238–249 (2009)

Development of Geometric Formulation of Elasticity

Odysseas Kosmas[(✉)] and Andrey Jivkov

School of Mechanical, Aerospace and Civil Engineering,
The University of Manchester, Manchester, UK
odysseas.kosmas@manchester.ac.uk

Abstract. Recently, new techniques have been presented that discretize continuous elasticity variables as cochains over a primal mesh, representing the solid, and an appropriately defined dual one. Discrete strain and stress can be then thought of as a vector-valued 1-form (or vector-valued 1-cochain on the primal mesh) and covector-valued 2-form (or vector-valued 2-cochain on the dual mesh) respectively. The governing equations can be formulated by requiring energy balance and invariance under time-dependent rigid translations and rotations of the ambient space. To obtain those, we project the discrete stress into normal and tangential components and formulate the boundary value problem with a system of two matrix equations. This allow for treating both classical and coupled-stress (micro-polar) elasticity. The link between discrete strains and stresses is provided by a material discrete Hodge star operator, which we define to include geometric and physical factors, such as lengths, areas, and moduli of elasticity and rigidity. The performance of the proposed formulation is demonstrated by a simple example.

Keywords: Discrete exterior calculus · Geometric mechanics · Mixed method
Primal and dual mesh · Discrete hodge star

1 Introduction

Focusing on problems of elasticity, for the cases when the ambient space is Euclidean, different methods have been presented in order to derive the (discrete) linear systems of equations that can be then solved by existing numerical methods. In order to study those discretization methods for partial differential equations (PDEs) the use of exterior calculus must be invoked. With that, one can define vector calculus to higher dimensions and to smooth manifolds [1]. But in order for someone to define the discretized quantities needed, Discrete Exterior Calculus (DEC) must be introduced. This way numerical methods for solving PDEs on simplicial complexes (triangle, tetrahedral or higher dimensional simplicial meshes) can be obtained [5, 8]. Using DEC for discretization of differential operators has been formulated and then employed by an increasing number of authors in recent years to create mimetic operators [4], develop multigrid solvers [3] and for the solution of various mechanics and physics problems [6, 9].

© Springer International Publishing AG, part of Springer Nature 2019
E. E. Gdoutos (Ed.): ICTAEM 2018, SI 5, pp. 262–267, 2019.
https://doi.org/10.1007/978-3-319-91989-8_58

In this work, we follow the techniques that have been presented lately, which discretize continuous elasticity variables as cochains over a primal mesh that represents the solid [5, 8], for the cases where the primal and dual meshes are orthonormal. Therefore we also define discrete strain and stress as a vector-valued 1-form and covector-valued 2-form respectively. However, we split (project) the discrete stress into two components, a normal and a tangential one. Doing so, the governing equations, which can be established by requiring energy balance invariance under time-dependent rigid translations and rotations of the ambient space, refer directly to these components. As a result, the equations have now a more general applicability than the one in [9] which can be considered as a constrained system, since to that the authors force the discrete stress to be normal to each face of the dual cell. Specifically, the equations can be used to represent both classical and coupled-stress elasticity. In the following, we first discuss the case of orthonormal primal and dual mesh, where the discrete Hodge star operator is a geometric diagonal matrix. This is then complemented by material parameters to capture the elastic behaviour of solids.

2   Basics of DEC with Reference to Elasticity

In linearized elasticity the basic unknown is the displacement field, which is a vector field on the reference configuration of the elastic body, Fig. 1. For the general case, in order to fully represent a topological space we consider that it can be a decomposition of simple pieces i.e. complexes K, which, in $\mathbb{R}^3$ involve lengths, areas and volumes. To define calculus on those complexes, a sign notation must be introduced. That can be though of as a reference orientation of any simplex σ^k, see for example [8]. The resulting oriented simplicial complex σ^k forms then a manifold, which, for the present, we define as a completely well-centered tetrahedral mesh, i.e. one in which the circumcenters are contained within the corresponding simplices [5, 8, 9].

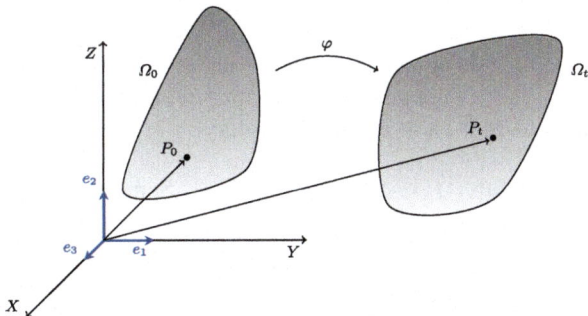

Fig. 1. Initial and deformed configuration of a continuum

Furthermore, to use those simplicial complexes as data structures, we need to define simplicial maps that approximate continuous maps, see [8]. To that end, we also use the k-chains and k-cochains on K, namely we define a function c with $c(-\sigma^k) = -c(\sigma^k)$ as a p-chain of K for the above oriented k-simplices, while we will use $C_k(K)$ for the space of p-chains. From the other hand, homomorphisms of $C_k(K)$ to $\mathbb{R}$ are called p-cochains of K, denoted $C^k(K; \mathbb{R})$ or $\text{Hom}(C_k(K), \mathbb{R})$. The dual p-cochains are denoted $C^k(\star K; \mathbb{R})$, where $\star$ is the dual complex of K. The discretization map from space of smooth p-forms to p-cochains is called the de Rham map [5]

$$R : C^k(K) \rightarrow C^k(K; \mathbb{R}) \tag{1}$$

That can be similarly defined for the dual one using $\star K$ to the later expression.

For elasticity problems, in order to discretize the operators needed, we first define the discrete exterior derivative d^k. That can be using the coboundary operator, which as a metric independent operator for K, defines d^k as [5, 7]

$$d^k : C^k(K) \rightarrow C^{k+1}(K) \tag{2}$$

The above operator acts on differential k-forms, while the sequence

$$0 \leftarrow C^n(K) \xleftarrow{d^{n-1}} \cdots \xleftarrow{d^k} C^k(K) \xleftarrow{d^{k-1}} \cdots \xleftarrow{d^{k0}} C^0(K) \leftarrow 0 \tag{3}$$

is called the cochain inducted by d^k. Similarly one can define discrete exterior derivative over a dual mesh or dual discrete exterior derivate

$$d^k : C^k(\star K) \rightarrow C^{k+1}(\star K) \tag{4}$$

Discrete operators can be determined by the incidence structure of the given simplicial mesh. The grad, curl and divergence can be defined using incidence matrices M_i^j which encode the connectivity information of a simplicial complex. We also use $\sigma^0(t)$ to describe a zero cell (at any time t) and its position vector in the Euclidean ambient space and $\sigma^j(t)$ for $j \in \mathbb{Z}$ to describe the j-th cell. Using those we can define the incidence matrices [7]

$$M_i^j = \begin{cases} 1 & \text{if } \sigma^i \text{ is coherent with the induced orientation of } \sigma^j \\ -1 & \text{if } \sigma^i \text{ is not coherent with the induced orientation of } \sigma^j \\ 0 & \text{if } \sigma^i \text{ is not on the boundary of } \sigma^j \end{cases} \tag{5}$$

for the cells σ^i and σ^j with $i, j \in \mathbb{Z}$ and $0 \le i < j \le 1$. Same way we define incidence matrices for the dual mesh by

$$M_j^i = \begin{cases} 1 & \text{if } \star \sigma^i \text{ is coherent with the induced orientation of } \star \sigma^j \\ -1 & \text{if } \star \sigma^i \text{ is not coherent with the induced orientation of } \star \sigma^j \\ 0 & \text{if } \star \sigma^i \text{ is not on the boundary of } \star \sigma^j \end{cases} \tag{6}$$

For the case when the primal and the dual mesh are orthonormal it can be shown that $M_i^j = \left(M_j^i \right)^T$ [7].

Given the definition of an inner product and corresponding metric tensor, the discrete Hodge star can be also defined as a pure diagonal matrix with components

$$H_{ij}^k = \frac{|\star\sigma_i^k|}{|\sigma_i^k|}\delta_{ij} \tag{7}$$

This operator relates p-cochains on the primal complex with (n-p)-cochains on the dual complex scaled by the inverse metric tensor, for further details see [5, 8, 9] and references therein.

3   Geometric Formulation of Elasticity

Assuming that a discretized continuum is modelled by the primal k-simplex embedded in an oriented Euclidean ambient space as described above. We further identify a zero cell $\sigma^0(t)$ (at any time t) with its position vector in the Euclidean ambient space. The time dependent simplicial mapping $\phi_t : k \rightarrow \phi_t(k)$ can be regarded as the discrete deformation map $\sigma_i^0(t) = \phi_t(\sigma_i^0), \forall\sigma_i \in k^{(0)}$.

Here we define the discrete traction as a 2-cochain on the dual mesh. More precisely we define it as a covector at the barycentre $B(\star\sigma^1)$ of σ^1, see Fig. 2. Given an orientation to primal cells for our oriented complex, discrete stress $\mathfrak{t}$ associates a 2-cochain to each oriented $\star\sigma^1$ (where $\star\sigma^1$ we consider the boundary of $\star\sigma^0$ that for the case of tetrahedral meshes consisting of triangles, Fig. 2). For the shared surface $\star\sigma^1$ between two neighbor $\star\sigma^0$ traction is considered to change sign between those dual cells [5, 9].

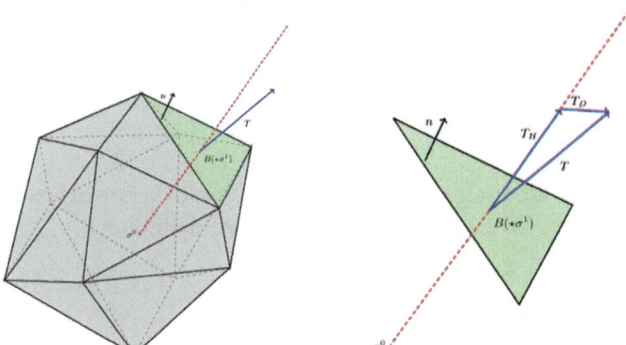

Fig. 2. The dual cell $\star\sigma^0$ around primal point σ^0 and the traction at the barycentre of $\star\sigma^1$ (left) and the decomposition of the traction vector T to a normal and a tangential one on $\star\sigma^1(t)$ (right)

Doing so, all balance laws can be obtained by postulating balance of energy and its invariance under rigid translations and rotations of the ambient space [5]. In order to derive those equations for the dual cell we decompose the traction vector on two components TH at the direction of the line connecting $\star\sigma^0$ with $B(\star\sigma^1)$ and T_D that follows from the equation

$$T = T_H + T_D \tag{8}$$

see Fig. 2.

If we use the incidence matrices described in (5) and (6), the balance of linear and angular momentum in $\mathbb{R}^3$ can be written in matrix form

$$\begin{bmatrix} RM_1^0 & I \\ R^{\perp}M_1^0 & 0 \end{bmatrix} \begin{bmatrix} T \\ B \end{bmatrix} = \begin{bmatrix} 0 \\ 0 \end{bmatrix} \tag{9}$$

where the matrix T is the discrete stress, B is the discrete body force and I the identity matrix. Furthermore matrices R and $R^{\perp}$ are the projection matrices of T to normal and tangential direction of $\star\sigma^1$. If we further denote U the matrix of displacements the discrete stress can be written as

$$T = H^1 M_1^0 U \tag{10}$$

The system of equations (9) can be solved in order to obtain the unknown displacements and stresses (or unknown body forces when a boundary point is fixed). Finally, in order to capture the elastic behavior of solids, we define the material discrete Hodge star H^k using the geometric part of (10) and a tensorial term that contains the modulus of elasticity E_a for all axial components and the modulus of rigidity G_b for the normal ones. So we have

$$H^k = \begin{bmatrix} \dfrac{|\star\sigma_i^k|}{|\sigma_i^k|} & 0 & 0 \\ 0 & \dfrac{|\star\sigma_i^k|}{|\sigma_i^k|} & 0 \\ 0 & 0 & \dfrac{|\star\sigma_i^k|}{|\sigma_i^k|} \end{bmatrix} \begin{bmatrix} E_a & 0 & 0 \\ 0 & G_b & 0 \\ 0 & 0 & G_b \end{bmatrix} \tag{11}$$

When barycenters are used to define the dual complex a correction factor has been proposed by [2] and could be also used to the above definition.

4   Bending of a Cantilever Beam

As a test example we present a classical solid mechanics problem, the bending of a prismatic cantilever beam. To our preliminary results we considered the beam to be completely fixed in all the directions on the left end while its cross section is 1×1 m and its length 10 m. We have used tetrahedral mesh as the primal one while the dual has been defined using circumcenter points of each simplex.

The loading of the right side consists of a point force of 9×10^6 N, see Fig. 3(left). We further assume that the solid behaves like a Hookean solid with Young's modulus $E_a = 5 \cdot 10^6$ N/m^2 and Poisson's ratio $v = 0.35$. Finally Fig. 3(right) presents the deflection of the beam in the z direction.

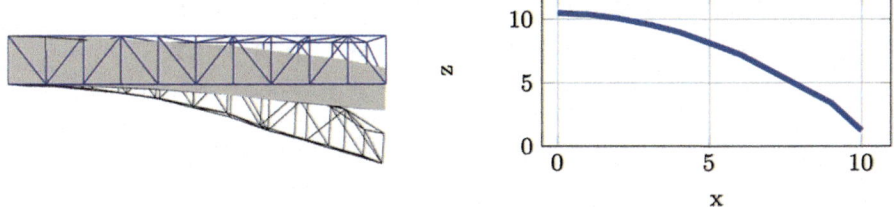

Fig. 3. Snapshots of the motion of a 3d cantilever beam (left) and deflection in the z direction (right)

5   Conclusions and Perspectives

The discrete formulation of elasticity proposed here gives the benefit of clear differentiation between the topological, geometric and physical contributions to the deformation behaviour of solids. This is the first step towards discrete modelling of more complex behaviour for engineering applications, including irreversible deformations and fracture, which is subject of ongoing work.

Acknowledgments. The authors appreciate highly the support of EPSRC via grant EP/N026136/1 "Geometric Mechanics of Solids".

References

1. Abraham, R., Marsden, J.E., Ratiu, R.: Manifolds, Tensor Analysis, and Applications, 2nd edn. Springer, New York (1988)
2. Auchmann, B., Kurz, S.: A geometrically defined discrete hodge operator on simplicial cells. IEEE Trans. Magn. **42**, 643–646 (2006)
3. Bell, N., Hirani, A.N.: PyDEC: software and algorithms for discretization of exterior calculus. ACM Trans. Math. Softw. **39**(1), 1–41 (2012)
4. Bochev, P.B., Hyman, J.M.: Principles of Mimetic Discretizations of Differential Operators, pp. 89–119. Springer, New York (2006)
5. Desbrun, M., Hirani, A.N., Leok, M., Marsden, J.E.: Discrete exterior calculus. arXiv Mathematics e-prints (2005)
6. Frauendiener, J.: Discrete differential forms in general relativity. Class. Quant. Gravity **23**(16), S369 (2006)
7. Grady, L., Polimeni, J.R.: Discrete Calculus - Applied Analysis on Graphs for Computational Science. Springer, London (2010)
8. Hirani, A.N.: Discrete exterior calculus. Ph.D. thesis, California Institute of Technology, Pasadena, CA, USA (2003)
9. Yavari, A.: On geometric discretization of elasticity. J. Math. Phys. **49**(2), 022901 (2008)

The Influence of Defects and Inclusions on Capacity for Work of Thin Plates

Nikita Morozov[(✉)], Yuri Petrov, and Boris Semenov

St. Petersburg State University, Universitetskaya Emb.,
St. Petersburg 199034, Russia
m.morozov@spbu.ru

Abstract. Some problems of strength and stability of thin constructions with defects and inclusions, which are of vital importance for nano- and microtechnology, are investigated. The attention is paid to the estimation of the number of defects influencing the capacity for work of nanosized plated and shells. The influence of surface effect is taken into account.

Keywords: Nanosized plates · Tension · Fracture · Stability · Inclusions
Cuts · Surface stresses

1 Introduction

When investigating the performance of structural elements, attention should be paid not only to their possible failure, but also to the loss of structural stability, which can occur with loads less destructive. This is typical for thin-walled structures.

Typically, structures fail due to buckling if loaded by compression. However, it is noted that the loss of stability can also be observed when the structure is stretched in the vicinity of various kinds of inhomogeneities: inclusions, cuts and etc. In the vicinity of these inhomogeneities, compressive stresses can arise under tension of the structure, causing a local buckling. This was confirmed both in experimental and theoretical studies [1–7].

This is especially true for structures of nanosized thickness in which the presence of such inhomogeneities can be due to both functional causes and defects when they are created.

It should be noted that in the case of nanoscale thicknesses of thin-walled elements, the situation becomes much more complicated, since the role of surface stresses that contribute significantly both to the effective elastic moduli and the bending stiffness of thin-walled elements increases. Therefore, a certain modification of the theory of plates and shells is needed. In particular, for plates it is shown that the bending stiffness of a plate of nanoscale thickness has the form

$$D = \frac{Eh^3}{12(1 - v^2)} + \mu_s h^2 + \frac{\lambda_s h^2}{2}$$

© Springer International Publishing AG, part of Springer Nature 2019
E. E. Gdoutos (Ed.): ICTAEM 2018, SI 5, pp. 268–272, 2019.
https://doi.org/10.1007/978-3-319-91989-8_59

That is, the relative change in the bending stiffness of such a plate is proportional to $O(h^{-1})$. Similar relationships are valid for effective elastic moduli and for the bending stiffness of shells. These corrections should be taken into account when estimating critical loads.

Within the framework of this work, the influence of the shape and orientation of elliptical cutouts, the relationship between the elastic moduli of the plate and inclusions on the destructive load and the ultimate load at which buckling is observed, will be investigated.

2   Results

2.1   Buckling of Plate with Elliptical Cutout

The problems of stability loss of a thin plate with an elliptical cut are considered. The dependence of critical loads on the eccentricity of the ellipse and the angle between the direction of tension and the minor axis of the ellipse is investigated.

From the analysis of the stressed state of a plate with an elliptical cut, it follows that for any angles of inclination of the ellipse relative to the direction of extension near the contour of the ellipse, there are regions of compression that can cause a loss of stability of the plate [8].

For analysis of the buckling problem a finite-element model of the plate with an elliptical cutout is constructed. The size of the plate is 20 μm × 20 μm, the thickness of the plate is 0.01 μm, the major axis of ellipse is $a = 1$ μm, the minor ellipse axes $b = 0.1$ μm, α is the angle between the minor axis of the ellipse and the direction of tension. The plate is stretched by a distributed force of 1 N/μm (see Fig. 1).

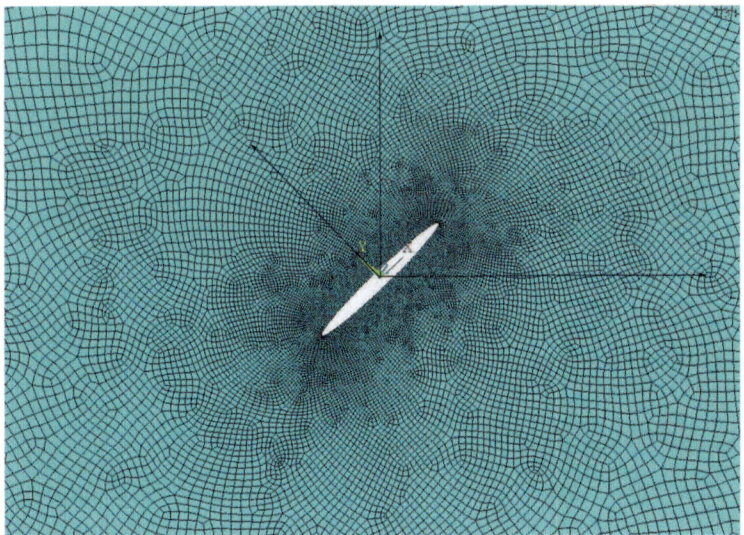

Fig. 1. Finite element model

The results of numerical simulation are given in Table 1.

Table 1. The dependence of the critical load, the maximum principal stress and the minimum principal stress on the angle between the minor axis of the ellipse and the direction of tension.

α	$p_{crit}\ N/\mu m$	$\sigma_{max}\ N/\mu m^2$	$\sigma_{min}\ N/\mu m^2$
0	0,33	1805,39	−102,22
15	0,37	1659,04	−131,38
30	0,53	1446,92	−204,3
45	1,01	1032,33	−318,89
60	2,96	732,7	−322,24
75	50	398,66	−202,76
90	312,39	120,24	−89,78

In Table 1 p_{crit} is the force at which local buckling of the plate occurs, σ_{max} is the maximum principal stress and σ_{min} is the minimum principal stress at the applied tensile force $1\ N/\mu m$.

From the analysis of the results of numerical simulation it follows that for a given geometry of a plate with an elliptical cut for small angles between the minor axis of the ellipse and the direction of tension, the loss of stability will occur at loads less than the destructive loads.

The first buckling form for the angle between the minor axis of the ellipse and the direction of tension equal 45° is shown in Fig. 2. The buckling forms for other angles are similar.

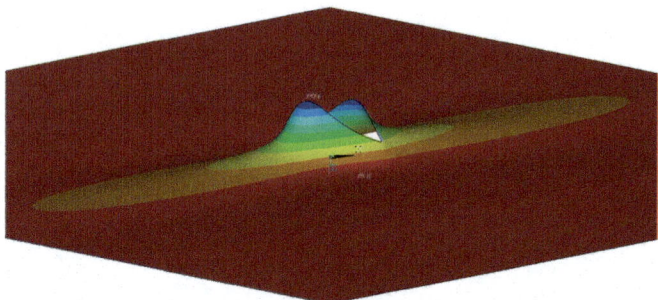

Fig. 2. First buckling form for the plate with elliptical cutout at tension in y direction for the angle between the minor axis of the ellipse and the direction of tension equal 45°.

2.2   Buckling of Plate with Inclusion

The numerical results and analysis in ANSYS showed that the loss of stability of plates with circular inclusion is possible if the elastic modulus of the inclusion is less than the modulus of elasticity of the plate $E_1 > E_2$ (inclusion softer than the plate) or when the elastic modulus of the inclusion larger than the modulus of elasticity of the plate

$E_1 < E_2$ (i.e. inclusion is stiffer). Note that the stability loss modes are different for the case $E_1/E_2 > 1$ and $E_1/E_2 < 1$. In the case when the inclusion is stiffer than the plate, as was noted above, the zones of compressive stresses lie along the x axis (they are shifted by 90° with respect to the case where the inclusion is softer than the plate).

We note that in the case of equal Young's moduli in the plate and inclusion, but of different Poisson's factors, regions of compressive stress appear under uniaxial tension. It means that in this case the plate can also lose the stability under uniaxial tension.

Figure 3 show the stress distribution before buckling and the buckling form of the plate in the case when $E_1 = E_2 = 2 \cdot 10^5$ MPa, Poisson's ratio of the plate is $v_1 = 0.01$ and Poisson's ratio of the inclusion is $v_2 = 0.49$.

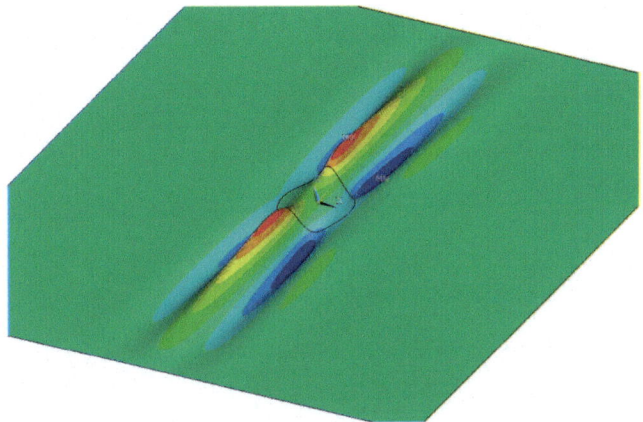

Fig. 3. Buckling form of plate with inclusion when $E_1 = E_2 = 2 \cdot 10^5$ MPa, Poisson's ratio of the plate is $v_1 = 0.01$ and Poisson's ratio of the inclusion is $v_2 = 0.49$

3   Conclusion

The problems of stability loss of a thin plate with an elliptical cut are considered. The dependence of critical loads on the eccentricity of the ellipse and the angle between the direction of extension and the semimajor axis of the ellipse is investigated.

It is shown that a sufficiently thin plate can lose its tensile stability when the Young's moduli coincide, and the Poisson's coefficients are different. A form of stability loss of an orthotropic plate with inclusion of the same material, but with turned orthotropic axes, is also constructed.

Acknowledgements. The authors would like to acknowledge the financial support of the Russian Foundation for Basic Research (RFBR) project No. 18-01-00884.

References

1. Dixon, J.R., Stranningan, J.S.: Stress distribution and buckling in thin sheets with central slits. In: Proceedings of the 2nd International Conference on Fracture, Brighton. Chapman and Hall, London (1969)
2. Dahl, Y.M.: On the local bending of an elongated plate with a crack. Izvestia Akademii Nauk. Mechanika tverdogo tela, No. 4, c. 135–141 (1978)
3. Guz, A.N., Dishel', M.S., Kuliev, G.G., Milovanova, O.B.: Fracture and Stability of Thin Bodies with Cracks. Naukova Dumka, Kiev (1981)
4. Rammerstorfer, F.G.: Buckling of elastic structures under tensile loads. Acta Mech. (2017). https://doi.org/10.1007/s00707-017-2006-1
5. Shimizu, S.: Tension buckling of plate having a hole. Thin Walled Struct. **45**(10–11), 827–833 (2007)
6. Bochkarev, A., Grekov, M.: Local instability of a plate with a circular nanohole under uniaxial tension. Dokl. Phys. **59**(7), 330–334 (2014)
7. Bauer, S., Kashtanova, S., Morozov, N., Semenov, B.: Stability of a nanoscale-thickness plate weakened by a circular hole. Dokl. Phys. **59**(9), 416–418 (2014)
8. Kachanov, M., Shafiro, B., Tsurkov, I.: Handbook of Elasticity Solutions. Kluwer Academic Publishers, Dordrecht (2003)

Limit of Ultra-high Strain Rates in Plastic Response of Metals

Alexander Mayer$^{(\boxtimes)}$, Vasiliy Krasnikov, and Victor Pogorelko

Chelyabinsk State University,
Bratiev Kashirinykh Street 129, 454001 Chelyabinsk, Russia
mayer@csu.ru

Abstract. Increase in strain rate requires an increase in density of lattice defects (dislocations) necessary for plastic relaxation of the increasing shear stress. Irradiation of metals by intensive femtosecond laser pulses creates compression waves with durations of tens of picoseconds and ultra-high strain rates up to an inverse nanosecond. At such strain rates, multiplication and motion of the initially existing dislocations become not enough for restriction of the shear stress, which can grow up to the limit of homogeneous nucleation of dislocation loops. We investigate regularities of the homogeneous nucleation of dislocations in metals with FCC, BCC and HCP lattices using the molecular dynamic simulations. Then we generalize these regularities in the form of continuum model of plasticity with nucleation and apply the model for simulation of shot compression pulses. It allows us to investigate the transition between the common mode of multiplication and the nucleation-controlled mode of plasticity.

Keywords: Dislocation plasticity · High strain rates
Homogeneous nucleation · Shock waves

1 Introduction

A broad range of strain rates is experimentally realized at present–from almost zero in quasi-static tests up to an inverse nanosecond in the experiments on thin foils irradiation by ultra-short laser pulses [1–3]. Increase in the strain rate on the shock wave front results in drastic increase in the elastic precursor amplitude [4], which can reach the amplitude of the main shock wave [5]. It is due to the fact that possible rate of plastic deformation is substantially restricted by the density of lattice defects, such as dislocations [6]. According to Orowan equation, the plastic strain rate is determined by multiplication of the Burgers vector amplitude, the dislocation velocity and the scalar density of dislocations. The dislocation velocity is restricted by the transverse sound velocity [7]; therefore, the maximal plastic strain rate becomes proportional to the dislocation density [6], which cannot growth arbitrarily fast by multiplication. As a result, the multiplication and motion of dislocations become not enough for restriction of the shear stress. On the other hand, growth of shear stress can initiate an alternative mechanism of homogeneous nucleation of dislocations [8, 9].

Here we investigate regularities of the homogeneous nucleation of dislocations in metals with FCC, BCC and HCP lattices using the molecular dynamic

© Springer International Publishing AG, part of Springer Nature 2019
E. E. Gdoutos (Ed.): ICTAEM 2018, SI 5, pp. 273–278, 2019.
https://doi.org/10.1007/978-3-319-91989-8_60

(MD) simulations. Then we generalize these regularities in the form of continuum model of plasticity with nucleation, which can be used for simulation of propagation of shot compression pulses in metals.

2 Molecular Dynamic Simulations

The MD simulations are performed with the help of LAMMPS [10] using the following inter-atomic potentials: angle-dependent (ADP) potential [11] for Cu and Al, embedded atom method (EAM) potentials [12], [13] and [14] for Ti, Mg and Fe, respectively. OVITO [15] is used for visualization and analysis of atom configurations; in particular, the dislocation extraction algorithm (DXA) [16] is used for searching of dislocations and calculation of the dislocation density.

Initially perfect single-crystal sample of parallelepiped shape with FCC (Cu and Al), HCP (Ti and Mg) or BCC (Fe) containing about a million atoms is considered with periodic boundary conditions on all faces. The simple shear is performed by 'deform' command of LAMMPs, which scales the atom coordinates. In the case of HCP lattices, the shear is in the basal plane; in other cases, it is in (001) plane. The temperature of the sample during the test is maintained by Nose-Hoover thermostat.

An example of the obtained results for Cu is shown in Fig. 1. The development of the dislocation structure starts from nucleation of couples of loops of Shockley partial dislocations. Subsequent growth and intersection of loops initiates the multiplication mechanism; the dislocation density sharply increases, which leads to fast decrease in shear stress. Further, both the dislocation density and the shear stress are stabilized by dynamic balance between multiplication and annihilation of dislocations.

3 Continuum Model of Dislocation Plasticity with Nucleation

We develop a continuum model of dislocation plasticity with the accounting of dis-location nucleation basing on the model proposed in [17] for the case of pore growth. The model can be used for simulation of the shock-wave processes in metals including the ultra-short shock waves with ultra-high strain rates. The nucleation of dislocations is described as a result of thermal fluctuations, which probability depends on the formation energy of the critical dislocation loop. Multiplication of dislocations plays an important role even in the case the process of the dislocation system development starts from the homogeneous nucleation. The dislocation multiplication is described on the basis of the assumption that a certain part of the plastically dissipated energy is spent on the formation of new dislocations. Annihilation of dislocations determines the final balance of the dislocation density; the annihilation is described basing on the standard approach. Plastic relaxation of stress is connected by Orowan equation with the dis-location velocity, which is determined by the acting shear stress.

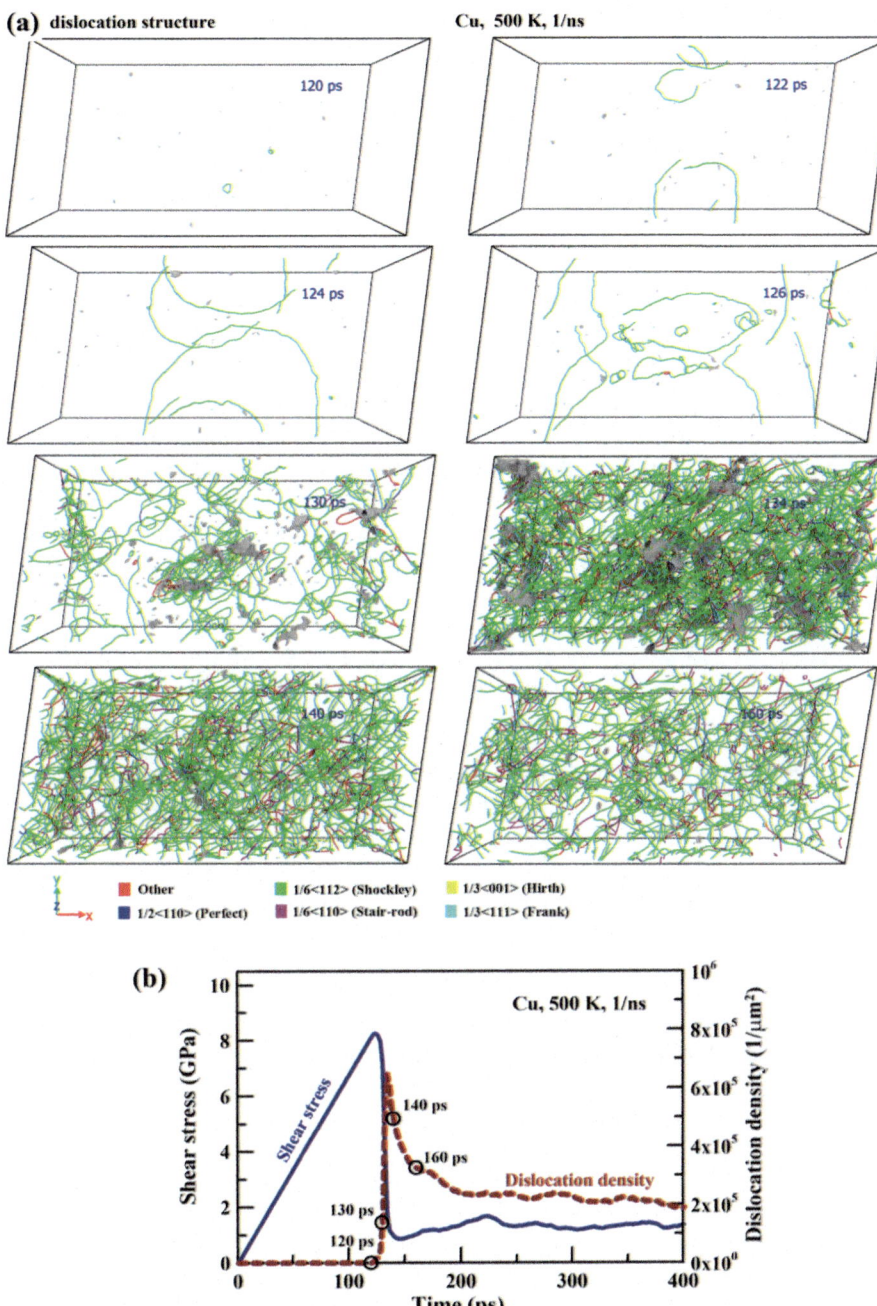

Fig. 1. (a) evolution of dislocation structure and (b) time dependences of the shear stress (solid blue curve) and dislocation density (doted red curve) at the pure shear of Cu single crystal (FCC lattice) with the strain rate of 1/ns at the temperature of 500 K. Circles in (b) mark some of time points shown in (a). Grey surfaces in (a) correspond to surface defects.

Evolution of the shear stress τ is determined by the Hook's law

$$\tau = 2G(\varepsilon - w), \tag{1}$$

where G is the shear modulus; ε is the geometric deformation due the external shear applied to the substance volume; w is the plastic deformation. In the case of the constant strain rate corresponding to the MD simulations, the geometric deformation is calculated as $\varepsilon = \dot{\varepsilon} \cdot t$, where t is the time; $\dot{\varepsilon}$ is the external strain rate. The plastic strain rate is determined from the Orowan equation

$$\frac{dw}{dt} = K(b\rho V), \tag{2}$$

where b is the modulus of the Burgers vector; ρ is the scalar density of dislocations; V is the dislocation velocity; K is factor less than 1, which takes into account the orientation of the slip systems with respect to the shear direction and the fact that not all dislocations are movable and not all movable segments lie in the active slip systems.

At low dislocation density and high level of shear stress, the dislocation velocity can be comparable with the transverse sound velocity c; therefore, the equation for the dislocation velocity accounts for the quasi-relativistic correction

$$V = \frac{b}{B}(\tau - Y)\left(1 - \left(\frac{V}{c}\right)^2\right)^{3/2}, \tag{3}$$

where B is the drag coefficient; Y is the static yield strength depending on the dislocation density according to the Taylor hardening law $Y = Y_0 + 0.5Gb\sqrt{\rho}$; Y_0 is the yield strength for perfect single crystal.

The kinetics equation for dislocation density expresses the balance between the rates of nucleation (Q_N), multiplication (Q_M) and annihilation (Q_A) of dislocations

$$\frac{d\rho}{dt} = Q_N + Q_M - Q_A. \tag{4}$$

The expression for nucleation rate follows form the thermal fluctuation mechanism of the process

$$Q_N = \frac{c}{a_c^4} \exp\left(-\frac{U_c}{kT}\right), \tag{5}$$

where $a_c = E/(b\tau)$ and $U_c = \pi E^2/(b\tau)$ are radius and energy of the critical loop, respectively; E is the formation energy per unit length of dislocation. The multiplication rate is determined by the energy dissipated during the plastic flow

$$Q_M = \left(\frac{dw}{dt}\right)\frac{\tau}{E}. \tag{6}$$

The annihilation rate is written by the standard way with the annihilation constant k

$$Q_A = kb\rho^2 V. \tag{7}$$

Figure 2 shows the comparison results of the continuum model and MD data for Cu single crystal at two temperatures and three strain rates. One can see that the model correctly describes evolution of both the shear stress and the dislocation density.

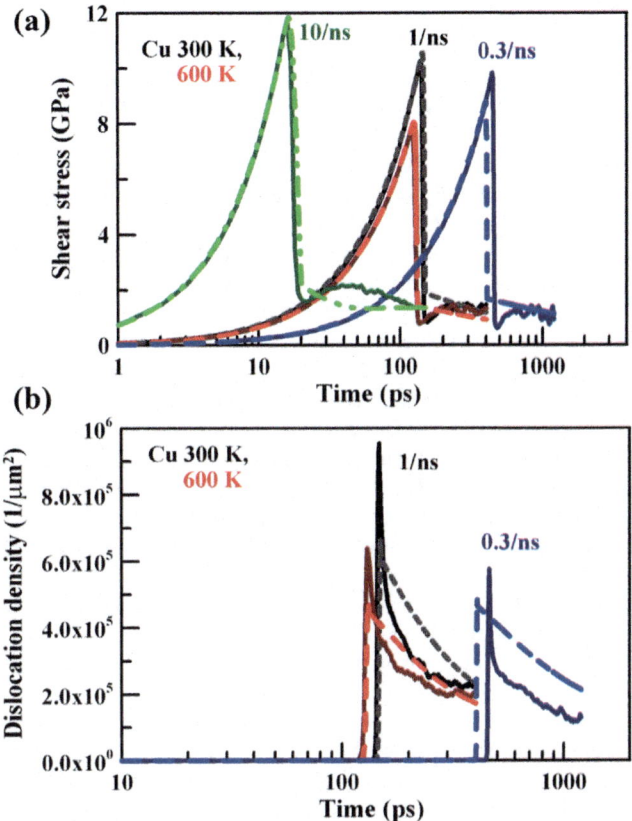

Fig. 2. Comparison of the continuum model results with MD data; shear of Cu single crystal with the strain rates of 10, 1 and 0.3/ns at the temperature of 300 K and with the strain rate of 1/ns at the temperature of 600 K (red curves): (a) shear stress and (b) dislocation density.

4   Conclusions

Basing on MD simulations, we formulate the continuum model of dislocation plasticity with accounting of the homogeneous nucleation of dislocation, which is necessary for investigation of the limit of ultra-high strain rates. The presented equation system (1)–(7) describes the case of uniform deformation realized in MD simulations, but it can be easily generalized on the case of non-uniform deformation for description of the shock wave processes in metals.

Acknowledgement. The work is supported by the Ministry of Education and Science of the Russian Federation, state task 3.2510.2017/PP.

References

1. Kanel, G.I., Zaretsky, E.B., Razorenov, S.V., Ashitkov, S.I., Fortov, V.E.: Unusual plasticity and strength of metals at ultra-short load durations. Phys. Usp. **60**(5), 490–508 (2017)
2. Matsuda, T., Sano, T., Arakawa, K., Sakata, O., Tajiri, H., Hirose, A.: Femtosecond laser-driven shock-induced dislocation structures in iron. Appl. Phys. Express **7**, 122704 (2014)
3. Agranat, M.B., Ashitkov, S.I., Komarov, P.S.: Metal behavior near theoretical ultimate strength in experiments with femtosecond laser pulses. Mech. Solids **49**(6), 643–648 (2014)
4. Smith, R.F., Eggert, J.H., Rudd, R.E., Swift, D.C., Bolme, C.A., Collins, G.W.: High strain-rate plastic flow in Al and Fe. J. Appl. Phys. **110**(12), 123515 (2011)
5. Mayer, A.E., Khishchenko, K.V., Levashov, P.R., Mayer, P.N.: Modeling of plasticity and fracture of metals at shock loading. J. Appl. Phys. **113**(19), 193508 (2013)
6. Borodin, E.N., Mayer, A.E.: Theoretical interpretation of abnormal ultrafine-grained material deformation dynamics. Model. Simul. Mater. Sci. Eng. **24**(2), 025013 (2016)
7. Krasnikov, V.S., Mayer, A.E.: Influence of local stresses on motion of edge dislocation in aluminum. Int. J. Plast. **101**, 170–187 (2018)
8. Bringa, E.M., Rosolankova, K., Rudd, R.E., Remington, B.A., Wark, J.S., Duchaineau, M., Kalantar, D.H., Hawreliak, J., Belak, J.: Shock deformation of face-centred-cubic metals on subnanosecond timescales. Nat. Mater. **5**, 805–809 (2006)
9. Norman, G.E., Yanilkin, A.V.: Homogeneous nucleation of dislocations. Phys. Solid State **53**(8), 1614–1619 (2012)
10. Plimpton, S.: Fast parallel algorithms for short-range molecular dynamics. J. Comput. Phys. **117**, 1–19 (1995). http://lammps.sandia.gov
11. Apostol, F., Mishin, Y.: Interatomic potential for the Al-Cu system. Phys. Rev. B **83**, 054116 (2011)
12. Mendelev, M.I., Underwood, T.L., Ackland, G.J.: Development of an interatomic potential for the simulation of defects, plasticity, and phase transformations in titanium. J. Chem. Phys. **145**(1), 154102 (2016)
13. Sun, D.Y., Mendelev, M.I., Becker, C.A., Kudin, K., Haxhimali, T., Asta, M., Hoyt, J.J., Karma, A., Srolovitz, D.J.: Crystal-melt interfacial free energies in HCP metals: a molecular dynamics study of Mg. Phys. Rev. B **73**, 024116 (2006)
14. Chamati, H., Papanicolaou, N.I., Mishin, Y., Papaconstantopoulos, D.A.: Embedded-atom potential for Fe and its application to self-diffusion on Fe(100). Surf. Sci. **600**, 1793 (2006)
15. Stukowski, A.: Visualization and analysis of atomistic simulation data with OVITO–the Open Visualization Tool. Model. Simul. Mater. Sci. Eng. **18**, 015012 (2010). http://www.ovito.org
16. Stukowski, A., Bulatov, V.V., Arsenlis, A.: Automated identification and indexing of dislocations in crystal interfaces. Model. Simul. Mater. Sci. Eng. **20**, 085007 (2012)
17. Krasnikov, V.S., Mayer, A.E.: Plasticity driven growth of nanovoids and strength of aluminum at high rate tension: Molecular dynamics simulations and continuum modeling. Int. J. Plast. **74**, 75–91 (2015)

Scattering of Waves by a Shear Band

Davide Bigoni[1], Domenico Capuani[2(✉)], and Diana Giarola[1]

[1] DICAM, University of Trento, Via Mesiano 77, 38123 Trento, Italy
[2] DA, University of Ferrara, Via Quartieri 8, 44121 Ferrara, Italy
domenico.capuani@unife.it

Abstract. The incremental behaviour of a prestressed, elastic, anisotropic and incompressible material is analyzed in the dynamic regime, under the plain strain condition. Dynamic perturbations of stress/deformation incident wave fields, caused by a shear band of finite length, formed inside the material at a certain stage of continued deformation, are investigated. At the base of the proposed dynamic perturbation approach is the time-harmonic infinite-body Green's function for incremental displacements obtained by Bigoni and Capuani [5] for small isochoric and plane deformation superimposed upon a nonlinear elastic and homogeneous strain. The integral representation relating the incremental stress at any point of the medium to the incremental displacement jump across the shear band faces, is obtained. Finally, a numerical procedure based on a collocation method is used to solve the boundary integral equation for incident wave scattering by a shear band.

Keywords: Shear band · Wave propagation · Pre-stress · Non-linear elasticity
Integral representation

1 Introduction

Localized deformations in the form of shear bands are known to be preferential near-failure deformation modes of ductile materials. The development of shear localization bands has been also shown to be possible in anisotropic composite materials consisting of random distributions of aligned rigid fibres of elliptical cross section in a soft elastomeric matrix [1].

When a ductile material is subject to severe strain, failure is preluded by an emergence of shear bands which initially nucleate in a small area, but quickly extend rectilinearly and accumulate damage, until they degenerate into fractures. Therefore, research on shear bands yields a fundamental understanding of the intimate rules of failure, so that it may be important in the design of new materials.

Modelling a shear band as a slip plane embedded in a highly prestressed material and perturbed by a mode II incremental strain, reveals that a highly inhomogeneous and strongly focussed stress state is created in the proximity of the shear band and aligned parallel to it. This evidence, together with the fact that the incremental energy release rate blows up when the stress state approaches the condition for ellipticity loss, may explain why shear bands grow rectilinearly and why they are a preferred mode of failure [2].

© Springer International Publishing AG, part of Springer Nature 2019
E. E. Gdoutos (Ed.): ICTAEM 2018, SI 5, pp. 279–288, 2019.
https://doi.org/10.1007/978-3-319-91989-8_61

Although it is expected that dynamic effects play an important role on shear band growth, most of the analyses conducted so far were limited to quasi-static conditions. The aim of the present paper is to investigate dynamic perturbations of stress/deformation incident wave fields, caused by a shear band of finite length, formed inside the material at a certain stage of continued deformation. The incremental behaviour of a prestressed, elastic, anisotropic and incompressible material is analyzed in the dynamic regime, under the plain strain condition. The integral representation relating the incremental stress at any point of the medium to the incremental displacement jump across the shear band faces, is obtained, and a numerical procedure, based on a collocation method, is used to solve the boundary integral equation for incident wave scattering by a shear band. The proposed approach provides a basis for the analysis of propagation of disturbances near the boundary of loss of ellipticity. Depending on the level of prestress and anisotropy, wave patterns are shown to emerge, with focussing of signals in the direction of shear bands. Varying the direction of the dynamic perturbation excites different wave patterns, which tend to degenerate to families of plane waves parallel to the shear bands, when the elliptic boundary is approached.

2   Constitutive Equations

The incremental behaviour of an infinite, incompressible, nonlinear elastic material, homogeneously deformed under the plane strain condition, is considered. According to Biot [3], the constitutive relations between the nominal stress increment $\dot{t}_{ij}$ and the gradient of incremental displacement $v_{i,j}$ (a comma denotes partial differentiation) can be expressed in the principal reference system of Cauchy stress (here denoted by axes x_1 and x_2) as follows

$$\dot{t}_{ij} = \mathbb{K}_{ijkl} v_{l,k} + \dot{p}\delta_{ij} \tag{1}$$

where repeated indices are summed and range between 1 and 2, δ_{ij} is the Kronecker delta, $\dot{p}$ is the incremental in-plane hydrostatic stress and $\mathbb{K}_{ijkl}$ are the instantaneous moduli. These moduli possess the major symmetry $\mathbb{K}_{ijkl} = \mathbb{K}_{klij}$ and are functions of the principal components of Cauchy stress, σ_1 and σ_2, describing the pre-stress, and of two incremental moduli μ and μ_* (which can depend arbitrarily on the current stress and strain) corresponding to shearing parallel to, and at $45°$ to, the principal stress axes. The non-null components are:

$$\mathbb{K}_{1111} = \mu_* - \frac{\sigma}{2} - p, \quad \mathbb{K}_{1122} = \mathbb{K}_{2211} = -\mu_*, \quad \mathbb{K}_{2222} = \mu_* + \frac{\sigma}{2} - p$$
$$\mathbb{K}_{1212} = \mu + \frac{\sigma}{2}, \quad \mathbb{K}_{1221} = \mathbb{K}_{2112} = \mu - p, \quad \mathbb{K}_{2121} = \mu - \frac{\sigma}{2} \tag{2}$$

with

$$\sigma = \sigma_1 - \sigma_2, \quad p = \frac{\sigma_1 + \sigma_2}{2}. \tag{3}$$

Equation (1) is complemented by the incompressibility constraint for incremental displacement v_i

$$v_{i,i} = 0. \tag{4}$$

Constitutive Eqs. (1)–(4) describe a broad class of material behaviours, including all possible elastic incompressible materials which are isotropic in an initial state, but also materials which are orthotropic with respect to the current principal stress directions. The latter situation has interesting practical applications in the field of fibre-reinforced elastic materials.

3   The Boundary Value Problem

Let S be a shear band of finite length, formed inside the infinite medium at a certain stage of continued deformation. A shear band of finite length can be seen as a very thin layer of material across which the normal component of incremental displacement and of nominal traction remain continuous, but the incremental nominal tangential traction vanishes, while the corresponding displacement is not prescribed. Therefore, it is possible to model such a shear band as a weak surface whose faces can freely slide, but are constrained to remain in contact. Note that this slip surface is different from a crack since it can carry normal tractions, so that only under special symmetry conditions on the prestress state it may behave as a crack subjected to shear parallel to it (the so-called "mode II" loading in fracture mechanics).

In Fig. 1 a shear band of total length $2l$ is represented together with a local reference system $(\hat{x}_1, \hat{x}_2)$ centered on the shear band, with $\hat{x}_1$-axis aligned parallel to the shear band, and rotated at an angle $\theta = \theta_0$ (taken positive when anticlockwise) with respect to the principal reference system (x_1, x_2) introduced for the constitutive Eq. (1).

Introducing the jump operator for a generic function f, smooth on two regions labeled "+" and "−", and discontinuous across the surface S, as

$$[\![f]\!] = f^+ - f^- \tag{5}$$

where $f^\pm$ denote the limits approached by function f at the faces of the discontinuity surface, the boundary conditions at the shear band surface S can be written as

$$[\![\hat{v}_2]\!] = 0, \quad [\![\hat{t}_{22}]\!] = 0, \quad \hat{t}_{21} = 0 \tag{6}$$

with $\hat{v}_i$, $\hat{t}_{ij}$ being incremental displacement and incremental stress components in the local reference system.

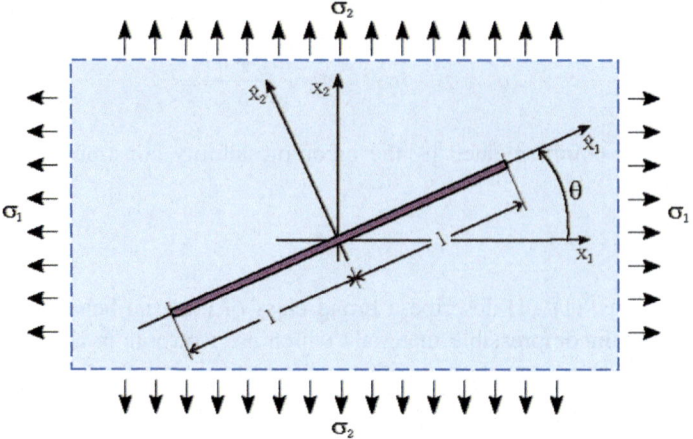

Fig. 1. Shear band of finite length ($2l$) and principal Cauchy stress components, σ_1 and σ_2.

In the hypothesis of time harmonic motion with circular frequency Ω, a wave characterized by an incremental displacement field $v_i^{inc}(\mathbf{x})e^{-i\Omega t}$ travels through the medium and is incident upon the shear band. Then, a scattered incremental displacement field $v_i^{sc}(\mathbf{x})e^{-i\Omega t}$ is generated by the interaction of the incident wave with the shear band such that the total incremental displacement field $v_i(\mathbf{x})e^{-i\Omega t}$ is given by

$$v_i = v_i^{inc} + v_i^{sc} \tag{7}$$

The scattered field v_i^{sc} must satisfy the radiation condition at infinity and conditions of finiteness of energy near the shear band edge. Outside the shear band, the incremental displacement field satisfies the equations of motion

$$(2\mu_* - \mu)v_{1,11} + \left(\mu - \frac{\sigma}{2}\right)v_{1,22} = -\dot{\pi}_{,1} - \rho\Omega^2 v_1$$

$$(2\mu_* - \mu)v_{2,22} + \left(\mu + \frac{\sigma}{2}\right)v_{2,11} = -\dot{\pi}_{,2} - \rho\Omega^2 v_2 \tag{8}$$

where ρ is the mass density and $\dot{\pi}$ is the increment of in-plane nominal hydrostatic stress

$$\dot{\pi} = \frac{\dot{t}_{11} + \dot{t}_{22}}{2} - \frac{\sigma}{2}v_{1,1} \tag{9}$$

Introducing the stream function $\psi(\mathbf{x})$ as

$$v_1 = \psi_{,2}, \quad v_2 = -\psi_{,1} \tag{10}$$

and the dimensionless prestress parameter

$$k = \frac{\sigma}{2\mu}, \tag{11}$$

Equation (8) can be combined to give

$$(1+k)\psi_{,1111} + 2\left(2\frac{\mu_*}{\mu} - 1\right)\psi_{,1122} + (1-k)\psi_{,2222} = -\frac{\varrho\Omega^2}{\mu}(\psi_{,11} + \psi_{,22}). \tag{12}$$

Equation (12) provides the regime classification, which is the same as for the quasi-static case (see Bigoni and Capuani [4]). The results in this paper will be restricted to the elliptic regime (E), defined through the condition that scalars γ_1 and γ_2

$$\left.\begin{array}{c}\gamma_1\\\gamma_2\end{array}\right\} = \frac{1 - 2\mu_*/\mu \pm \sqrt{\Delta}}{1+k}, \quad \Delta = k^2 - 4\frac{\mu_*}{\mu} + 4\left(\frac{\mu_*}{\mu}\right)^2 \tag{13}$$

are either both real and negative in the elliptic imaginary regime (EI) or a conjugate pair in the elliptic complex regime (EC). Note that Δ is positive in (EI) and negative in (EC).

A consequence of the above discussion is that the emergence of weakly discontinuous surfaces in the medium corresponds to failure of ellipticity, in the present context as in the quasi-static case. This occurs in a continuous loading path (starting from E) either when $k = 1$ (so that $\gamma_1 = 0$) or when $\Delta = 0$ (so that $\gamma_1 = \gamma_2$). The former case defines the elliptic-imaginary/parabolic boundary, while the latter the elliptic-complex/hyperbolic boundary.

4   Integral Representation

With reference to any given point $\mathbf{y}$ outside the shear band, the scattered field can be given the following integral representation in terms of jumps of incremental displacements and incremental stress across the discontinuity surface

$$v_g^{sc}(\mathbf{y}) = -\int_S \left([[t_{ij}]] n_i v_j^g(\mathbf{x}, \mathbf{y}) - t_{ij}^g(\mathbf{x}, \mathbf{y}) n_i [[v_j]] \right) dl_x \tag{14}$$

where $\mathbf{n}$ is the unit normal at every point of S, $v_j^g(\mathbf{x}, \mathbf{y})$ is the incremental displacement of the time-harmonic Green function for the infinite prestressed medium found by Bigoni and Capuani [5], and $t_{ij}^g(\mathbf{x}, \mathbf{y})$ is the associated nominal stress increment. In Eq. (14), due to the continuity of the incident incremental field $(v_i^{inc}, t_{ij}^{inc})$, jumps of the total incremental field $([[v_i]], [[t_{ij}]])$ coincide with jumps of the scattered incremental field $([[v_i^{sc}]], [[t_{ij}^{sc}]])$.

Owing to boundary conditions (6),

$$[[\dot{t}_{ij}]]n_i = 0 \tag{15}$$

so that Eq. (14) reduces to

$$v_g^{sc}(\mathbf{y}) = \int_S \dot{t}_{ij}^g(\mathbf{x}, \mathbf{y})n_i[[v_j]]dl_x. \tag{16}$$

The gradient of incremental displacement can be evaluated from (16) as

$$v_{g,k}^{sc}(\mathbf{y}) = -\int_S \dot{t}_{ij,k}^g(\mathbf{x}, \mathbf{y})n_i[[v_j]]dl_x, \tag{17}$$

and the incremental stress can be deduced from constitutive Eq. (1):

$$\dot{t}_{lm}^{sc}(\mathbf{y}) = -\mathbb{K}_{lmkg}\int_S \dot{t}_{ij,k}^g(\mathbf{x}, \mathbf{y})n_i[[v_j]]dl_x + \dot{p}\delta_{lm}. \tag{18}$$

In order to determine the incremental displacement jump $[[v_i]]$, Eq. (18) is to be written for $\mathbf{y}$ approaching a point of S, where the boundary conditions (6) are prescribed. Denoting by $\mathbf{s}$ the unit tangent vector at any point of S, boundary condition (6$_3$) can be rewritten as

$$\mathbf{n} \cdot \dot{\mathbf{t}}^{sc}\mathbf{s} = -\mathbf{n} \cdot \dot{\mathbf{t}}^{inc}\mathbf{s} \tag{19}$$

Hence, using the incremental stress representation (18) into boundary condition (19), leads to

$$\mathbf{n} \cdot \dot{\mathbf{t}}^{inc}\mathbf{s} = n_l s_m \mathbb{K}_{lmkg}\int_S \dot{t}_{ij,k}^g(\mathbf{x}, \mathbf{y})n_i[[v_j]]dl_x. \tag{20}$$

Equation (20) represents the boundary integral formulation for the boundary value problem at hand. The kernel of the integral Eq. (20) is hypersingular of order r^{-2} as $r \to 0$, r being the distance between field point $\mathbf{x}$ and source point $\mathbf{y}$:

$$r = |\mathbf{x} - \mathbf{y}| = \sqrt{(x_1 - y_1)^2 + (x_2 - y_2)^2} \tag{21}$$

Therefore, the integral on right-hand side of (20) is specified in the finite-part Hadamard sense.

Equation (20) can be given a more explicit expression by introducing the relations for the incremental displacement components in both the principal reference system (x_1, x_2) and the local reference system $(\hat{x}_1, \hat{x}_2)$:

$$\mathbf{v} = \mathbf{Q}\widehat{\mathbf{v}}, \quad [\mathbf{Q}] = \begin{bmatrix} \cos\theta_0 & \sin\theta_0 \\ -\sin\theta_0 & \cos\theta_0 \end{bmatrix} \tag{22}$$

so that, due to boundary condition (6_1),

$$[[v_j]] = Q_{j1}[[\widehat{v}_1]] = s_j[[\widehat{v}_1]]. \tag{23}$$

By using Eqs. (23), (20) can be given the final form

$$\mathbf{n} \cdot \overset{inc}{\mathbf{t}}\,\mathbf{s} = n_l s_m \mathbb{K}_{lmkg} \int_S t^g_{ij,k}(\mathbf{x}, \mathbf{y}) n_i s_j [[\widehat{v}_1]] dl_x \tag{24}$$

showing that the solution of the problem is given by a linear integral equation in the unknown scalar function $[[\widehat{v}_1]]$, i.e. the jump of tangential incremental displacement across the shear band faces.

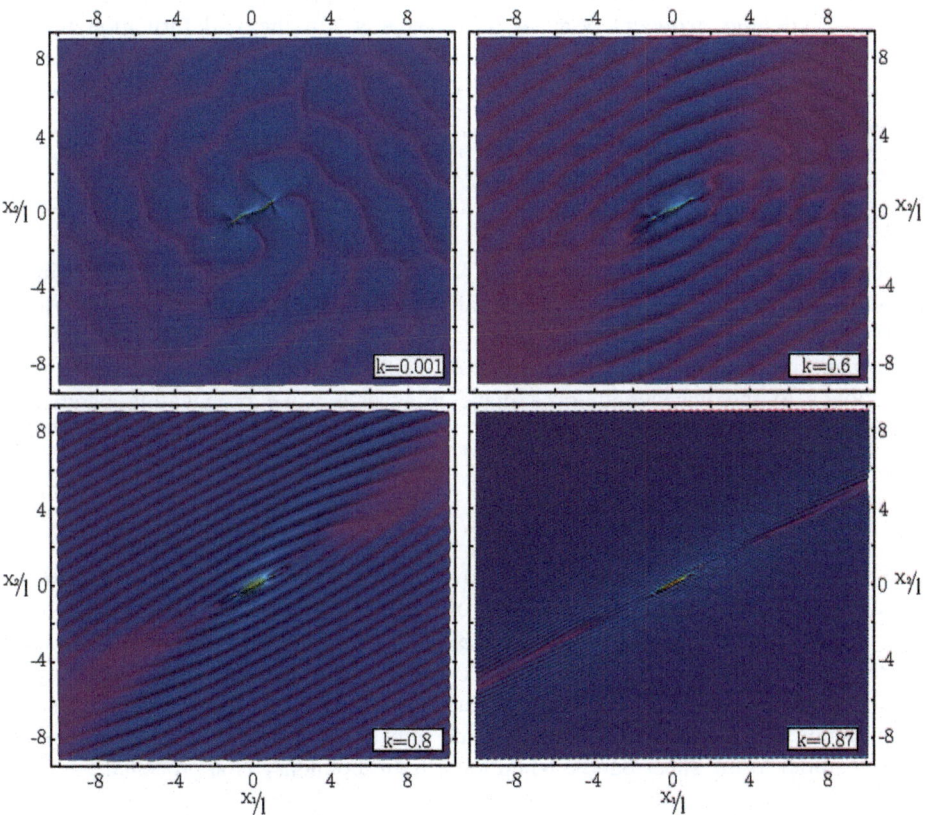

Fig. 2. Real part of scattered incremental deviatoric strain at different prestress levels.

In the particular case $\theta_0 = 0$, Eq. (24) becomes

$$t_{21}^{inc} = \int_S \left[(\mu - p)\dot{t}_{21,1}^2(\mathbf{x}, \mathbf{y}) + \left(\mu - \frac{\sigma}{2} \right)\dot{t}_{21,2}^1(\mathbf{x}, \mathbf{y}) \right] [\![\hat{v}_1]\!] dl_x. \tag{25}$$

5 Numerical Examples

The boundary integral Eq. (24) is solved by a collocation method. The shear band is divided into Q line elements and a linear variation of the incremental displacement jump $[\![\hat{v}_1]\!]$ is assumed within each line element, with the exception of two line elements situated at the shear band tips, where a square root variation of the incremental displacement jump $[\![\hat{v}_1]\!]$ is considered to take into account the singularity at the shear band tip. In particular, the shear band is discretized with $Q = 100$ line elements to obtain the numerical results shown in this section.

A ductile low-hardening metal, modelled within the J_2-deformation theory [6, 7], with the hardening exponent $N = 0.4$, is considered. In the J_2-deformation theory, which is particularly suited to analyze the loading branch of the constitutive response of ductile metals, the prestress parameter k of Eq. (11), and the orthotropy parameter $\xi = \mu_*/\mu$, are given by the relations

$$k = \frac{\lambda^4 - 1}{\lambda^4 + 1}, \quad \xi = \frac{N(\lambda^4 - 1)}{2(\ln \lambda)(\lambda^4 + 1)} \tag{26}$$

where N is the hardening exponent, and λ is the logarithmic stretch representing a prestrain measure. In the case of $N = 0.4$, failure of ellipticity occurs at a prestress level $k = 0.8753$.

A plane incident wave field is considered, with a wavelength $\lambda_1 = 2\pi l$, where λ_1 corresponds to the wavelength of a plane transverse wave propagating parallel to x_1-axis with propagation speed c, i.e.

$$c = \sqrt{\frac{\mu(1+k)}{\rho}}, \quad \lambda_1 = 2\pi \frac{c}{\Omega} \tag{27}$$

The material dynamic response is shown in terms of level sets of incremental deviatoric strain. Level sets of the real part of incremental deviatoric strain are reported in Fig. 2 for the scattered field and in Fig. 3 for the total field. Depending on the level of prestress and anisotropy, wave patterns are shown to emerge, with focussing of signals in the direction of shear bands. Varying the direction of the dynamic perturbation excites different wave patterns, which tend to degenerate to families of plane waves parallel to the shear bands, when the elliptic boundary is approached.

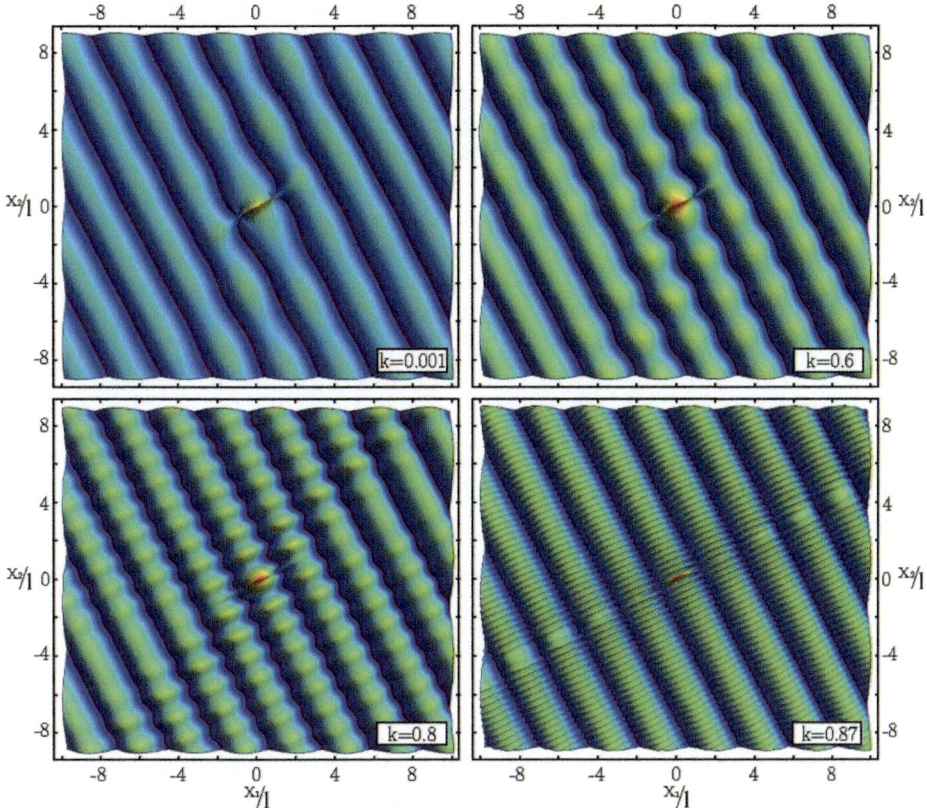

Fig. 3. Real part of total incremental deviatoric strain at different prestress levels.

Acknowledgements. Financial support from the ERC advanced grant ERC-2013-ADG-340561-INSTABILITIES is gratefully acknowledged.

References

1. Avazmohammadi, R., Ponte Castañeda, P.: Macroscopic constitutive relations for elastomers reinforced with short aligned fibers: instabilities and post-bifurcation response. J. Mech. Phys. Solids **97**, 37–67 (2016). https://doi.org/(10.1016/j.jmps.2015.07.007)
2. Bigoni, D., Dal Corso, F.: The unrestrainable growth of a shear band in a prestressed material. Proc. Roy. Soc. A **464**, 2365–2390 (2008). https://doi.org/10.1098/rspa.2008.0029
3. Biot, M.A.: Mechanics of Incremental Deformations. Wiley, New York (1965)
4. Bigoni, D., Capuani, D.: Green's function for incremental nonlinear elasticity: shear bands and boundary integral formulation. J. Mech. Phys. Solids **50**, 471–500 (2002). https://doi.org/10.1016/S0022-5096(01)00090-4
5. Bigoni, D., Capuani, D.: Time-harmonic Green's function and boundary integral formulation for incremental nonlinear elasticity: dynamics of wave patterns and shear bands. J. Mech. Phys. Solids **53**, 1163–1187 (2005). https://doi.org/10.1016/j.jmps.2004.11.007

6. Hutchinson, J.W., Neale, K.W.: Finite strain J_2-deformation theory. In: Carlson, D.E., Shield, R.T. (eds.) Proceedings of the IUTAM Symposium on Finite Elasticity, pp. 237–247. Martinus Nijhoff, The Hague (1979)
7. Argani, L., Bigoni, D., Capuani, D., Movchan, N.V.: Cones of localized shear strain in incompressible elasticity with prestress: Green's function and integral representations. Proc. Roy. Soc. A **470**, 20140423 (2014). https://doi.org/10.1098/rspa.2014.0423

Finite Element Simulation of a Lead-Core Bearing Device Mechanical Response to a Near-Fault Ground Motion

Todor Zhelyazov[1]($\boxtimes$), Rajesh Ruphakety[2], and Simon Olaffson[2]

[1] Technical University of Sofia, 8 Kliment Ohridski Blvd, 1000, Sofia, Bulgaria
elovar@yahoo.com
[2] Earthquake Engineering Research Center,
University of Iceland, Austurvegur 2a, 800, Selfoss, Iceland
{rajesh, simon}@hi.is

Keywords: Finite element analysis · Strong ground motion
Acceleration time series records · Time-History analysis

The mechanical response of a lead-core bearing device to a real, near-fault, strong ground motion is simulated by performing a finite element analysis.

An explicit 3-D finite element model of the bearing device is built (see Fig. 1). Taking into account some of the existing symmetries, a half-model space is considered.

The following material models are used: linear, elastic and isotropic model (with the assumption that the yield stress will not be reached) – for the steel elements (i); Neo-Hookean model – for the rubber layers (ii); bilinear model with isotropic hardening - for the lead-core (iii) [1–3].

The strong ground motion is modeled by defining an array that contains the discrete values of the x-, y- and z- components of the support displacements. The elements of the array are obtained on the basis of the recorded acceleration time-series [4, 5] (see Fig. 2).

An approximate solution is obtained by decimating the acceleration time-series. The time domain is divided into substeps of uniform length. The x-, y- and z- components of the acceleration are assumed constant within each substep. The x-, y- and z- components of the velocity and of the acceleration are successively defined by integration of the recorded acceleration time-series. A time-history finite element analysis is then performed.

Three strategies to investigate the mechanical response of the lead-core bearing device seen as a component of a base-isolated structure (i.e. a bridge) are taken into consideration: the bottom side of the bearing device is fixed (e.g. all nodes that belong to the bottom side are constrained) and displacement that model the ground motion are applied to nodes that belong to the top surface as described above (i); point masses are associated to keypoints that belong to one (top or bottom) surfaces of the bearing devices, whereas support displacements are applied to nodes belonging to the other side by using the array defined on the basis of the recorded acceleration time-series (ii); the bearing device is modeled along with a part of the superstructure (iii).

E. E. Gdoutos (Ed.): ICTAEM 2018, SI 5, pp. 289–290, 2019.
https://doi.org/10.1007/978-3-319-91989-8_62

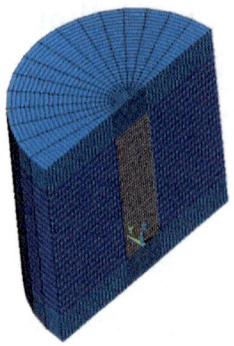

Fig. 1. Half-model space: finite element mesh; light blue - steel elements, dark blue - rubber elements and grey - lead-core.

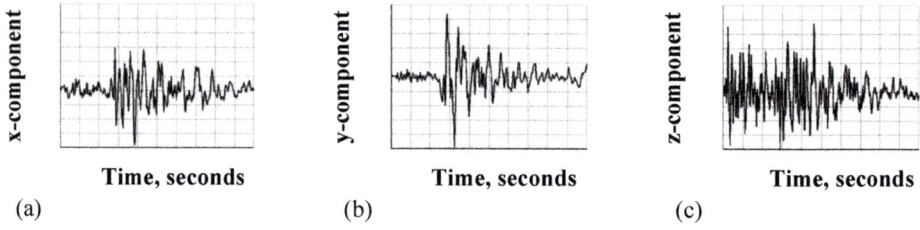

Fig. 2. Record of the near-fault strong ground motion; *x*- component of the accelerations (a); *y*-component of the acceleration (b); *z*- component of the acceleration.

This contribution is situated in the framework of research works aimed at formulating a finite element subroutine capable of reproducing the macro-response of a lead-core bearing device.

References

1. Weisman, J., Warn, G.P.: Stability of elastomeric and lead-rubber seismic isolation bearings. J. Struct. Eng. **138**(2), 215–233 (2012)
2. Ogden, R.W.: Nonlinear Elastic Deformations. Dover Publications, New York (1984)
3. Crisfield, M.A.: Non-linear Finite Element Analysis of Solids and Structures. Advanced Topics, vol. 2. Wiley, Chichester (1997)
4. Sigbjörnsson, R., Ólafsson, S., Snæbjörnsson, J.Th.: Macroseismic effects related to strong ground motion: a study of the South Iceland earthquakes in June 2000. Bull. Earthq. Eng. **5**(4), 591–608 (2007)
5. Sigbjörnsson, R., Snæbjörnsson, J.Th., Higgins, S.M., Halldórsson, B., Ólafsson, S.: A note on the Mw 6.3 earthquake in Iceland on 29 May. Bull. Earthq. Eng. **7**, 113–126 (2009)

Research on Optimization of Face Gear Grinding Machine Structure Design Based on Dynamic Performance of Machine Tool

Yanzhong Wang[1(✉)], Xiaomeng Chu[1], and Jingbo Guo[2]

[1] College of Mechanical Engineering and Automation,
Beihang University, Beijing 100191, China
yzwang63@163.com
[2] College of Mechanical Engineering, Shijiazhuang Tiedao University,
Shijiazhuang 050043, China

Abstract. A dynamic analysis method for face gear grinding machine is proposed to guide the structure optimization design of weak part. Firstly, dynamic analysis, harmonic response analysis and modal analysis of machine tool are implemented using ABAQUS. The analysis result indicates that the machine tool bed and spindle are key structures affecting the dynamic performance of the machine tool. Secondly, take the machine tool bed and spindle as optimization goal to optimum design their structures, respectively. Meanwhile, establish their optimal design model, thus the machine tool bed and spindle structure are optimized by harmonic response analysis and modal analysis. Finally, the analysis results demonstrate that the proposed method can improve dynamic performance of the machine tool effectively. On the premise of the machine tool bed and spindle mass do not increase, all the six nature frequencies of the whole machine tool are increased to different degrees after the optimum design.

Keywords: Dynamic analysis · Face gear · Harmonic response analysis
Modal analysis

1 Introduction

With the improvement of modern machining and manufacturing to machining accuracy, stability and reliability of the machine tool, the optimization of the machine tool structure has been gradually to improve the dynamic performance. The stiffness of the machine tool has a great influence on the machining performance of the whole machine. Traditional machine tool designer's stiffness design is mostly based on experience, which stays in the design stage of static stiffness. Recent years. The personnel of machine tool design and research gradually realize that dynamic optimization design is an indispensable means in the research and development of high-end CNC machine tools. In the study of dynamic optimization design of machine tools, most scholars optimize the single structural parts directly and lack of consideration for the dynamic performance of the whole machine [1–4].

Ohta et al. carry out simulation calculation and experimental verification of 16 modes of rolling guide [5]. Weck et al. propose that the optimization goal of structural

© Springer International Publishing AG, part of Springer Nature 2019
E. E. Gdoutos (Ed.): ICTAEM 2018, SI 5, pp. 291–302, 2019.
https://doi.org/10.1007/978-3-319-91989-8_63

parts is to improve stiffness and reduce the flexibility of acceleration [6]. Yigit et al. explicitly propose that the average degree of flexibility in the range of the frequency range should be regarded as one of the indicators to evaluate the dynamic performance of the machine tool [7]. Mi et al. use ANSYS to model a horizontal machining center, and analyze the influence of its whole mode and precompression on dynamic stiffness [8]. Zaghbani et al. use the working mode test method to compare the machine modes under different rotating speed and working conditions, and find that the natural frequency of the machine parts decreases when the rotational speed is increased [9]. Because the most part of the machine tool structure adopts the reinforcement plate and the reinforcing rib structure, it can improve the stiffness of the bearing and reduce the quality of the parts. Therefore, it is also an important method to optimize the structure of machine parts by using different reinforcement plates [10].

The study on the optimization of machine tool performance is to determine the weak parts which have great influence on the performance of the machine, and then optimize the research on the weak parts. Huang puts forward a method to identify the static stiffness of the whole machine through finite element analysis [11]. Yu identifies the weak link of the engine under different excitation frequency through the machine tool dynamic stiffness analysis, and uses the weak parts as optimization object, adapts the size optimization algorithm to improve the dynamic stiffness of machine tool [12]. By the structure optimization design of a machine tool column, Liu proposes a structural design method based on topology optimization, selection and layout of reinforcement plate and optimization of size [13].

This paper analyzes the dynamic characteristics of the face gear grinding machine by the finite element method, and determines the parts that affect the dynamic performance of the whole machine and then optimize the design. The aim is to improve the natural frequency, reduce the resonance amplitude, and improve the dynamic performance, as well as optimize the design of the machine tool bed and spindle. Analysis results show that the method significantly improves the optimization efficiency and greatly improves the dynamic performance of the machine.

2 Dynamics Analysis and Optimization Target

2.1 Whole Machine Model Simplified

Figure 1 illustrates the configuration of the face gear grinding machine. The three CNC axes are assigned labels [X, Y, Z] for three linear movements and [A, B, C] for three rotation motion. Moving along the X direction of the guide track can achieve the radial feed motion of the cutter; moving in the Y direction of the guide track can achieve the axial feed motion of the cutter; moving in the Z direction of the guide rail can achieve the additional motion of the cutter relative to the face gear; The A direction rotation can realize the cutter oscillation along the axis of the virtual slotting cutter. The B direction rotation can realize the split tooth movement of the face gear; The C direction is the high speed rotary motion of the cutter itself.

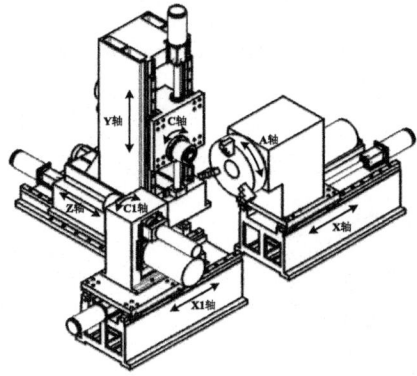

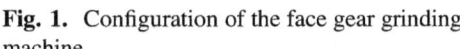

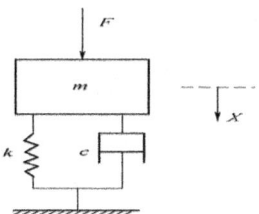

Fig. 1. Configuration of the face gear grinding machine.

Fig. 2. Whole machine vibration system diagram.

The machine is regarded as a vibration system composed of mass m, spring k and damping c, which is shown in Fig. 2. When the system is stimulated by an external drive F, it will generate a small displacement x, the oscillation equation is expressed as:

$$mx'' + cx' + kx = F \tag{1}$$

The modal parameters of the structure are independent of the external excitation, and the influence of the structure damping can be obtained by the undamped natural frequency of the system.

$$\omega_n = \sqrt{\frac{k}{m}} \tag{2}$$

When the system is subjected to an external force $F = F_0 \sin \omega_0 t$, the amplitude of the structure at the resonance point is expressed as:

$$X = \frac{F_0}{c \cdot \omega_n} \tag{3}$$

From formula (2) and (3), on the premise that the mass does not increase, increasing the static stiffness of the structure or improving the natural frequency of the structure can reduce the amplitude of the resonance point, thereby improving the dynamic performance of the machine tool.

2.2 Whole Machine Dynamic Analysis

2.2.1 Performance Analysis of Face Gear Grinding Machine

In order to ensure the face gear grinding machine has enough stiffness and machining precision, the overall structure of the machine must be analyzed.

According to the overall structure of the machine, the finite element analysis is carried out. Firstly, the 3d simplified model is established and the grid is divided, as shown Fig. 3.

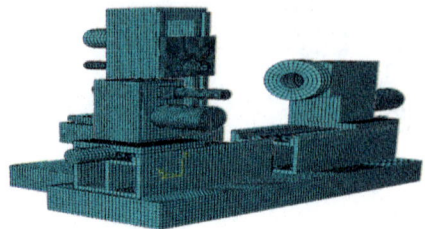

Fig. 3. Simulation configure of the face gear grinding machine.

2.2.2   Model Analysis

The purpose of modal analysis of machine tool is to get the natural frequency and mode of the machine. According to ABAQUS software analysis, the first 6 frequency and the mode of vibration mode are shown in Table 1.

Table 1. The first 6 order natural frequencies.

Order	1	2	3	4	5	6
Frequency	13.3	22.9	41.4	51.2	82.7	85.6

As show in Fig. 4, It can be seen that the whole machine of low-order modes behave the vibration mode of lathe bed and the spindle. Therefore, the lathe bed and spindle are the structural parts with the greatest influence on the low-order vibration mode of the whole machine.

2.2.3   Harmonic Response Analysis of the Whole Machine

In order to reflect the dynamic performance of machine tool in different spindle speed. We apply the simple harmonic force in three directions at the tool place, and the force amplitude is 2000 N. According to the dynamic characteristics of the machine tool obtained by the modal analysis, the frequency of the harmonic force is set in the external force frequency band of the spindle. Then analyze on the excitation vibration of whole machine, and the results as shown in Fig. 5.

As shown in Fig. 5, it can be seen that the harmonic response values are the maximum in each direction at 20 Hz, the magnitude of the harmonic response in the X direction and Y direction are smaller, and the magnitude of the harmonic response in the Z direction is lager.

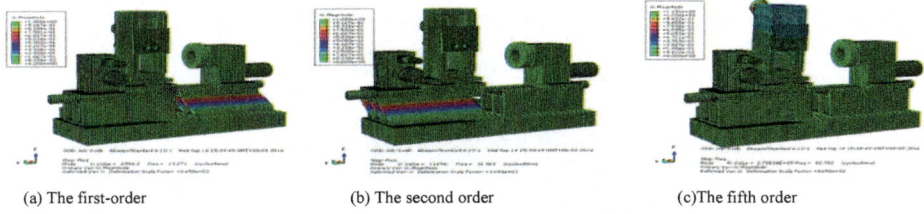

(a) The first-order	(b) The second order	(c)The fifth order

Fig. 4. The results of modal analysis of Grinding machine of face-gear.

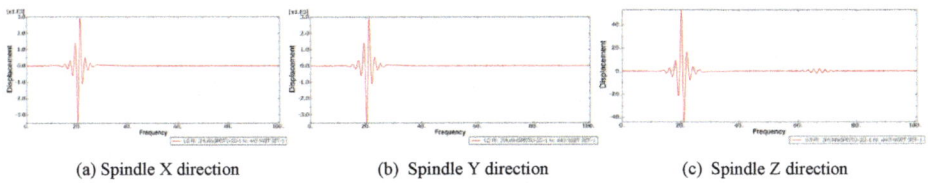

(a) Spindle X direction	(b) Spindle Y direction	(c) Spindle Z direction

Fig. 5. The result of the vertex harmonic response analysis of Machine tool.

2.3 Identify the Optimization Objective of Bed and Spindle

According to the above analysis, the bed and spindle have the greatest influence on the low-order mode of the machine. The low order natural frequency of the whole machine has the greatest influence on the dynamic performance of the machine. Since each natural frequency of the structure corresponds to a mode of vibration, we can determine that the bed body and the column structure are the key structural components that influence the dynamic performance of the machine tool. By optimizing the bed and spindle structure, it can reduce the blindness of the dynamic performance optimization and thus improve the efficiency of optimization.

3 Structural Design of Key Parts for Machine Tools

3.1 Structural Design of Bed

Machine tool lathe bed, as key components of the grinding machine, is used for supporting pillar, guide rail, etc. in order to meet the precision grinding of Face-gear, it is required that gear grinding machine for NC forming has good static and dynamic characteristics. Cast iron is the first choice for column and base material with its good wear resistance, heat resistance, shock absorption and easy casting to complex shape. The base structure of the three parts of the machine is the same, as shown in the Fig. 6.

3.1.1 The Modal Analysis of the Bed

The modal analysis of the bed is performed using ABAQUS software, and the first 6 natural frequencies and the first 6 vibration modes are obtained from modal analysis, which are shown in Table 2 (Fig. 7).

Fig. 6. Structure diagrams of Machine tool bed.

Table 2. The natural frequency and amplitude of the first 6 order of the bed.

Order	1	2	3	4	5	6
Frequency	1844.0	3063.7	4242.0	5189.7	5332.5	5478.4
Amplitude	1.004	1.034	1.000	1.000	1.203	1.161

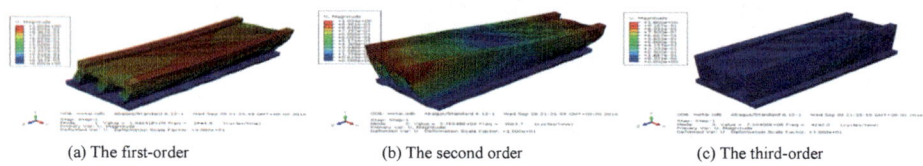

(a) The first-order   (b) The second order   (c) The third-order

Fig. 7. The first 3-order vibration mode after optimizing to bed.

3.1.2 Harmonic Response Analysis of Bed

According to harmonic response analysis, the range of first six order natural frequency of lathe bed, which is not in the range of external vibration frequency caused by grinding. Therefore, here just a brief analysis of the lathe bed frequency range of harmonic response. Choose the response on the outside of the bed surface as the object, and the harmonic response spectral lines in the X, Y, Z direction are shown in the Fig. 8.

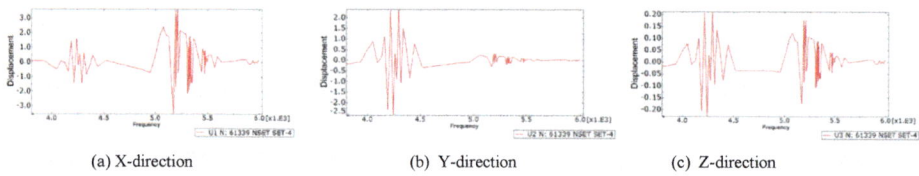

(a) X-direction   (b) Y-direction   (c) Z-direction

Fig. 8. The results of the harmonic response analysis on the upper lateral center of the bed guide rail.

As can be seen from the Fig. 8, the harmonic response values are larger in each direction at 4240 Hz and 5200 Hz, the magnitude of the harmonic response in the X direction and Y direction are the maximum, and the magnitude of the harmonic response in the Z direction is the smallest.

3.2   Work Piece System Design

The work piece system consists of a rotating shaft and a moving shaft, which mainly completes the dividing of the surface gear and the feeding motion in the deep direction of the tooth. The rotary motion is mainly driven by the servo motor belt and the synchronous belt wheel. The moving shaft is mainly driven by the servo motor and the ball screw, which is shown in the Fig. 9.

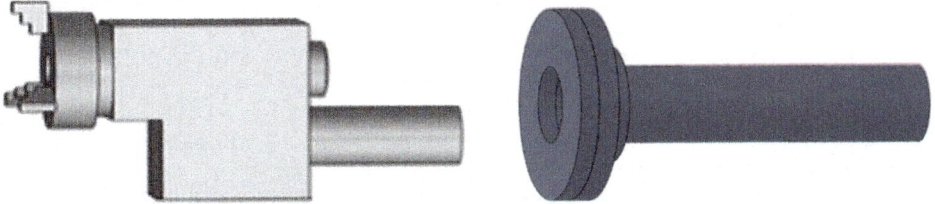

Fig. 9. Work piece spindle system diagram.   **Fig. 10.** The structure diagram of work piece spindle.

Analyze the main shaft part of the work piece. Figure 10 illustrates the structure diagram of work piece spindle.

3.2.1   The Modal Analysis of the Work Piece Spindle

The modal analysis of the work piece spindle is performed using ABAQUS software, and the first 6 natural frequencies and the first 6 vibration modes are obtained from modal analysis, which are shown in Table 3 (Fig. 15).

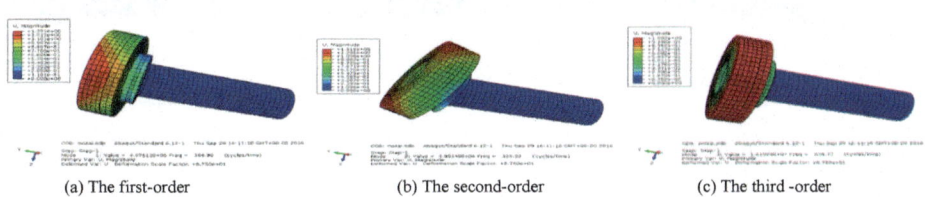

(a) The first-order   (b) The second-order   (c) The third -order

Fig. 11.   The first 3-order vibration mode after optimizing to the work piece spindle.

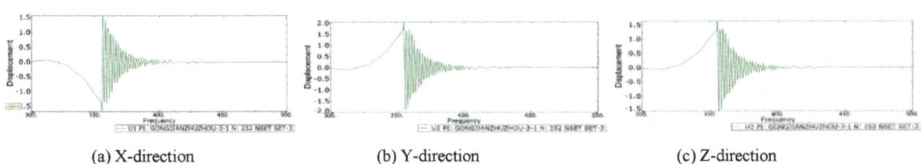

(a) X-direction   (b) Y-direction   (c) Z-direction

Fig. 12.   The results of the harmonic response analysis on the outside point of the work piece spindle.

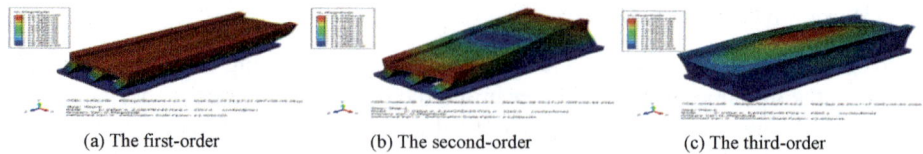

(a) The first-order   (b) The second-order   (c) The third-order

Fig. 13. The first 3-order vibration mode after optimizing to the bed body.

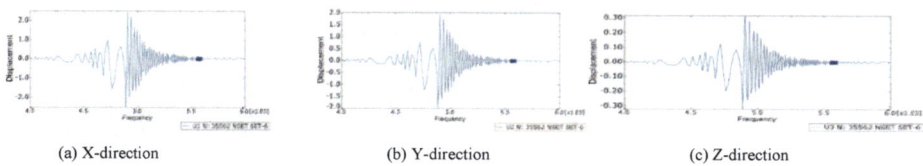

(a) X-direction   (b) Y-direction   (c) Z-direction

Fig. 14. The results of the harmonic response analysis on the upper lateral center of the bed guide rail.

Table 3. The natural frequency and amplitude of the first 6 order of the work piece spindle.

Order	1	2	3	4	5	6
Frequency	354.9	355.2	638.8	894.1	902.6	902.6
Amplitude	1.321	1.315	1.002	1.002	1.008	1.008

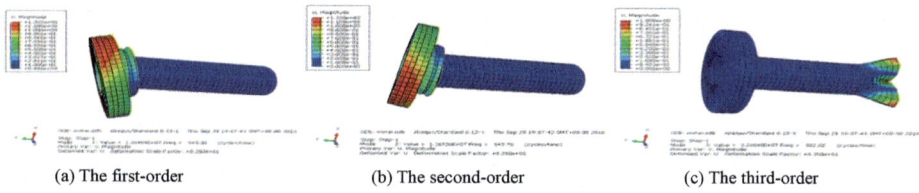

(a) The first-order   (b) The second-order   (c) The third-order

Fig. 15. The first 3-order vibration mode after optimizing to the work piece spindle.

As can be seen from the Fig. 11, the first order vibration is the bending of the X-axis, and the amplitude is larger. The second order vibration is torsion of the Z axis and the amplitude is larger. The third order vibration is the bending of the Z axis. The fourth order vibration is the twist around the Y-axis. The fifth and sixth orders vibration are twisted around the Y-axis.

3.2.2   Harmonic Response Analysis of the Work Piece Spindle

According to harmonic response analysis, the range of first 6 order natural frequency of the work piece spindle, which is not in the range of external vibration frequency caused by grinding. Therefore, here just a brief analysis of the lathe bed frequency range of harmonic response. Choose the response on the outside of the bed surface as the object, and the harmonic response spectral lines in the X, Y, Z direction are shown in the Fig. 12.

As can be seen from the Fig. 12, the harmonic response values are the maximum in each direction at 350 Hz, the magnitude of the harmonic response in the X direction and Z direction are smaller, and the magnitude of the harmonic response in the Y direction is lager.

3.3 The Optimal Structural Design of Bed and Spindle

According to the above analysis, the machine tool bed and spindle structure are optimized. Based on the analysis results, in order to further improve the rigidity and anti-torsion characteristics of the bed, it is necessary to increase the wall thickness of the bed structure and increase the number of reinforcement plates on the basis of the original structure. In order to reduce the inertia of the spindle of the machine tool, the thickness of the spindle head of the work-piece needs to be reduced.

4 The Verification of Results After Optimization

4.1 The Results Verification of the Bed Body After Optimization

4.1.1 The Results Verification of the Bed Body After Optimization
The modal analysis of the bed body is performed using ABAQUS software, and the first 6 natural frequencies and the first 6 vibration modes are obtained from modal analysis, which are shown in Table 4.

Table 4. The natural frequency and amplitude of the first 6 order of the bed body after optimization.

Order	1	2	3	4	5	6
Frequency	2353.6	3240.0	4880.1	5519.9	5603.6	7285.3
Amplitude	1.001	1.037	1.000	1.181	1.173	1.015

As can be seen from the Fig. 13, the first order vibration mode is the torsion around the X-axis. The second order vibration mode is the torsion around the Z axis. The third order vibration is the bending around the X-axis. The fourth order vibration is the bending around the X-axis and the torsion around the Z axis, and the amplitude is larger. The fifth order vibration is the bending around the X-axis and the torsion around the Y-axis. The sixth order vibration is the bending around the X-axis.

4.1.2 Harmonic Response Analysis of Bed Body
After optimization, the harmonic response spectral lines in the X, Y, Z direction are shown in the Fig. 14.

As can be seen from the Fig. 14, the harmonic response values are the maximum in each direction at 4880 Hz, the magnitude of the harmonic response in the X direction and Y direction are the maximum, and the magnitude of the harmonic response in the Z direction is the smallest.

4.2   The Optimization Results Verification of the Work Piece Spindle

4.2.1   The Modal Analysis of the Work Piece Spindle

The modal analysis of the work piece spindle has been performed using ABAQUS software, and the first 6 natural frequencies and the first 6 vibration modes are obtained from modal analysis, which are shown in Table 5.

Table 5. The natural frequency and amplitude of the first 6 order of the work piece spindle after Optimization.

Order	1	2	3	4	5	6
Frequency	543.1	543.7	902.6	902.6	920.0	989.2
Amplitude	1.207	1.200	1.008	1.008	1.002	1.000

4.2.2   Harmonic Response Analysis of the Work Piece Spindle

According to harmonic response analysis, the range of first 6 order natural frequency of the work piece spindle is 354–902 Hz, which is not in the range of external vibration frequency caused by grinding. Therefore, here just a brief analysis of the lathe bed frequency range of harmonic response, in this case, we take 0–1000 Hz, and the frequency is divided into 20. Choose the response on the outside of the bed surface as the object.

After the harmonic response analysis of the work piece spindle has been performed, the harmonic response spectral lines in the X, Y, Z direction can be obtained and they are shown in the Fig. 16.

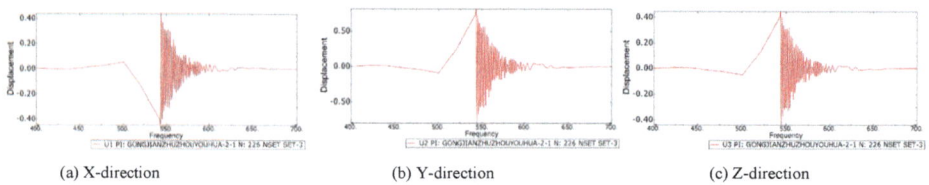

| (a) X-direction | (b) Y-direction | (c) Z-direction |

Fig. 16. The results of the harmonic response analysis on the outside point of the work piece spindle after optimization.

As can be seen from the Fig. 16, the harmonic response values are the maximum in each direction at 540 Hz, the magnitude of the harmonic response in the X direction and Z direction are smaller, and the magnitude of the harmonic response in the Y direction is lager. Therefore, after optimization, the natural frequency of the bed is increased, the resonance amplitude is reduced, and the dynamic performance is improved to meet the design requirements (Table 6).

4.3 The Whole Machine Optimization Result Verification

Table 6. The first 6 order natural frequencies.

Order	1	2	3	4	5	6
Original frequency	13.3	22.9	41.4	51.2	82.7	85.6
Optimizing frequency	15.6	30.3	50.6	70.5	100.8	107.9
Change volume (%)	17.3 ↑	32.3 ↑	22.2 ↑	37.1 ↑	21.9 ↑	26.1 ↑

5 Conclusions

According to the modal analysis of the machine tool, it is determined that the bed body and spindle have the greatest influence on the low-order mode of the whole machine. Based on the harmonic response analysis, it is found that the low order natural frequency has the greatest influence on the dynamic performance of the whole machine. Therefore, the machine tool bed and spindle are the key structural parts which have the greatest influence on the dynamic performance of CNC machine tool. The modal analysis and harmonic response analysis of lathe bed and spindle of the original machine tool have been finished, respectively. The first six order natural frequency of machine tool lathe bed and the amplitude of delaying X, Y, Z directions can be obtained, thus puts forward the optimized design scheme of machine tool bed and spindle. After optimization, the modal analysis and harmonic response analysis of lathe bed and spindle of CNC lathe bed and spindle is carried out. The amplitude of X, Y and Z directions decreases respectively. Furthermore, the optimization results of the machine tool are verified, and the first six order natural frequencies are improved to different degrees.

References

1. Peng, W.: The structure dynamic optimum design of machine tool column based on sensitivity analysis. Modular Mach. Tool Autom. Manuf. Tech. **3**, 29–31 (2006)
2. Luo, H., Chen, W.F., Ye, W.H.: Sensitivity analysis and multi-objective optimization design of a machine column. Mech. Sci. Technol. Aerosp. Eng. **28**(4), 487–491 (2009)
3. Zheng, X.M.: Optimization design for bed structure of XH7910 machining center based on sensitivity analysis. Mach. Des. Manuf. (2012)
4. Guo, L., Zhang, H., Wang, J.S., et al.: Theoretical modeling of unit structure method with verifications. J. Mech. Eng. **47**(23), 135–143 (2011)
5. Ohta, H., Hayashi, E.: Vibration of linear guideway type recirculating linear ball bearings. J. Sound Vibr. **235**(4), 847–861 (1999)
6. Weck, M., Brecher, C.: Werkzeugmaschinen: Konstruktion und Berechnung (2006)
7. Yigit, A.S., Ulsoy, A.G.: Dynamic stiffness evaluation for reconfigurable machine tools including weakly non-linear joint characteristics. Proc. Inst. Mech. Eng. Part B J. Eng. Manuf. **216**(1), 87–101 (2002)
8. Mi, L., Yin, G.F., Sun, M.N., et al.: Effects of preloads on joints on dynamic stiffness of a whole machine tool structure. J. Mech. Sci. Technol. **26**(2), 495–508 (2012)

9. Zaghbani, I., Songmene, V.: Estimation of machine-tool dynamic parameters during machining operation through operational modal analysis. Int. J. Mach. Tools Manuf **49**(12), 947–957 (2009)
10. Kong, J.: The dynamic characteristics analysis and optimization of rail weld grinding machine lathe bed. J Hunan Univ. Arts Sci. **26**(1), 46–49 (2014)
11. Huang, D., Lee, J.J.: On obtaining machine tool stiffness by CAE techniques. Int. J. Mach. Tools Manuf **41**(8), 1149–1163 (2001)
12. Yu, C.L., Zhang, H., Wang, R.C., et al.: Study on method for weak link identification of dynamic stiffness of a machine tool and optimization design. J. Mach. Eng. **49**(21), 11–17 (2013)
13. Liu, C., Tan, F., Wang, L.P., Cai, Z.Y., et al.: Research on optimization of column structure design for dynamic performance of machine tool. J. Mach. Eng. **52**(3), 161–168 (2016)

Review of Thermal, Turbulent and Misalignment Effects on Hydrodynamic Journal Bearings

D. Sivakumar[1(✉)], Suresh Nagesh[2], and K. N. Seetharamu[1]

[1] PES University, Bengaluru 560085, India
dsivakumardesign@gmail.com
[2] CORI Lab, PES University, Bengaluru 560085, India

Abstract. As there is an increasing demand for high performance turbo-machinery hydrodynamic bearings are expected to run at high speed and high load operating conditions with low power loss and oil flow. Bearings play a critical role to achieve machine's overall reliability level. This paper reviews various analytical and experimental studies that have been done on hydrodynamic bearing performance published in various journals. The analytical study brings challenges in formulating the boundary conditions consideration of thermal effects on governing equations effect of turbulence and consideration of re-circulating oil with incoming oil in inlet oil groove etc. Experimental studies pose the challenges on the measurement of performance parameters such as temperature distribution oil film distribution etc. From the survey, many researchers found that thermal and turbulent effects had the considerable impact on predicting accurate static and dynamic performance of the hydrodynamic journal bearing.

Keywords: Fluid film · Hydrodynamic · Journal bearing · Turbulence
Thermal effect · Film thickness · Dynamic

1 Introduction

As there is an increasing demand for high performance turbo-machinery, hydrodynamic bearings are expected to run at high speed and high load applications with low power loss and oil flow. Thermal, turbulent and misalignment effects play important role for performance of the journal bearings. Numerous excellent studies have been carried out to study these effects. This paper surveys those technical papers and reports. It includes analytical and experimental works carried out on these effects by many researchers. The analytical study brings challenges on formulating the boundary conditions, consideration of thermal effects on governing equations, effect of turbulence and consideration of re-circulating oil with incoming oil in inlet oil groove etc., Experimental studies pose the challenges on measurement of performance parameters such as temperature distribution, oil film distribution etc.

2 Thermal and Turbulent Effects

One of the earliest studies was made by Elrod and Ng [1] on turbulent oil film lubrication. They followed the concept of "law of wall" in which the eddy viscosity is assumed as a function of wall shear stress and distance away from the wall. On the wall surface, the flow is considered as a laminar and there is no contribution from eddy viscosity. The eddy viscosity increases when the position moves from the wall to the core of the film. The specific formula was used to quantify the eddy viscosity in this reference. Safar and Szeri [2] made study on thermo hydrodynamic lubrication on Laminar and Turbulent regimes. Effective viscosity was used to calculate the performance of finite journal bearings. Temperature and local shear dependant viscosity is calculated through iterations of momentum and energy equations. They extended this study by considering turbulent flow regime. Assuming that bearing only conducts heat in the radial direction, this study was performed.

Szeri and Sugunami [3] attempted to build hydrodynamic lubrication model suitable for both laminar and turbulent regions. Here theoretical predications were validated with experimental data also. They predicted that the constant temperature boundary condition at shaft-film interface is suitable for laminar bearing operations. However they have recommended that the insulated shaft boundary condition was more appropriate for turbulent flow regime since theoretical data was very close with experimental data.

Frene et al. [5] compared the theoretical and experimental thermo hydrodynamic performance of finite length journal bearings. They had taken into account of heat transfer between the film along with shaft and the bush, Cavitation and lubricant recirculation. Good agreement is observed between theoretical and experimental results when actual operating bearing clearances are used. Thermal deformations along with differential thermal expansion between journal and bearing must therefore be considered in both theoretical and experimental studies. In another attempt [6], they found that heat re-circulates from hottest point to the groove area in the fluid due to convection, in the bush due to conduction and in the shaft due to shaft rotation. Also it is found that most of the heat was evacuated by fluid flow. Khonsari [7] made comprehensive survey on thermal effects in hydrodynamic bearings. This paper surveyed the theoretical and experimental work done on thermal effects.

Booser and Wilcock [8] studied about temperature fade, a drop in bearing temperature beyond the point of minimum film thickness commonly observed in hydrodynamic journal bearings. Two factors are found to be significant in predicting fade, namely heat transfer to the shaft considered as a recuperator with no net heat loss, and a reduction in the assumed viscous heat generation in the cavitation zone. From the study it is found that factors such as feed pressure in the groove, sub-ambient pressure in the cavitation zone, backflow from the feed groove, and end leakage from the feed groove having little or no influence on temperature fade.

Mittwollen et al. [11] elaborated governing equations for the performance prediction of oil film journal bearings in high speed and high load turbo-machinery. The lubricant temperature and viscosity were admitted to vary in all three dimensions and a possible reverse flow was considered for Laminar-turbulent flow-transition. Also, thermal groove mixing and film rupture were taken into account with refined models. Computer code was built to compare the performance of various multi-lobe bearing designs in the region of high thermal loads. It was predicted that maximum film temperature as the failure criterion in severe operating conditions. Flow transition and hot oil carry-over also have considerable effects.

Khonsari [12] presents transient thermo-hydrodynamic behaviour of the lubricating film in journal bearings. The analysis results indicated that the lubricant film reaches a steady-state temperature in a fraction of a second. The time needed to reach steady state is a strong function of the speed, eccentricity, and the inlet conditions. Fitzgerald and Neal [13] carried out experimental study to improve the temperature distribution data and heat transfer data available for journal bearings. It was found that temperature variation on circumferential direction was strongly dependant on speed and load. More variation was observed in shorter bearing than in a long bearing.

Another experimental study was carried out Simmons and Dixon [14] on journal pad bearing to study the effect of load direction, Preload, and clearance ratio. It was found that Load between pad configurations reduces bearing temperature significantly than load on configuration. Also it was observed that laminar to turbulent transition helps in cooling of unloaded region of oil film than in the hydrodynamic film. Fillon et al. [15] made experimental study on gearboxes to understand the thermal effects of journal bearings performance. It was observed that shifting the pivot from the central position to 55% position leads to decrease in maximum temperature. Also when the load is applied at 50° from the trailing edge of a lobe instead of applying at middle of the lobe, bearing temperature and power loss decreases significantly.

Taylor et al. [16] carried out experimental study the thermal effects on circular and elliptical journal bearings. It was found that the thermal behaviour of the journal bearings was affected significantly by rotational speed and the maximum bush temperature, power loss and flow rate increased exponentially with speed. Also it was noticed that the significant temperature gradient the circumference of the bearings. However, the similar trend was not observed in axial direction.

Zhang et al. [18] studied theoretical and experimental THD analyses of high speed heavily loaded journal bearings. Numerical solutions are developed including effects of thermal deformation, mass conserving cavitation and turbulent effects. It is observed that the radial temperature gradient is bigger in the cavitation region than in the full film region. Nicholas et al. [20] studied about full adiabatic thermo hydro dynamic analysis for pressure dam bearings. The thermal effects have significant influence on the predicted bearing performance such as journal eccentricity, dynamic coefficients, and rotor stability.

Bouyer et al. [21] analysed the influence of global and local thermal effects and also mechanical and thermal deformations on bearing performance. Local thermal effects are important in the case of a highly loaded bearing as these effects are concentrated within a small zone of the bearing. The thermo Elasto hydro dynamic study, included deformations due to pressure also. From the analysis results it is found that the thermal effects have a strong influence on the minimum film thickness, the maximum pressure, and the dissipated power.

Yuan et al. [22] studied the suitability of different turbulent models for journal bearings from both their static and dynamic performances. Based on experimental results and theoretical results for different turbulent models, they attempted to comment on the turbulent model which is appropriate for analysis of the static and dynamic performance of large power unit journal bearings. The theoretical results in Ng-Pan turbulent model are very close to the test results for journal velocity up to 55 m/s. However, Philip Aoki model would be the most suitable turbulent model for journal velocity of 75 m/s.

Takabi and Khonsari [23] made comprehensive survey on the nature of thermally induced seizure (TIS) bearing failure mode. This survey indicates that seizure can occur in loaded, unloaded, lubricated, un- lubricated bearings. By increasing more oil flow and by cooling of the shaft along with decreasing the cooling rate of the housing can significantly reduce TIS in bearings.

3   Misalignment Effects

Pinkus et al. [4] made comprehensive study on misaligned journal bearings. It was found that an orientation of journal misalignment normal to the groove is more advantageous with a single-axial groove bearing. Long L/D ratios were undesirable for misaligned bearings. For a given journal slope, longer bearings will always yield smaller values of minimum film thickness and they will become inoperable (onset of contact) at misalignments at which shorter bearings will continue to operate satisfactorily.

Vijayaraghavan et al. [9] studied the effect of different degrees of cavitation for different bearing L/D ratios for various misalignment factors. Since thermal effects were not considered in this study, another study [10] was made and found that for small amount of misalignment, the performance of the bearing was not varied significantly. However if misalignment factor is greater than 4, static and dynamic characteristics of the bearings were changed drastically.

Prabhu [17] made experimental investigation study on bearing performance of misaligned cylindrical and three lobe journal bearings. It was found that the coefficient of friction is increased with an increase in horizontal misalignment. Also, the minimum film thickness decreases with an increase in horizontal misalignment. Misalignment has relatively greater effect on the cylindrical bearing than three lobe bearing.

Fillon et al. [19] made experimental investigation on performance of a plain journal bearing, when it is experienced misalignment. It is found that the largest misalignment has considerable effect on reduction in maximum pressure and the minimum film thickness. They also found that misalignment had significant effect in bearing performance when bearing operates at low Sommerfeld number.

Xu et al. [24] made comprehensive analysis on misaligned journal bearing including turbulent and thermo hydrodynamic effects. The results indicate that thermal and turbulence effects had a significant effect on the lubrication of misaligned journal bearings when they operate at large eccentricity ratio large. This paper recommended considering the effects of journal misalignment, turbulent and thermal effect into account in the design and analysis of journal bearings.

4 Summary Conclusions

This literature surveys shows that the amount of interest shown by various researchers on hydrodynamic bearings. With today's usage of high speed computers, many researchers have been working on unexplored area of hydrodynamic journal bearings. One such area is accurately predicting the performance of heavily loaded high surface velocity journal bearings.

Some of the important conclusions of this literature survey are summarized below:

1. Under sever operating conditions, thermal and turbulent effects are playing major role in predicting the static and dynamic performance of the bearings. Thermo Elasto Hydro Dynamic (TEHD) model includes pressure, temperature and elasticity effects yield good predictions of bearing performance.
2. Maximum film temperature as the failure criterion in severe operating conditions. Flow transition and hot oil carry-over also have considerable effects.
3. When the load is applied at 50° from the trailing edge of a lobe instead of applying at middle of the lobe, bearing temperature and power loss decreases significantly.
4. The thermal behaviour of the journal bearings was affected significantly by rotational speed and the maximum bush temperature, power loss and flow rate increased exponentially with speed. Also it was noticed that the significant temperature gradient the circumference of the bearings.
5. Thermal effects have a strong influence on the minimum film thickness, the maximum pressure, and the dissipated power.
6. The radial temperature gradient is bigger in the cavitation region than in the full film region.
7. Ng-Pan turbulent model are the closest to the test results for journal velocity up to 55 m/s. But for 75 m/s elliptical journal bearing, the most suitable turbulent model is Philip Aoki model.
8. If misalignment factor is greater than 4, static and dynamic characteristics of the bearings were changed drastically.
9. Misalignment has a much greater effect on the cylindrical bearing than three lobe bearing. The system damping factor increases with misalignment.
10. Turbulence has detrimental effect on the lubrication of misaligned journal bearings when the eccentricity or misalignment ratio is large.

References

1. Elrod, H., Ng, C.: A theory for turbulent fluid films and its application to bearings. J. Lubr. Technol. **89**, 346–362 (1967)
2. Safar, Z., Szeri, A.Z.: Thermohydrodynamic lubrication in laminar and turbulent regimes. J. Lubr. Technol. **96**, 48–56 (1974)
3. Suganami, T., Szeri, A.Z.: A thermohydrodynamic analysis of journal bearings. J. Lubr. Technol. **101**, 21–27 (1979)
4. Pinkus, O., Bupara, S.S.: Analysis of misaligned grooved journal bearing. J. Lubr. Technol. **101**(4), 503–509 (1979)

5. Boncompain, R., Frene, J.: A study of the thermo hydrodynamic performance of a plain journal bearing comparison between theory and experiments. J. Lubr. Technol. **105**, 423 (1983)
6. Fillon, I.V.I., Frene, J.: Analysis of thermal effects in hydrodynamic bearings. J. Tibol. **108**, 219 (1986)
7. Khonsari, M.M.: A review of thermal effects in hydrodynamic bearings. Part II: journal bearings. ASLE Trans. **30**(1), 26–33 (1987)
8. Booser, E.R., Wilcock, D.F.: Temperature fade in journal bearing exit regions. Tribol. Trans. **31**(4), 405–410 (1988)
9. Vijayaraghavan, D., Keith, T.G.: Effect of cavitation on the performance of a grooved misaligned journal bearing. Wear **134**(2), 377–397 (1989)
10. Vijayaraghavan, D., Keith, T.G.: Analysis of a finite grooved misaligned journal bearing considering cavitation and starvation effects. J. Tribol. **112**(1), 60–67 (1990)
11. Mittwollen, N., Glienicke, J.: Operating conditions of multi-lobe journal bearings under high thermal loads. J. Tribol. **112**(2), 330–338 (1990)
12. Khonsari, M.M., Wang, S.H.: Notes on transient THD effects in a lubricating film. Tribol. Trans. **35**(1), 177–183 (1992)
13. Fitzgerald, M.K., Neal, P.B.: Temperature distributions and heat transfer in journal bearings. ASME **130**, 122–130 (1992)
14. Simmons, J.E.L., Dixon, S.J.: Effect of load direction, Preload, Clearance ratio and oil flow on the performance of 200 mm journal pad bearing. Tribol. Trans. **37**(2), 227–236 (1994)
15. Bouchoule, C., Fillon, M., Nicolas, D., Barresi, F., Thermal effects in hydrodynamic journal bearings of speed increasing and reduction gearboxes. In: Proceedings of the Twenty-Fourth Turbomachinery Symposium, pp. 85–95 (1995)
16. Ma, M.T., Taylor, C.M.: An experimental investigation of thermal effects in circular and elliptical plain journal bearings. Tribol. Int. **29**(1), 19–26 (1996)
17. Prabhu, B.S.: An experimental investigation on the misalignment effects in journal bearings. Tribol. Trans. **40**(2), 235–242 (1997)
18. Zhang, C., Yi, Z., Zhang, Z.: THD analysis of high speed heavily loaded journal bearings including thermal deformation, mass conserving cavitation, and turbulent effects. J. Tribol. **122**(3), 597–602 (2000)
19. Bouyer, J., Fillon, M.: An experimental analysis of misalignment effects on hydrodynamic plain journal bearing performances. J. Tribol. **124**(2), 313–319 (2002)
20. He, M., Allaire, P., Cloud, C.H., Nicholas, J.: A pressure dam bearing analysis with adiabatic thermal effects. Tribol. Trans. **47**(1), 70–76 (2004)
21. Bouyer, J., Fillon, M.: On the significance of thermal and deformation effects on a plain journal bearing subjected to severe operating conditions. J. Tribol. **126**(4), 819–822 (2004)
22. Ji, F., Guo, Y., Yuan, X., Yang, L., Zhao, W.: Turbulent model analysis and experimental research for lubrication performance of large power units journal bearing. In: ICIEA 2009 4th IEEE Conference, pp. 206–210 (2009)
23. Takabi, J., Khonsari, M.M.: On the thermally-induced seizure in bearings: a review. Tribol. Int. **91**, 118–130 (2015)
24. Xu, G., Zhou, J., Geng, H., Lu, M., Yang, L., Yu, L.: Research on the static and dynamic characteristics of misaligned journal bearing considering the turbulent and thermohydrodynamic effects. J. Tribol. **137**(2), 024504 (2015)

Autowave Plasticity. Localization and Collective Modes

Lev B. Zuev[⊠]

Institute of Strength Physics and Materials Science SB RAS,
Tomsk 634055, Russia
lbz@ispms.tsc.ru

Abstract. The autowave theory of plastic flow of materials is suggested. It is shown that this is based on the elastic-plastic invariant, which connects the elastic and plastic parameters of deforming solids. The consequences from the invariant describe correctly the principal features of plastic flow.

Keywords: Deformation · Plasticity · Elasticity · Plastic flow
Dislocation · Autowave

1 Introduction

The processes responsible for plastic deformation and failure in crystalline solids have been systematically investigated during several decades. Today it is our firm belief that the inhomogeneous plastic flow behavior is one of the most fascinating and intriguing aspects of the plastic deformation. The localization of plastic deformation would occur at different scale levels and would show up clearly from the yield limit, σ_y, to the ultimate strength, σ_B. The intermittent flow behavior has a huge variety of manifestations: slip lines, cell and streaky substructures, Lüders fronts, dislocation avalanches, space-time oscillations of plastic deformation, etc. These phenomena are widely covered in the literature. In order to address the plastic flow instability and necking, the occurrence in the deforming material of localized strain zones was considered in connection with its work hardening. Mechanical testing of materials was performed using various methods; the dislocation substructure of deformed materials was investigated. Thus, abundant experimental evidence was obtained, which enabled setting the above plastic flow non-homogeneities against the dynamic strain aging of material and the Portevin-Le Châtelier effect.

The most significant finding of our studies is that on a macro-scale level, the plastic deformation would develop non-homogeneously, with the deforming medium becoming stratified spontaneously into alternating non-deforming and defor-ming zones also called active nuclei or stratums. The active deformation nuclei are ordered in space and with time and spaced at intervals $\lambda \approx 10^{-2}$m. In the past few years the present conception was employed by a lot of workers for addressing macro-localization processes in the deforming medium. To mention just a few: the theoretical works by Aifantis and Bell as well as the experimental observations described.

© Springer International Publishing AG, part of Springer Nature 2019
E. E. Gdoutos (Ed.): ICTAEM 2018, SI 5, pp. 309–312, 2019.
https://doi.org/10.1007/978-3-319-91989-8_65

The present study is aimed to extend our knowledge of plastic flow localization by widening the range of materials tested, which differ in crystal lattice type, structure, grain size and chemical composition as well as deformation mechanisms (dislocation glide, twinning and phase transformation-induced plasticity).

Thus, the examination of the above data points to the overall conclusion that the development of plastic flow occurs as a regular space-time process. However, we still lack in many instances the most elementary information, which is crucial on several accounts. There is the great difficulty for explaining of plasticity localization features above. It is the significant difference between scale of inhomogeneity of plastic flow ($\lambda \approx 10^{-2}$m) and dislocation scale (the Burgers vector $b \approx 10^{-10}$m). Apparently, it would be impossible to explain these phenomena remaining in the traditional frame-works of dislocation theory. On this reason we propose new approach to problem of plastic flow localization in solids.

2 The Principal Relations

Most of traditional dislocation models for solids plasticity are based on the Taylor-Orowan equation for dislocation plastic deformation rate

$$\partial \varepsilon / \partial t = b \rho_m V_{disl}, \tag{1}$$

where ρ_m is the density of dislocations moving at rate V_{disl} under the action of the applied stress, σ.

However, the equation for plastic flow must take into account all the components of plastic deformation. It can be achieved using the autowave approximation for plastic flow description

$$\partial \varepsilon / \partial t = \alpha b \rho_m V_{disl} + D_\varepsilon \partial^2 \varepsilon / \partial x^2. \tag{2}$$

It is understandable that Eqs. (1) and (2) are interrelated. Then, provided the dislocations are homogeneously distributed in the crystal volume, transformation of the first term in the right side of Eq. (2) would yield $\partial \varepsilon / \partial x = d^{-1} \cdot b/d \approx b \rho_m$ (here d is the distance between dislocations; d; the ratio b/d has the meaning of shear strain for a dislocation path d and the value $d^{-2} \approx \rho_m$ is the density of mobile dislocations). For the case of diffusion-type flow, $D_\varepsilon \approx L_{disl} \cdot V_{disl}$ (here $L_{disl} \approx \alpha x$ is a dislocation path and $V_{disl} = const$ is dislocation velocity).

Apparently, first term in right-hand side of Eq. (2) coincides with Eq. (1). Hence, the Taylor-Orowan equation might be regarded as a special case of Eq. (2). It is assumed that Eq. (1) finds limited use; it is generally employed for description of elementary relaxation event, which occurs as a jump-wise deformation process. From this view-point, Eq. (2) can be regarded as an extended version of dislocation dynamics equation, which accounts for both the hydrodynamic and the diffusion type of deformation flow, $\phi(\sigma, \varepsilon) = b \rho_m V_{disl}$ and $D_\varepsilon \partial^2 \varepsilon / \partial x^2$, respectively. In the framework of traditional dislocation approaches one can address a homogeneous distribution of dislocations forming no substructure elements (ensembles). Against this general background, new approaches to problems of the work hardening theory are being developed, some of these involving long-range stress fields to account for the deformation.

A majority of modern physical approaches to plasticity are based still on the dislocation theory. Therefore, we need to find a relation which might be useful for addressing autowaves in question in the frame of dislocation models. Consider the physical interpretation. The member in its right side takes account of two types of plastic flow occurring simultaneously in the deforming solid. One type of plastic flow is termed as a 'hydrodynamic' one; this occurs in material volume due to the activation of local shears on the deformation fronts for $\dot{\varepsilon} \sim V_{disl}$. The member $D_\varepsilon \partial^2 \varepsilon / \partial x^2$ in Eq. (2) denotes the other type of plastic flow that is diffusion-like in character. In the case of hydrodynamic flow type, the deformation is apparently focused on the Lüders front. The contribution of the diffusion-like type of plastic flow decreases slowly as soon as the Lüders front travels a certain distance, $x \sim \sqrt{t}$. In the latter case, the dislocation sources immediately ahead of the Lüders front would be activated. The above suggests that new fronts would emerge in a random fashion at a distance $\sim \lambda$ from the front of hydrodynamic deformation flow already in existence.

3   Elastic-Plastics Invariant

We suggest a link between plastic flow macro-parameters and crystal lattice characteristics. For this purpose, two products are matched, i.e. $\lambda \cdot V_{aw}$ and $\chi \cdot V_t$, which characterize plastic flow and elastic deformation, respectively. The quantities χ and V_t are interplanar spacing of crystal lattice and transverse ultrasound wave velocity, respectively. Numerical analysis was performed using experimentally obtained values λ and V_{aw} and hand-book values χ and V_t. The experimental data allow one to write the equality

$$\lambda \cdot V_{aw} / \chi \cdot V_t \approx const = \hat{Z}, \qquad (3)$$

which holds true for all studied materials. The value $\hat{Z}$ was averaged over nineteen materials to give $\langle \hat{Z} \rangle = 0.48 \pm 0.25 \approx 1/2$. This result constitutes both a formal and a physical proof that elastic and plastic processes are closely related in the course of deformation. Therefore, Eq. (3) has been labeled as *'elastic-plastic strain invariant'*. As it is shown in this paper, the consequences from Eq. (3) describe the principal features of plastic flow.

4   Conclusions

(i)   At all its stages the plastic flow would exhibit a localization behavior, with specific patterns emerging in the deforming solid. Thus localization is taken to be an integral attribute of the plastic flow.

(ii) The emergent patterns correspond to different types of localized plastic flow autowaves. The autowave propagation rate is found to be inversely proportional to the work hardening coefficient.

(iii) Dispersion relation of quadratic form is derived for autowaves in question, which describes nonlinear processes in the defect system of the deforming medium. The autowave equation is found to be related to the dislocation dynamics equation (the Taylor-Orowan equation).

(iv) It is shown that the latter equation is a special case of the autowave equation; its use restricts the applicability of traditional dislocation approaches for adequate description of plastic flow features.

Acknowledgement. *This study was supported by the Russian Scientific Fund (Grant No. 16-19-10025).*

Microstructure-Based Modelling and Digital Image Correlation Measurement of (Residual) Strain Fields in Austenitic Stainless Steel 316L During Tension Loading

V. Herrera-Solaz[1(✉)], L. Patriarca[2], J. Segurado[3],
and M. Niffenegger[1]

[1] Nuclear Energy and Safety Research Department, Paul Scherrer Institute,
5232 Villigen, Switzerland
vicente.herrerasolaz@psi.ch
[2] Department of Mechanical Engineering, Politecnico di Milano,
20156 Milan, Italy
[3] IMDEA Materials Institute, C/Eric Kandel 2, Getafe, Madrid, Spain

Abstract. The present work aims to compare the residual strain fields of the austenitic stainless steel 316L samples under specific loading conditions, obtained by different techniques. Experimentally, Digital Image Correlation (DIC) technique is used to track the (residual) strain maps during the specimen loading. Microstructure-based Finite Element (FE) models based on two different approaches for the material behavior (Crystal Plasticity (CP) and SNF) provide the numerical results. The quality of the outcomes reinforces the use of a combination of these techniques to study the residual strain locations in the microstructure, as well as its evolution during the deformation process. Whereas the results based on CP and SNF showed good agreement, some minor differences between calculated and measured (DIC) strains could be observed.

Keywords: DIC · Crystal plasticity finite element · EBSD · Microstructure

1 Material and Experimental Procedure

Flat tensile specimens made from hot-rolled austenitic stainless steel AISI 316L with different average grain sizes (165, 341, 687 and 1590 μm) having a thickness of 4 mm and a dog-bone shape were tested in load-controlled mode. DIC technique was used to measure the real-time deformations (In-situ DIC) and, upon specimen unloading, the residual strain maps (ex-situ DIC) at higher resolutions [1]. Ex-situ DIC gives 10 times higher image resolution than the in-situ since it employs an optical microscope instead of a CCD camera with optical lens.

© Springer International Publishing AG, part of Springer Nature 2019
E. E. Gdoutos (Ed.): ICTAEM 2018, SI 5, pp. 313–314, 2019.
https://doi.org/10.1007/978-3-319-91989-8_66

2   Numerical Modelling

Pseudo 3D Microstructure-based FE models have been computed by Abaqus following the procedure prescribed in [2]. The grain geometry and orientations were measured by EBSD and they were successively used to define the model. The material behavior was modelled considering two different microstructure-based approaches. Firstly, a CP model was implemented by means of the CAPSUL code [3] to obtain more precise strain distributions at the grain size scale. On the other hand, an in-house SNF [4] code, that simplifies the material behavior, was employed to optimize the computational time providing an acceptable accuracy of the results.

3   Results and Discussion

Experimental results coming from DIC measurements were compared with those obtained from the numerical simulations with the CAPSUL and SNF codes. In general, the agreement with regard to the locations of strain concentrations and their mean values was quite good in the majority of the cases, as can be seen in Fig. 1. The fact that the results from all the analyses show remarkable similarities makes them credible and useful for the study of (residual) local stresses at the microstructure level, and consequently they could help to a better understanding of the material damage process.

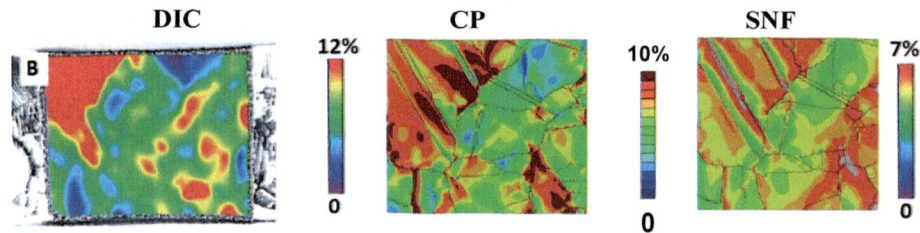

Fig. 1. Residual strain fields obtained with DIC, CP and SNF methodologies in a sample with an average grain size of 1590 μm after yielding and unloading.

References

1. Patriarca, L., Filippini, M., Beretta, S.: Digital image correlation-based analysis of strain accumulation on a duplex γ-TiAl. Intermetallics **75**, 42–50 (2016)
2. Sistaninia, M., Niffenegger, M.: Prediction of damage-growth based fatigue life of polycrystalline materials using a microstructural modeling approach. Int. J. Fatigue **66**, 118–126 (2014)
3. Cruzado, A., LLorca, J., Segurado, J.: Modeling cyclic deformation of inconel 718 superalloy by means of crystal plasticity and computational homogenization. Int. J. Solids Structu. **122-123**, 148–161 (2017)
4. Herrera-Solaz, V., Niffenegger, M.: Application of hysteresis energy criterion in a microstructure-based model for fatigue crack initiation and evolution in austenitic stainless steel. Int. J. Fatigue **100**, 84–93 (2017)

A Study on the Effect of Component Geometry on Fatigue Property

Bo Qin and Liyang Xie[(⊠)]

Key Laboratory of Vibration and Control of Aero-Propulsion System of Ministry
of Education, Northeastern University, Shenyang 110819, China
lyxie@me.neu.edu.cn

Stress concentration, size and surface state were believed to be the main factors for the fatigue endurance strength and fatigue life of a component contrasting to a smooth standard test specimen with the same nominal stress. Especially, there are complicated interactions among these factors.

To predict component endurance strength and fatigue life according to the respective properties of a standard specimen, it is very important to know the roles of the factors such as stress concentration caused by component geometry, the size of the critical zone or critical section of the component, as well as the surface state determined by component manufacturing process and finishing.

Assuming that these factors play their roles independently, traditional method to reflect the effects of these factors is as simple as using a complex modification factor as

$$K_a = \frac{K_f}{\varepsilon \beta}$$

Where, K_f stands for fatigue stress concentration factor, ε stands for size factor, and β stands for surface factor, or a little bit more comprehensive one as

$$K_a = \left(\frac{K_f}{\varepsilon} + \frac{1}{\beta} - 1 \right)$$

Both of these two models can not reflect the coupling effect of the three factors, and either is applied for both of component fatigue endurance strength modification and component fatigue life modification.

Obviously, there will be strong interactions among the three factors. For instance, components with different geometries, thus different stress gradients at their respective critical zones, will show different degrees of size effects even though these components have the same size at critical sections. Similarly, stress concentration will also mitigate the effect of surface state since not all the component surface will contribute to its fatigue property. On the other hand, it is not seems reasonable to modify component S-N (cyclic stress – fatigue life) curve by means of such a complex factor.

The present paper studies the mechanism by which the three factors interact of each other, and develop parameters, methods and models to express the effects of the factors on component fatigue endurance strength and fatigue life.

© Springer International Publishing AG, part of Springer Nature 2019
E. E. Gdoutos (Ed.): ICTAEM 2018, SI 5, pp. 315–316, 2019.
https://doi.org/10.1007/978-3-319-91989-8_67

First of all, a stress state factor is applied to take the place of traditional stress concentration factor, and fatigue stress is defined to characterize the stress level in a small element at the critical zone. The reasons are as the following:

(1) Local high stress at component critical point can be easily computed by finite element method nowadays, it is not necessary to use stress concentration factor to calculated local stress according to nominal stress. Moreover, there is no nominal stress for many components of complex geometry by which local high stress can be calculated in such a way.

(2) For most of situations, the same factor causing stress concentration (or stress gradient) also causes multi-axial stress state, that has also effect on fatigue endurance strength and fatigue life.

(3) The stress on a small size element of the critical zone, instead of the highest stress at the critical point, is responsible to component fatigue behavior. Therefore, the stress gradient at the critical point is a fundamental parameter to reflect stress concentration effect.

On the size factor, the present paper uses push-pull cyclic stress based size factor, instead of the traditional rotating-bending stress based size factor. The latter mixed size effect and stress gradient effect. It is the material defects in high stress zone that affects component property, therefore only size with higher stress making sense. Surface state effect will also be considered with reference to component geometry and stress gradient, since only the surface with higher stress is the potential fatigue area.

Fatigue stress is a stress describing the average stress level on a small three-dimensional element

$$\sigma_{fat} = \iint_{x,y,z \in \Delta V} f(x,y,z)dxdydz$$

Where, x, y and z are the coordinates originate from the highest stress point, ΔV is the element volume.

Component fatigue endurance strength modification model is

$$S_{fat} = S\frac{\sqrt{\beta}\sqrt[3]{\varepsilon}}{K_s}$$

Where, K_s stands for multi-axial stress state factor, ε stands for size factor, and β stands for surface factor.

On component S-N curve to be modified from smooth standard test specimen, this paper suggests to modify the material fatigue life constants m and C, respectively, instead of using an equivalent stress like traditional method. The difference between this paper's modification and the traditional modification to the S-N curve of a standard test specimen is obvious. It is easy to understand that the new modification is better than the traditional modification since the latter obtains a component S-N curve parallel to the material (smooth standard test specimen) S-N curve.

Anatomic and Biomechanical Study to Guide in the Choice of Atlantoaxial Fixation Pattern

Chuang Liu$^{(\boxtimes)}$ ⓘ and Yunhui Yan ⓘ

School of Mechanical Engineering and Automation, Northeastern University,
Shenyang 110819, Liaoning, People's Republic of China
l_c68@163.com

Abstract. Placing posterior atlas screws is technically demanding, and a misplaced screw can result in serious medical consequences. However, the current surgical approach and choice of internal fixation technique are usually decided according to the doctor's personal experience and preferences. The purpose of this paper was to set up an integrated solution for screw placement of posterior atlas fixation, using 3D interactive visualization anatomic measurement and biomechanical simulation analysis. The atlas posterior arch screw demonstrated the lowest risk factor for damaging nearby tissues. The three mainstream posterior internal fixations can provide similar stability for the odontoid fractures. Atlas pedicle screw generated the lowest stresses on screws showing it stabilizing capability. The three fixations have the ability to offer enough stability for atlantoaxial instability caused by odontoid fractures, and each with its own set of merits and demerits. The results of this study can serve as process for customized surgical planning.

Keywords: Anatomic measurement · Biomechanical simulation
Atlas fixation

1 Introduction

Atlantoaxial vertebra is the junction of the cranial and cervical spine where the rotation of the head and neck is mainly concentrated in the atlantoaxial joint. The unique mode of connection and complex biomechanics of the atlantoaxial vertebrae make the treatment of atlantoaxial instability a challenge in the field of spinal surgery (Fig. 1).

There have been numerous posterior internal fixation techniques after years of development, the biomechanical stability of the pedicle screw fixation technique is the most effective fixation method at present. However, the anatomy of the atlanto vertebral and its surrounding structures is very personalized. That has great influence on the selection of internal fixations and operative procedures, resulting in the failure of scheduled operation plan, and serious complications. Placing posterior atlas screws is technically demanding, and a misplaced screw can result in injury to the vertebral artery, spinal cord, or internal carotid artery. However, the current preoperative planning of spinal surgery is mainly based on preoperative clinical imaging data to measure bone morphology parameters, and decide the type of surgical procedure based on the doctor's experience and personal preferences. This treatment lacks the objective

© Springer International Publishing AG, part of Springer Nature 2019
E. E. Gdoutos (Ed.): ICTAEM 2018, SI 5, pp. 317–318, 2019.
https://doi.org/10.1007/978-3-319-91989-8_68

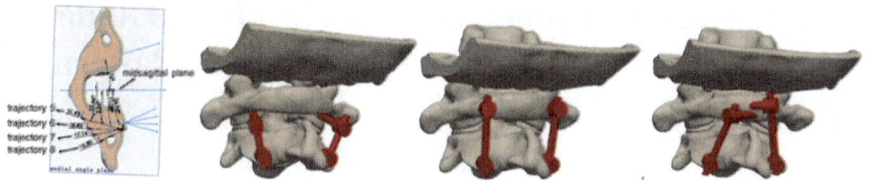

Fig. 1. Screw trajectory-optimizing and the three mainstream methods of Atlas vertebral internal fixation.

quantitative risk assessment of the placement of screw and the bone mass analysis, and the biomechanical simulation evaluation before the operation.

Atlas vertebral internal fixation is a very complicated operation, and the difficulty and key problem of this operation is screw placement for Atlas vertebral. The special anatomical structure and special pathological structure of different patients have increased the risk and uncertainty of the operation. According to those differences among patients, personalized analysis will be able to avoid the risk and find the most suitable fixing method, finally will greatly improve the safety of atlantoaxial fixation and stability and open a new epoch of accurate treatment. Therefore, this paper studied on 3D interactive visualization anatomic measurement and biomechanical simulation analysis of dominant methods for Posterior Atlas Fixation, to set up an integrated solution for screw placement of posterior atlas fixation, and discuss the optimization selection of internal fixation methods.

Acknowledgements. The authors would like to thank the National Natural Science Foundation of China (51374063) and the Fundamental Research Funds for the Central Universities (N141008001, N150308001).

References

1. Hu, Y., Dong, W.X.: An anatomic study to determine the optimal entry point, medial angles, and effective length for safe fixation using posterior C1 lateral mass screws. Spine **40**, 191–198 (2015)
2. Melcher, R.P., Puttlitz, C.M.: Biomechanical testing of posterior atlantoaxial fixation techniques. Spine **27**, 2435–2440 (2002)
3. Tan, M., Wang, H.: Morphometric evaluation of screw fixation in atlas via posterior arch and lateral mass. Spine **28**, 888–895 (2003)
4. Badhiwala, J.H., Nassiri, F.: Does transection of the C2 nerve roots during C1 lateral mass screw placement for atlantoaxial fixation result in a superior outcome? a systematic review of the literature and meta-analysis. Spine **42**, 1067–1076 (2017)
5. Brolin, K., Halldin, P.: Development of a finite element model of the upper cervical spine and a parameter study of ligament characteristics. Spine **29**, 376–385 (2004)

Challenges and Progress in Residual Stress Evaluation and Analysis at the Nanoscale

Alexander M. Korsunsky[(⊠)] [iD]

MBLEM, Department of Engineering Science,
University of Oxford, Oxford OX1 3PJ, UK
alexander.korsunsky@eng.ox.ac.uk

Abstract. Residual Stresses are an aspect of material state that is complex in many different respects. As any stress quantity, it has multi-component tensor nature, so that at each point up to 6 components need to be determined. Moreover, in contrast with 'live' stresses, residual stresses (RS) cannot always be determined by tracking their evolution from an initial state. Furthermore, RS are scale-dependent, so that if the resolution (gauge volume) of the measurement is changed, so does the perceived stress state. In the context of RS 'measurement', conventional classification into Type I, II and III (macro-, micro- and nano-scale stresses) is often used, but the intricate relationship between these quantities remains insufficiently well understood. Finally, the interaction of RS with material processing and service behaviour introduces further associated complexities, and deserves additional discussion.

Keywords: Residual stress · FIB-DIC · Micro-ring-core milling
Synchrotron X-ray micro-diffraction

1 Introduction

Residual stresses (RS) determine the performance and durability of many different natural and engineered systems, ranging from plants, seashells, bone and dental tissue, to electronic parts and electrochemical systems (batteries, fuel cells) to structural components and assemblies used in transport, power generation and civil construction. Residual stresses affect the stability and evolution of systems from the initial intended structural and mechanical state that may result in the degradation of internal bonds between parts of the system that may result in fracture or voiding; excessive deformation or distortion, etc.

Over many decades of mechanical and materials engineering practice and research, recognition of the importance of RS and control over them has increased steadily. One of the reasons for this is that with the miniaturization of components and devices and the attendant improvement in our capability to image the details at greater resolution it became gradually clearer that degradation processes that limit the lifetime and performance of systems begin at the finest atomic and molecular scales, progress through the nano- and micro-scale, and culminate in failure modes that manifest themselves at millimeter to meter to kilometer ranges.

© Springer International Publishing AG, part of Springer Nature 2019
E. E. Gdoutos (Ed.): ICTAEM 2018, SI 5, pp. 319–321, 2019.
https://doi.org/10.1007/978-3-319-91989-8_69

Accordingly, RS evaluation and modelling advanced apace, with new principles and experimental configurations being proposed to allow analysis of different materials types (metals, ceramics, polymers, composites, coatings, poly- and single crystals, glasses, etc.) under varying conditions. In a recently published monograph [1] the author presented a systematic approach to RS analysis and simulation by pulling together theoretical and experimental approaches and drawing on examples to illustrate their validity.

2 Aspects and Advances in the State-of-the-Art

In the current state-of-the-art in experimental residual stress analysis at the nanoscale, two principal techniques dominate: the use of nano-focused beams of electrons or X-rays that define a nano-scale interaction gauge volume from which information is extracted using scattering or spectroscopy. A highlight in the recent advances in the use of electron nanobeam diffraction is presented in [2], where the authors report the mapping of both 'live' and residual stresses around dislocations in stainless steel samples. The method relies on the use of moderately convergent beam to obtain diffraction patterns that can be interpreted in terms of strain.

When crystalline materials are studies, sharp diffraction patterns are obtained with distinct peaks whose centre position can be directly interpreted in terms of interplanar lattice spacing. Either single crystal (Laue) or powder diffraction (Debye-Scherrer) modes can then be used, depending on the grain structure of the object. In contrast, no such sharp peaks arise when glassy or amorphous materials are investigated. The smooth intensity variation can be interpreted using inverse Fourier transform to obtain the X-ray Pair Distribution Function (XPDF) in which distinct peaks correspond to atomic neighbours and coordination shells. In [3] this approach has been used to detect the variation in strain as a function of length scale within amorphous silica. It was discovered that for atomic neighbours lying within less than ~ 2 nm from a given atom, strain value and the mechanism of its accommodation differs from the long range (macroscopic) regime: instead of changing the bond length in response to applied stress, it is the inter-bond angles that undergo change. This highlights the length scale for short range disorder in fused silica.

An alternative approach to beam interaction techniques described above is the use of mechanical relief caused by material removal to evaluate stress. In this case the experimental approach is to detect the strain change caused by the load redistribution. Nevertheless, particle beams remain the sharpest tool available, so in the recently reported study [4], combined use was made of focused ion and electron beams to perform micro-ring-core milling and Digital Image Correlation (DIC) analysis. Data was collected for a series of experiments performed using different core diameters. Crucial to this advance made in residual stress analysis was the use of *eigenstrain* as a means of modelling the deformation state of the sample. By adopting eigenstrain approach to the analysis it became possible to deconvolve the residual stress variation with position below the sample surface (depth profiling) with the resolution down to

50 nm. The results were validated by comparison with nanofocus synchrotron X-ray diffraction. The fundamental advantage of the new approach lies in the ability to perform residual stress evaluation directly on bulk sample surface, without the need for lamella preparation.

3 Conclusion

The illustrations of recent advances in experimental residual stress analysis establish the current frontier in this field, and also indicate the directions for further development and exploration expected to occur in the coming decade(s).

References

1. Korsunsky, A.M.: A Teaching Essay on Residual Stresses and Eigenstrains, 1st edn. Butterworth-Heinemann (2017). ISBN 9780128109908
2. Pekin, T.C., Gammer, T., Ciston, J., Ophus, C., Minor, A.M.: In situ nanobeam electron diffraction strain mapping of planar slip in stainless steel. Scr. Mater. **146**, 87 (2018)
3. Lunt, A.J.G., Chater, P., Korsunsky, A.M.: On the origins of strain inhomogeneity in amorphous materials. Sci. Rep. **8**, 1574 (2018)
4. Korsunsky, A.M., Salvati, E., Lunt, A.J.G., Sui, T., Mughal, M.Z., Daniel, R., Keckes, J., Bemporad, E., Sebastiani, M.: Nanoscale residual stress depth profiling by Focused Ion Beam milling and eigenstrain analysis. Mater. Des. **145**, 55 (2018)

Symposium on: "Mechanics of Amor-phous and Nanocrystalline Metals," by Jamie J. Kruzic

Shear Bands in Metallic Glasses: Atomic Transport, Propagation – and Relaxation Behavior

Gerhard Wilde[(⊠)]

Institute of Materials Physics, University of Münster, Münster, Germany
matphysik@uni-muenster.de

Abstract. Plastic deformation of metallic glasses, if exerted at low homologous temperatures with respect to the glass transition temperature, is mostly localized in plate-like mesoscopic defects, so-called shear bands. Although the occurrence of shear bands is well known and often determines the mechanical performance of the material, their actual physical properties remain fairly unknown. Additionally, it is widely accepted that localized regions, so-called shear transformation zones, undergo plastic yielding through a shear transformation at low strains. Yet, how those localized regions are distributed, in what way they are structurally distinct from the surrounding matrix and how they couple to evolve into shear bands with macroscopic lengths at higher strains is also unclear. This contribution addresses those coupled questions and issues by combining complementing experimental methods and – for selected situations – results from atomistic simulations [1–11]. Here, experimental data on the rate of atomic diffusion within shear bands have been obtained using the radiotracer method on post-deformed specimens. Those measurements that had been performed at temperatures below T_g on samples after they underwent plastic deformation, showed a drastic increase of the atomic mobility inside the shear bands. In fact, as indicated in Fig. 1, the diffusion coefficient inside the shear bands was found to be more than six orders of magnitude larger than the diffusion coefficient of the undeformed matrix of the same material.

Keywords: Metallic glasses · Shear bands · Atomic transfer
Glass transition temperature

Additionally, novel TEM-based methods served to experimentally determine the local specific volume as well as the local degree of medium-range order and the local chemical composition quantitatively. For this reason, HAADF-STEM (High Angle Annular Dark Field-Scanning Transmission Electron Microscopy) and EELS (Electron Energy Loss Spectroscopy) measurements have been combined so that from both simultaneously collected signals, the local mass density can be obtained. Additionally, this method has also been utilized to obtain arrays of nanobeam diffraction pattern over defined nanoscale scan areas inside and outside shear bands. From those series of nanobeam diffraction pattern, so-called "fluctuation electron microscopy/FEM" can yield information on the local medium range order at precidely defined locations inside the sample. Figure 2 shows the obtained density differences along a shear band axis and also the FEM result based on the nanobeam diffraction series.

© Springer International Publishing AG, part of Springer Nature 2019
E. E. Gdoutos (Ed.): ICTAEM 2018, SI 5, pp. 325–328, 2019.
https://doi.org/10.1007/978-3-319-91989-8_70

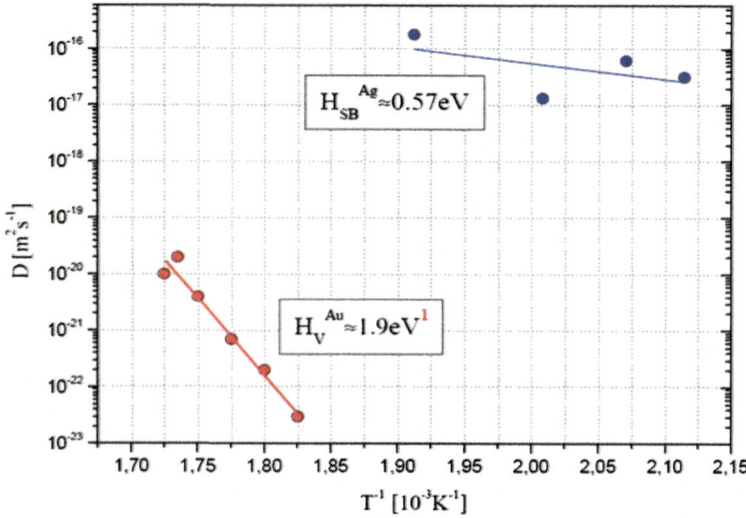

Fig. 1. Diffusion coefficients of radioactive Ag and Au isotopes in $Pd_{40}Ni_{40}P_{20}$ bulk metallic glass. The diffusion coefficients in the shear band (blue) were measured after deformation by cold rolling. Remarkably, the activation enthalpy for diffusion inside the shear bands amounts to only one third of the activation enthalpy for bulk diffusion. This large difference indicates large modifications of the atomic structure inside the shear bands.

One striking result is given by the fact that strong variations of the mass density are observed along single shear bands. Those variations have an average "wavelength" of about 100–150 nm. Yet, even more striking seems the observation that in addition to regions that show the expected, reduced density inside shear bands also segments with clearly increased density exist. In other words, the activation of the shear band has created regions inside the shear band that have a lower and others that have a higher mass density as compared to the un-deformed matrix far away from any shear band.

Moreover, local strain fields at shear bands have also been analyzed and the impact of shear deformation on the low-temperature heat capacity anomaly known as the "Boson peak" has been addressed. Relaxation experiments indicated an unexpected temporal evolution of the shear bands, including so-called cross-over behavior. The experimental results are discussed with respect to the underlying mechanism during the early stages of shear band activation and their temporal evolution as well as on the properties characterizing these "defects" in deformed metallic glasses.

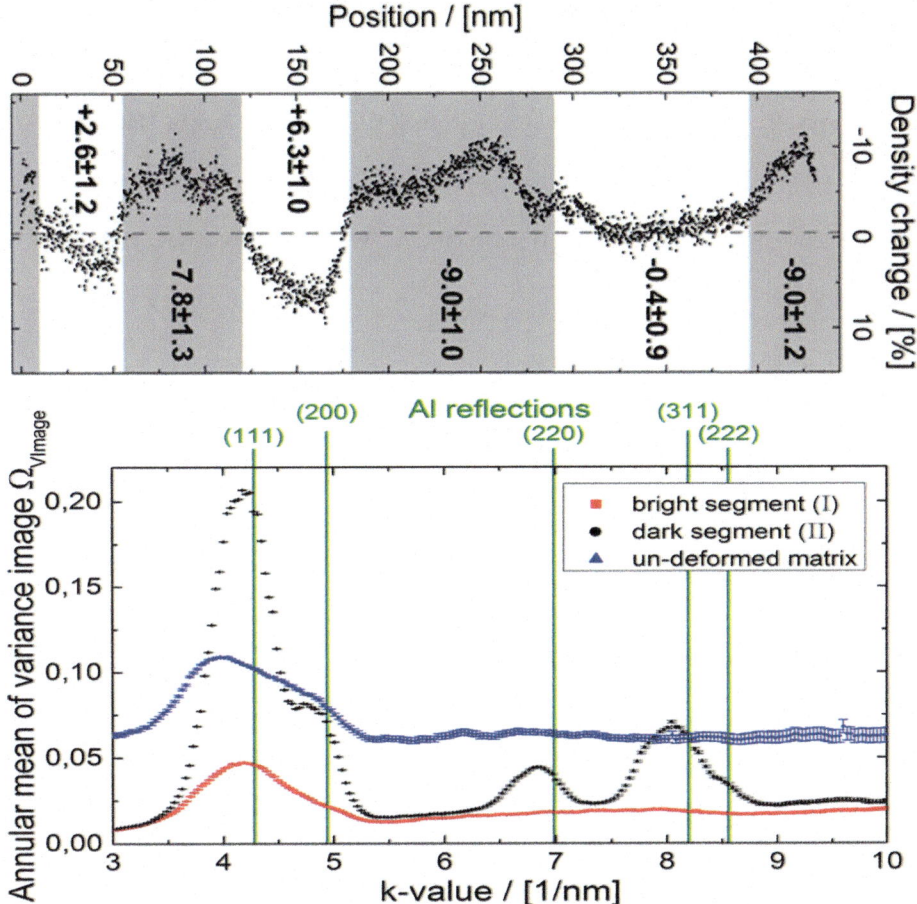

Fig. 2. (upper) mass density distribution along a single shear band. (lower) result of FEM obtained on the same shear band in regions of different density and also obtained in the un-deformed matrix.

References

1. Bokeloh, J., Reglitz, G., Divinski, S., Wilde, G.: Phys. Rev. Lett. **107**, 235503 (2011)
2. Wilde, G., Rösner, H.: Appl. Phys. Lett. **98**, 251904 (2011)
3. Rösner, H., Peterlechner, M., Kübel, C., Schmidt, V., Wilde, G.: Ultramicroscopy **142**, 1 (2014)
4. Bünz, J., Brink, T., Tsuchija, K., Meng, F., Wilde, G., Albe, K.: Phys. Rev. Lett. **112**, 135501 (2014)
5. Mitrofanov, Y.P., Peterlechner, M., Divinski, S.V., Wilde, G.: Phys. Rev. Lett. **112**, 135901 (2014)
6. Mitrofanov, Y.P., Peterlechner, M., Binkowski, I., Zadorozhnyy, M.Y., Golovin, I.S., Divinski, S.V., Wilde, G.: Acta Mater. **90**, 318 (2015)

7. Binkowski, I., Schlottbom, S., Leuthold, J., Ostendorp, S., Divinski, S.V., Wilde, G.: Appl. Phys. Lett. **107**, 221902 (2015)
8. Schmidt, V., Rösner, H., Peterlechner, M., Wilde, G., Voyles, P.M.: Phys. Rev. Lett. **115**, 035501 (2015)
9. Nollmann, N., Binkowski, I., Schmidt, V., Rösner, H., Wilde, G.: Scripta Mater. **111**, 119 (2016)
10. Binkowski, I., Shrivastav, G.P., Horbach, J., Divinski, S.V., Wilde, G.: Acta Mater. **109**, 330 (2016)
11. Hieronymus-Schmidt, V., Rösner, H., Zaccone, A., Wilde, G.: Phys. Rev. B **95**, 134111 (2017)

Plastic Deformation and Failure Mechanisms in Notched Nano-Scale Metallic Glass Specimens

R. Narasimhan[1]([✉]), Tanmay Dutta[1], and I. Singh[2]

[1] Department of Mechanical Engineering, Indian Institute of Science,
Bangalore 560012, India
narasi@iisc.ac.in
[2] Discipline of Mechanical Engineering, Indian Institute of Technology Indore,
Simrol, Indore 453552, India

Abstract. Finite element analysis (FEA) of tensile loading of nano-scale double edge notched (DEN) metallic glass (MG) specimens are conducted using a non-local plasticity model. The effects of notch acuity on the plastic deformation response of the samples are studied. Transition is observed from localized plastic deformation by single shear band (SSB) to necking of the ligament and further to double shear bands (DSBs) with increase in notch acuity. These results corroborate with molecular dynamics (MD) simulations and are rationalized from spread of plastic zone in the ligament.

Keywords: Metallic glasses · Failure mechanisms · Notched specimens

1 Introduction

Under tensile loading, bulk MG specimens fail by localized deformation occurring due to formation of a dominant shear band. On the other hand, nano-scale MG specimens exhibit homogeneous plastic deformation followed by necking. Singh and Narasimhan [1] demonstrated that single edge notched (SEN) nano-scale tensile specimens may experience SSB or necking depending on ligament width. Although MD simulations of DEN specimens have also shown transition from shear banding to necking, the mechanics aspects of the deformation process are not well understood. This is expected to be different from SEN specimens owing to the interaction between the plastic zones developing at the two notch roots. In order to address this issue, a combined MD - FEA of tensile loading of DEN specimens has been undertaken in this work.

2 Constitutive Model

A finite deformation based non-local plasticity model [1, 2] with free volume as an internal variable is employed in the FEA. This model has a length scale l_c which scales with the width of a shear band and characterizes non-local back stress associated with flow defects. Also, complementary MD simulations are performed using Cu-Zr MG specimens [2].

© Springer International Publishing AG, part of Springer Nature 2019
E. E. Gdoutos (Ed.): ICTAEM 2018, SI 5, pp. 329–330, 2019.
https://doi.org/10.1007/978-3-319-91989-8_71

3   Plastic Deformation Response and Failure Behavior

Plane strain simulations of DEN nano-scale samples under tensile loading are carried out keeping the uncracked ligament width B_0 fixed, while varying the normalized notch root radius $2R/B_0$. Also, in FEA, normalized internal length scale l_c/B_0 is varied. Figure 1(a) shows DSB from FEA with $l_c/B_0 = 3.2$ and $2R/B_0 = 0.5$, while necking and SSB are observed in Fig. 1(b) for $2R/B_0 = 0.83$ and in Fig. 1(c) for $2R/B_0 = 2.5$, respectively. The normalized load-displacement curves are displayed in Fig. 1(d) with necking and failure stages marked with 'o' and '×', respectively. The specimen with large $2R/B_0$ behaves akin to an unnotched one leading to failure by SSB (Fig. 1(c)) and rapid load drop (Fig. 1(d)). For moderately blunt notch ($2R/B_0 = 0.83$), plastic zone size r_p engulfs the full ligament as seen in Fig. 1(e) leading to necking (Fig. 1(b)). By contrast, for acute notches, r_p saturates as noticed from Fig. 1 (e) paving the way for failure by DSB (Fig. 1(a)). MD simulations show qualitatively similar behavior [2].

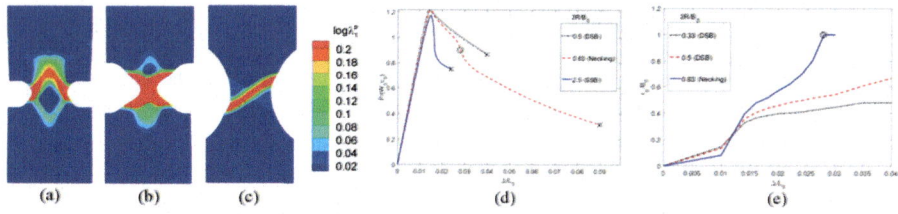

Fig. 1. Contours of log λ_1^p for MG specimens with $l_c/B_0 = 3.2$: (a) $2R/B_0 = 0.5$ at $\Delta/L_0 = 0.04$; (b) $2R/B_0 = 0.83$ at $\Delta/L_0 = 0.09$; (c) $2R/B_0 = 2.5$ at $\Delta/L_0 = 0.024$; (d) Normalized load-displacement curves for different $2R/B_0$ and (e) Evolution of r_p/B_0 with Δ/L_0.

4   Conclusions

Mode of failure in nano-scale notched MG specimens under tensile loading transitions from SSB to necking and further to DSB on increasing notch acuity. The specimen with moderately blunt notch exhibits necking and large failure strain.

References

1. Singh, I., Narasimhan, R.: Notch sensitivity in nanoscale metallic glass specimens: insights from continuum simulations. J. Mech. Phys. Solids **86**, 53–69 (2016)
2. Dutta, T., Chauniyal, A., Singh, I., Narasimhan, R., Thamburaja, P., Ramamurty, U.: Plastic deformation and failure mechanisms in nano-scale notched metallic glass specimens under tensile loading. J. Mech. Phys. Solids **111**, 393–413 (2018)

On the Fracture Behavior of Bulk Metallic Glasses

Bernd Gludovatz[1]([✉]) [iD], Jamie J. Kruzic[1] [iD],
and Robert O. Ritchie[2,3] [iD]

[1] UNSW Sydney, Sydney, NSW 2052, Australia
b.gludovatz@unsw.edu.au
[2] Lawrence Berkeley National Laboratory, Berkeley, CA 94720, USA
[3] University of California, Berkeley, CA 94720, USA

Abstract. The excellent combination of high strength and low stiffness make bulk-metallic glasses (BMGs) candidate materials for many structural applications. Their fracture toughness, however, can vary markedly, and their ductility, particularly in tension/compression, is rather limited, although in bending they are quite ductile. Here, we report on a systematic study on Zr and Pd-based glasses to investigate the influence of sample size on the fracture toughness of BMGs. Results show that with decreasing sample size the fracture behavior changes from brittle failure with low fracture toughness, via a semi-brittle failure regime, to fully ductile fracture and non-catastrophic failure with sub-critical crack growth, i.e., *R*-curve behavior.

Keywords: Bulk Metallic Glasses · Fracture toughness · Sample size

1 Introduction

High strength, low stiffness and near net-shape castability make bulk-metallic glasses (BMGs) candidate materials for many structural applications. A major drawback for their use in engineering service, however, are highly variable fracture toughness values. Due to the often limited dimensions of cast BMGs, standard fracture-toughness tests are generally performed on samples with dimensions comparable to their critical bending thickness which is defined as the dimension below which a glass can achieve the relevant number of shear bands to demonstrate significant bending ductility [1] (Fig. 1a). To date, however, it is not clear how BMGs would behave in fracture toughness tests evaluated on samples with dimensions that are either below, above, or comparable to their critical bending thickness.

2 Results

In this study, we have investigated the influence of sample size on the fracture toughness of BMGs using three-point bending samples of different sizes of Zr and Pd-based glasses. Sample dimensions were chosen with respect to the critical bending thickness of each glass; for the Zr-glass well above and in the range of the critical

© Springer International Publishing AG, part of Springer Nature 2019
E. E. Gdoutos (Ed.): ICTAEM 2018, SI 5, pp. 331–332, 2019.
https://doi.org/10.1007/978-3-319-91989-8_72

(a)   **(b)**

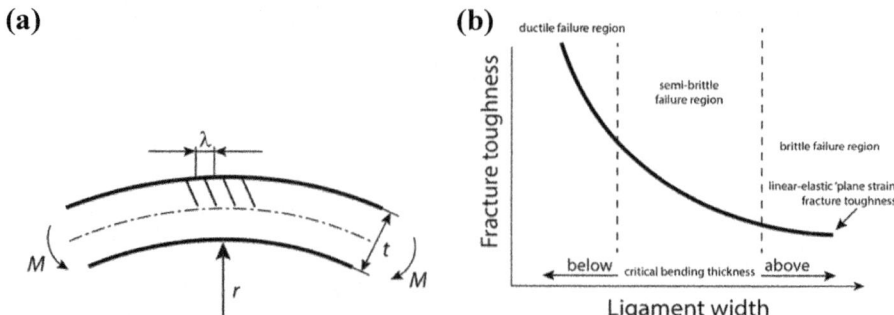

Fig. 1. (a) Below a critical thickness, t, the spacing of shear bands, λ, in BMG samples decreases with increasing bending moment, M, and decreasing radius, r. (b) BMG samples show a strong dependence of their fracture toughness on the ligament width, b. (Figures taken from [2].)

bending thickness, and for the Pd-glass in the range of the critical bending thickness and below. Results suggest that BMGs show a transition in fracture behavior that is linked to their sample size (Fig. 1b). Samples with ligament widths larger than the critical bending thickness show brittle failure characteristics and low fracture toughness values with only small variations in the results; in this regime, the linear-elastic plane-strain K_{Ic} fracture toughness of the material can be determined. Samples tested with dimensions comparable to the critical bending thickness yield highly variable fracture toughness results whereas samples that are below the critical bending thickness can show fully ductile failure characteristics, less variability in the results, and non-catastrophic failure with subcritical crack growth and R-curve behavior.

3   Conclusions

The fracture properties of bulk metallic glass have been investigated with the aim of understanding the large variability in fracture toughness values in BMGs. Results show that both fracture behavior and the obtained fracture toughness values are likely sample size dependent when testing samples with dimensions comparable or smaller than the critical bending thickness of a glass.

References

1. Conner, R.D., Johnson, W.L., Paton, N.E., Nix, W.D.: Shear bands and cracking of metallic glass plates in bending. J. Appl. Phys. **94**, 904–911 (2003)
2. Gludovatz, B., Granata, D., Thurston, K.V.S., Löffler, J.F., Ritchie, R.O.: On the understanding of the effects of sample size on the variability in fracture toughness of bulk metallic glasses. Acta Mater. **126**, 494–506 (2017)

Synthesis and Properties of Bulk Metallic Glass Composites

Lisa Krämer[1], Marlene Kapp[1], Verena Maier-Kiener[2],
Karoline Kormout[1], Yannick Champion[3], and Reinhard Pippan[1(✉)]

[1] Erich-Schmid Institute of Materials Science,
Austrian Academy of Sciences, Leoben, Austria
reinhard.pippan@oeaw.ac.at
[2] Department of Physical Metallurgy and Materials Testing,
Montanuniversität Leoben, Leoben, Austria
[3] Univ. Grenoble Alpes, CNRS, Grenoble INP, SIMaP, 38000 Grenoble, France

Abstract. The aim of this study is to show that new types of bulk metallic glass composites (BMGCs) can be produced via severe plastic deformation (SPD). The initial materials are mixtures of a Zr-metallic glass (MG) and crystalline Cu powders that were mixed and then consolidated, cold welded together and refined by high pressure torsion (HPT). Four different compositions (Zr-MG Xwt% Cu, X = 20, 40, 60, 80) were produced as well as single phase Zr-MG samples as reference. To investigate the influence of the degree of deformation and the ratio of the two phases on the evolution of the microstructure and mechanical properties, scanning electron microscopy (SEM), X-ray diffraction (XRD), hardness and microcompression measurements were used.

Keywords: Bulk metallic glass · Severe plastic deformation
Mechanical properties

Bulk metallic glasses (BMGs) have advantages over crystalline metals including high hardness, high elastic energy storage and high corrosion resistivity, but also some major drawbacks. BMGs are very brittle and especially show poor ductility in tensile testing. Changing the microstructure is a common way to tune properties of materials. A way to realize this is to produce composites (BMGCs) containing an additional amorphous or crystalline phase. Instead of coarse composite as in [1] we used as initial materials are a mixtures of a Zr-metallic glass (MG) and crystalline Cu powders. They were mixed and then consolidated, the powder particle were welded together by the heavy plastic shear deformation induced by high pressure torsion (HPT). Four different compositions (Zr-MG Xwt% Cu, X = 20, 40, 60, 80) were produced as well as single phase Zr-MG samples as reference.

The microstructure of the HPT deformed Cu + Zr-MG mixture and their mechanical properties depend strongly on the composition and on the applied strain [2, 3]. Figure 1 shows the scanning electron microscope (SEM) micrographs of 2 compositions processed to different shear strains. At low strains, a coarse bulk composite is formed, in which the elongated phases start to refine and the hardness increases strongly (see [3]). With increasing the applied strain, a continuous thinning of the

© Springer International Publishing AG, part of Springer Nature 2019
E. E. Gdoutos (Ed.): ICTAEM 2018, SI 5, pp. 333–335, 2019.
https://doi.org/10.1007/978-3-319-91989-8_73

lamellae can be found for 20 and 40wt% of Cu, whereas, the lamellas tend to break down at higher contents of Cu. For 60 and 80wt% Cu, the length of the lamellae decreases more than their thickness (for this details see [3]). In addition, a mixing of the two phases takes place and at very high strains, a single phase BMG is obtained for all composition but not for Zr-MG 80wt% where a nearly single phase supersaturated solid solution nanocrystalline Cu alloy has been formed at very high applied strains. The XRD results and hardness evolution is described in [3] and the micromechanical experiments in [4].

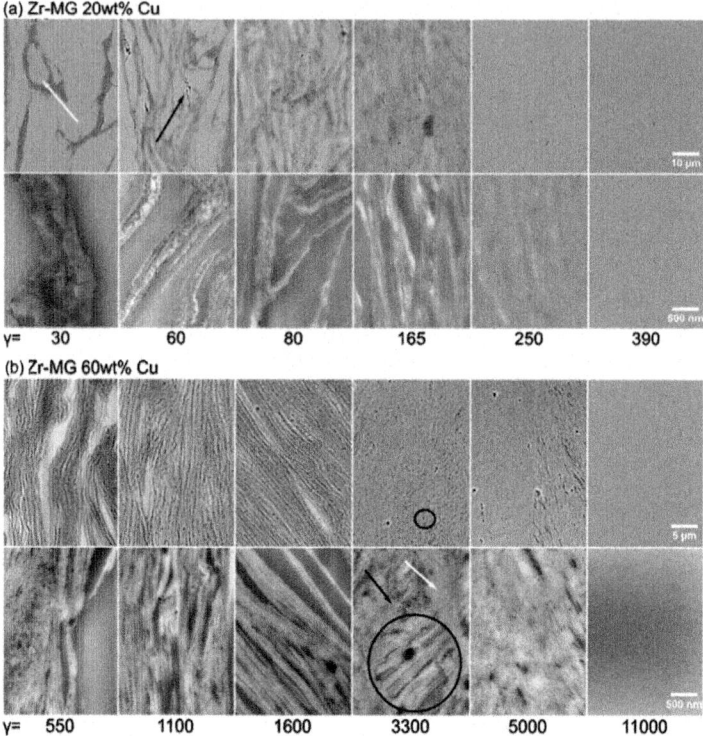

Fig. 1. Illustration of SEM micrographs of (a) Zr-MG 20wt% Cu and (b) Zr-MG 60wt% Cu as a function of the applied shear strain γ. The top and bottom row of micrographs are taken at different magnifications. The deformation increases from the left to the right and the microstructure changes from Zr-MG particles glued together by crystalline Cu to a lamellar structure. These lamellae refine with increasing applied strain until a complete mixing of the two phases occurs. Take notice that different strains are necessary for complete mixing for different contents of Cu.

References

1. Sauvage, X., Champion, Y., Pippan, R., Cuvilly, F., Perriere, L., Akhotova, A., Renk, O.: Structure and properties of a nanoscaled composition modulated metallic glass. J. Mater. Sci. **49**, 5640–5645 (2014)
2. Kraemer, L., Kormout, K., Setman, D., Champion, Y., Pippan, R.: Production of bulk metallic glasses by severe plastic deformation. Metals (Basel) **5**, 720 (2015)
3. Kraemer, L., Champion, Y., Pippan, R.: From powders to bulk metallic glass composites. Sci. Rep. **7** (2017). Article Number: 6651
4. Kraemer, L.: (in preparation)

Surface Behavior of Metallic Glasses Under Irradiation

Kang Sun, X. L. Bian, and Gang Wang[⊠]

Shanghai University, Shanghai 200444, China
g.wang@shu.edu.cn

Abstract. Structural inhomogeneity and defects caused in the forming process in metallic glasses lead to some weak areas (such as free volumes and liquid-like zones) embedding into the glassy phase, which are associated with the plastic flow. The energy introduced by an ion irradiation can change the number and distribution of free volumes. The irradiation at low dosage can change the structure by increasing the density of free volume, thus resulting in a decrease in the strength. The irradiation at high dosage makes nanocrystallization occurring in the metallic glass. The analysis of relaxation spectrum suggests that the relaxation strength increases, and then decreases with increasing the irradiation dosage. The results of this study highlight the fact that a large amount of free volumes can be achieved through a low-dose ion irradiation, which can modify the mechanical behavior of metallic glasses.

Keywords: Metallic glasses · Ion irradiation · Nanoindentation

1 First Section

1.1 A Subsection Sample

Please note that the first paragraph of a section or subsection is not indented. The first paragraphs that follows a table, figure, equation etc. does not have an indent, either.

Subsequent paragraphs, however, are indented.

Sample Heading (Third Level). Only two levels of headings should be numbered. Lower level headings remain unnumbered; they are formatted as run-in headings.

Sample Heading (Forth Level). The contribution should contain no more than four levels of headings. The following Table 1 *gives a summary of all heading levels.*

Displayed equations are centered and set on a separate line.

$$x + y = z \tag{1}$$

Please try to avoid rasterized images for line-art diagrams and schemas. Whenever possible, use vector graphics instead (see Fig. 1).

For citations of references, we prefer the use of square brackets and consecutive numbers. Citations using labels or the author/year convention are also acceptable.

E. E. Gdoutos (Ed.): ICTAEM 2018, SI 5, pp. 336–337, 2019.
https://doi.org/10.1007/978-3-319-91989-8_74

Table 1. Table captions should be placed above the tables.

Heading level	Example	Font size and style
Title (centered)	**Lecture Notes**	14 point, bold
1st-level heading	**1 Introduction**	12 point, bold
2nd-level heading	**2.1 Printing Area**	10 point, bold
3rd-level heading	**Run-in Heading in Bold.** Text follows	10 point, bold
4th-level heading	*Lowest Level Heading.* Text follows	10 point, italic

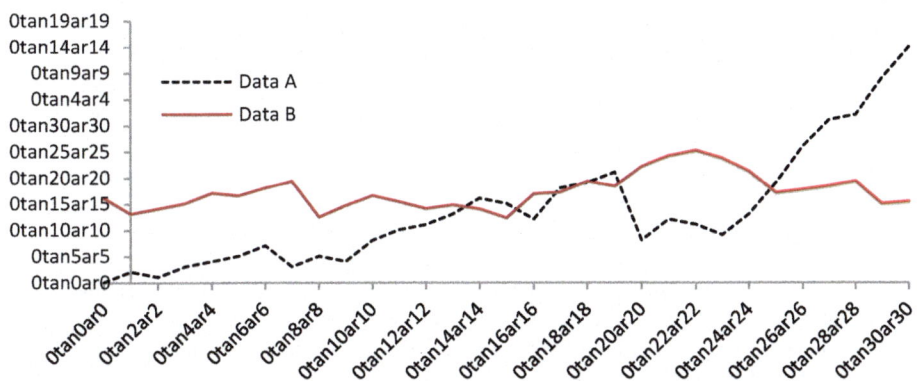

Fig. 1. A figure caption is always placed below the illustration. Short captions are centered, while long ones are justified. The macro button chooses the correct format automatically.

The following bibliography provides a sample reference list with entries for journal articles [1], an LNCS chapter [2], a book [3], proceedings without editors [4], as well as a URL [5].

References

1. Author, F.: Article title. Journal **2**(5), 99–110 (2016)
2. Author, F., Author, S.: Title of a proceedings paper. In: Editor, F., Editor, S. (eds.) CONFERENCE 2016. LNCS, vol. 9999, pp. 1–13. Springer, Heidelberg (2016)
3. Author, F., Author, S., Author, T.: Book Title, 2nd edn. Publisher, Location (1999)
4. Author, F.: Contribution title. In: 9th International Proceedings on Proceedings, pp. 1–2. Publisher, Location (2010)
5. LNCS Homepage. http://www.springer.com/lncs. Accessed 21 November 2016

Shear Banding in Bulk Metallic Glass Matrix Composites with Dendrite Reinforcements

Stephen R. Niezgoda[1]([✉]), Michael P. Gibbons[1], Wolfgang Windl[1], and Katharine M. Flores[2]

[1] The Ohio State University, Columbus, OH 43085, USA
niezgoda.6@osu.edu
[2] Washington University in St. Louis, St. Louis, MO 64130, USA

Abstract. Bulk metallic glass matrix composites (BMGMCs) with metallic dendrite reinforcements combine the excellent strength, hardness, and elastic strain limit of amorphous metallic glass with a ductile crystalline phase to achieve extraordinary toughness with minimal degradation in strength. In order to explore the mechanical interactions between the amorphous and crystalline phases a full-field micromechanical model, which couples a free volume based constitutive model for the matrix with crystal plasticity, has been implemented in an elastic-viscoplastic Fast-Fourier Transform (FFT) solver. The findings indicate that in BMGMCs, local inhomogeneities in the glass phase are less influential on the mechanical performance than the contrast in individual phase properties. Due to the strong contrast in mechanical properties, heterogenous stress fields develop, contributing to regionally confined free volume generation, localized flow and softening in the glass. In these softened regions, plastic flow rapidly localizes into shear bands.

Keywords: Bulk metallic glass · Shear bands · Yield phenomena
Micromechanical simulation

One strategy for mitigating the poor failure resistance in BMGs is to incorporate crystalline phases which dissipate energy and accommodate plastic strain by dislocation slip, effectively blocking long-range shear band propagation. This has led to considerable interest in bulk metallic glass-matrix composites (BMGMCs) with crystalline reinforcements [1]. Computational modeling offers a way to accelerate the design of BMGMCs by exploring the effect of microstructural parameters and individual phase properties on the overall composite mechanical performance. Micromechanical deformation simulations also provide important insight on the nucleation and interaction of shear bands with the crystalline dendrites, as well as how the microstructure and individual phase properties can be tailored to optimize the mechanical performance of the composite.

Prior continuum and atomistic-scale modeling efforts have worked to establish robust constitutive models for BMG and BMGMC deformation, introduce fundamental variables to describe the local state of the material, and develop mathematical frameworks for efficiently simulating the micromechanical deformation associated with shear banding [2, 3]. Here we implement a free volume based constitutive model for the matrix coupled with crystal plasticity within an efficient fast Fourier-transform based

© Springer International Publishing AG, part of Springer Nature 2019
E. E. Gdoutos (Ed.): ICTAEM 2018, SI 5, pp. 338–340, 2019.
https://doi.org/10.1007/978-3-319-91989-8_75

micromechanical solver [4] for the simulation of quasi-static deformation of BMGMCs, in terms of both the macroscopic mechanical response as well as the local characteristics of shear band operation. We show that in contrast to shear band nucleation in monolithic BMGs, where it is reasonably assumed the nucleation sites arise from chemical and structural fluctuations at the atomic scale, in metallic glass composites, the spatial distribution and contrast in individual phase properties are the dominating factors controlling where, when, and how shear bands are formed. We show that shear bands in metallic glass composites are reproducible in numerical simulations without the introduction of intrinsic flaws or an artificial nucleation sources, provided there exists sufficient contrast in properties between the dendrites and amorphous matrix. We then link experimentally observed oscillatory stress-strain behavior to the concurrent localization of strain in shear bands and the corresponding relaxation of the glass-matrix in the surrounding regions. Results from the simulation showing the development of shear bands along with the result evolution of stress-strain and accumulated strain energy is shown in Fig. 1.

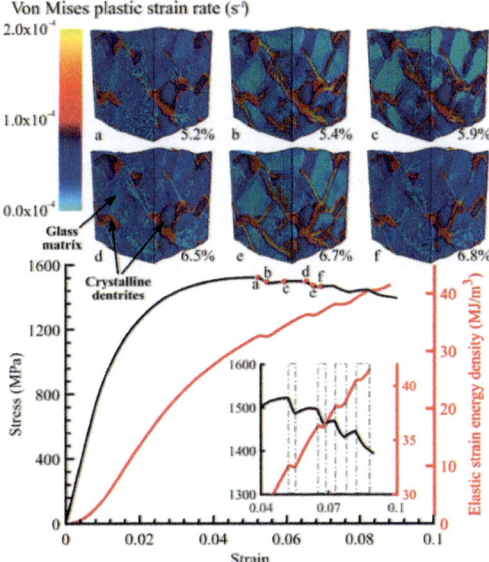

Fig. 1. 3D deformation maps (top) colored by the local plastic strain-rate at overall strain levels corresponding to the marked positions on the stress-strain curve (bottom). The elastic strain energy density highlight its relationship to shear band formation and the associated oscillations in the stress-strain curve.

References

1. Hofmann, D.C., Suh, J.Y., Wiest, A., Lind, M.L., Demetriou, M.D., Johnson, W.L.: Development of tough, low-density titanium-based bulk metallic glass matrix composites with tensile ductility. Appl. Phys. Sci. **105**(51), 20136–20140 (2008)
2. Zhao, P., Li, J., Wang, Y.: Heterogeneously randomized STZ model of metallic glasses: Softening and extreme value statistics during deformation. Int. J. Plast. **40**, 1–22 (2013)
3. Anand, L., Aslan, O., Chester, S.A.: A large-deformation gradient theory for elastic–plastic materials: strain softening and regularization of shear bands. Int. J. Plast. **30**, 116–143 (2012)
4. Lebensohn, R.A., Kanjarla, A.K., Eisenlohr, P.: An elasto-viscoplastic formulation based on fast Fourier transforms for the prediction of micromechanical fields in polycrystalline materials. Int. J. Plast. **32**, 59–69 (2012)

Origin of Embrittlement of Metallic Glasses

Marios D. Demetriou$^{(\boxtimes)}$, Glenn R. Garrett, and William L. Johnson

Glassimetal Technology, Pasadena, CA 91107, USA
marios@glassimetal.com

Abstract. Owing to their glassy nature, metallic glasses demonstrate a toughness that is extremely sensitive to the frozen-in configurational state. This sensitivity gives rise to the so-called "annealing embrittlement", which is often severe and in many respects limits the technological advancement of these materials. Here, equilibrium configurations (i.e. "inherent states") of a metallic glass are established around the glass transition, and the configurational properties along with the plane-strain fracture toughness are evaluated. An association between the intrinsic toughness and the inherent state properties is attempted in an effort to identify the fundamental origin of embrittlement of metallic glasses. The established correlations reveal a one-to-one correspondence between a decreasing toughness and an increasing shear modulus, which is robust and continuous over a broad range of inherent states. This annealing embrittlement sensitivity is shown to vary substantially between metallic glass compositions, and appears to correlate well with the fragility of the metallic glass.

Keywords: Amorphous metal · Annealing embrittlement sensitivity
Fragility

Unlike conventional oxide glasses, metallic glasses can be considerably tougher often approaching values typical of engineering metals. Specifically, the toughness of metallic glasses can vary from values as high as 150–200 MPa·m$^{1/2}$, comparable to low carbon steels, to as low as 2–3 MPa·m$^{1/2}$, which is more typical of oxide glasses. But like any glass, metallic glasses tend to embrittle when annealed around the glass transition, as evidenced by a measurable drop in their toughness and their capacity to deform plastically. This phenomenon, typically referred to as annealing embrittlement, has broad scientific and technological ramifications. From a technological perspective, annealing embrittlement is at least partly responsible for the large variability in toughness observed when the glass is quenched at different cooling rates, thereby limiting their mechanical performance and engineering reliability. From a physics perspective, the physical origin of annealing embrittlement is still a topic of debate and controversy that is far from being settled, as varying and contradictory theories have been proposed by various groups over the years.

Here, we seek to establish a fundamental correspondence between fracture toughness and the inherent state of the metallic glass that could reveal the underlying origin of embrittlement. High-purity fully-amorphous samples with size and geometry consistent with the E399 ASTM standard were produced in that work. The samples

© Springer International Publishing AG, part of Springer Nature 2019
E. E. Gdoutos (Ed.): ICTAEM 2018, SI 5, pp. 341–342, 2019.
https://doi.org/10.1007/978-3-319-91989-8_76

were fully relaxed at well-defined temperatures, T_R, near the glass transition temperature, T_g, and are carefully characterized for "inherent-state" properties and plane-strain fracture toughness, K_{IC}.

The "intrinsic" plane-strain fracture toughness K_{IC} is found to follow exponentially decaying functions of the equilibrium isoconfigurational shear modulus G. This relation appears to be continuous over a broad range of equilibrium configurational states of the glass. Such correlation reveals that the plane-strain fracture toughness of samples equilibrated at different temperatures is linked uniquely to properties characterizing the inherent state of the glass (e.g. the shear modulus). It is interesting to examine how this sensitivity to annealing embrittlement varies among various metallic glasses. The annealing embrittlement sensitivity can be defined as the relative drop in fracture toughness K_{IC} with respect to a relative drop in relaxation temperature.

Plotting fracture toughness vs. temperature data for metallic glass $Zr_{35}Ti_{30}$ $Cu_{8.25}Be_{26.75}$ gathered in the present work, along with data for $Pt_{57.5}Ni_{5.3}Cu_{14.7}P_{22.5}$ and $Zr_{44}Ti_{11}Cu_{10}Ni_{10}Be_{25}$ reveals that the three metallic glasses exhibit very different propensities for annealing embrittlement. Specifically, $Pt_{57.5}Ni_{5.3}Cu_{14.7}P_{22.5}$ and $Zr_{44}Ti_{11}Cu_{10}Ni_{10}Be_{25}$ demonstrating the highest and lowest annealing embrittlement sensitivity respectively and $Zr_{35}Ti_{30}Cu_{8.25}Be_{26.75}$ falling in between.

Interestingly, the order of annealing embrittlement sensitivity actually follows the order of the fragility of the three glasses. The fragility, defined by Angell as the steepness of the viscosity temperature dependence at T_g, is often used as a parameter characterizing the potential energy landscape morphology of a glass/liquid. By plotting the annealing embrittlement sensitivity vs fragility for the three metallic glasses, a one-to-one correspondence is revealed. Since fragility is understood to be controlled by the decrease of the shear modulus with increasing temperature, and considering the one-to-one correspondence between K_{IC} and G is demonstrated here, it is conceivable that fragility can be uniquely associated with the steepness of the fracture toughness drop with temperature, or equivalently with the annealing embrittlement sensitivity.

The outcome of this study could have significant scientific and technological implications. From a scientific perspective, this result suggests a thermodynamic origin of annealing embrittlement, but also a structural one. In Angell's original conception, fragility quantifies how rapidly the "structure" (i.e. the local order) of an equilibrium liquid develops as the liquid cools towards T_g. As such, the present result suggests that annealing embrittlement should be associated with the degree of order established in each equilibrium state at which the atomic structure of the glass is relaxed to. From a technological perspective, the present result suggests that more fragile glasses are more prone to embrittlement than stronger glasses. Specifically, it suggests that fragile glasses would tend to embrittle at low fictive temperatures and toughen at high fictive temperatures, while stronger glasses would have a more constant toughness with varying fictive temperature.

Crack Propagation in Bulk Metallic Glasses

Jamie J. Kruzic[(⊠)]

School of Mechanical and Manufacturing Engineering, UNSW Sydney,
Kensington, NSW 2052, Australia
j.kruzic@unsw.edu.au

Abstract. Bulk metallic glasses (BMGs) possess high strength, but their fracture toughness and fatigue crack growth resistance can vary widely. Fracture toughness and fatigue crack growth results for Zr-based BMGs show that fracture toughness is highly sensitive to the glassy micro- and/or nanostructure, while the fatigue crack growth resistance is relatively insensitive. Soft and heterogeneous glassy structures promote crack tip blunting and high fracture toughness, and can be promoted by thermo-mechanical treatments. During fatigue cycling, *in-situ* free volume generation occurs ahead of the crack tip that defines the local structure independent of the initial structural state. Accordingly, fatigue crack propagation rates depend little on the initial structure of the BMG. Finally, synchrotron XRD analysis of crack-tip strain fields revealed that the crack-tip deformation extends farther, and is more homogeneous, than visible shear bands suggest and the mechanism of fatigue crack propagation is identical to crystalline metals despite the different deformation mechanism.

Keywords: Bulk metallic glasses · Fracture toughness · Fatigue

1 Introduction

Bulk metallic glasses (BMGs) have excellent high tensile yield strengths (~ 1–5 GPa is common) and large elastic strain limits ($\sim 2\%$). However, damage tolerance properties, such as fracture toughness and fatigue crack growth resistance, can vary widely from excellent to poor and the current understanding of how to control these properties is limited [1], thus motivating the present work.

2 Results

Fracture toughness studies on $Zr_{63.78}Cu_{14.72}Ni_{10}Al_{10}Nb_{1.5}$ and $Zr_{52.5}Ti_5Cu_{18}Ni_{14.5}Al_{10}$ (at.%) BMGs revealed cold deformation, e.g., cold rolling and imprinting, and thermal cycling 70 times to cryogenic temperatures (77 K) results in increased fracture toughness values [2, 3]. This affect is attributed to the softer and more heterogeneous glassy structures that result after these treatments, that in turn promote crack blunting. While annealing a $Zr_{44}Ti_{11}Ni_{10}Cu_{10}Be_{25}$ BMG to reduce free volume in the glass structure severely lowers the fracture toughness, fatigue crack growth rates are unaffected [4]. This is attributed to a fatigue transformation zone of higher free volume that is formed in the cyclic plastic zone ahead of the crack tip, as revealed by positron

© Springer International Publishing AG, part of Springer Nature 2019
E. E. Gdoutos (Ed.): ICTAEM 2018, SI 5, pp. 343–344, 2019.
https://doi.org/10.1007/978-3-319-91989-8_77

annihilation spectroscopy (Fig. 1a) [5]. Furthermore, synchrotron x-ray studies of fatigue crack tips in a $Zr_{52.5}Ti_5Cu_{18}Ni_{14.5}Al_{10}$ BMG reveal a crack tip strain field corresponds well with crystalline metals and that extends farther, and is more homogenous, than the visible shear bands suggest. Results confirm that the mechanism of fatigue crack growth in BMGs is alternating crack tip blunting and re-sharpening arising from compressive stresses that evolve in the cyclic plastic zone (Fig. 1b) [6].

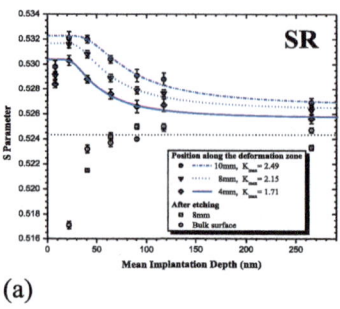
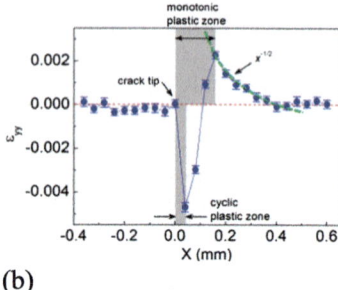

(a)   (b)

Fig. 1. (a) Elevated free volume in a fatigue crack cyclic plastic zone is revealed by the higher S-parameter measured via positron annihilation spectroscopy [5]. (b) Crack tip strain field showing the compressive and tensile residual elastic strains induced ahead of a crack tip (X = 0) [6].

3 Conclusions

Soft and heterogeneous glassy structures promoted by thermo-mechanical processing methods promote crack tip blunting and high fracture toughness in BMGs. In contrast, fatigue crack propagation rates depend little on the glassy structure of the BMG due to *in-situ* free volume generation that redefines the structure ahead of the crack tip during fatigue cycling. Finally, synchrotron XRD analysis of crack tip strain fields reveal that the mechanism of fatigue crack propagation is identical to crystalline metals.

References

1. Kruzic, J.J.: Bulk metallic glasses as structural materials: a review. Adv. Eng. Mater. **18**, 1308–1331 (2016)
2. Li, B.S., et al.: Designed heterogeneities improve the fracture reliability of a Zr-based bulk metallic glass. Mater. Sci. Eng. A **646**, 242–248 (2015)
3. Xie, S.H., Kruzic, J.J.: Cold rolling improves the fracture toughness of a Zr-based bulk metallic glass. J. Alloys Compd. **694**, 1109–1120 (2017)
4. Launey, M.E., et al.: Effect of free volume changes and residual stresses on the fatigue and fracture behavior of a Zr-Ti-Ni-Cu-Be bulk metallic glass. Acta Mater. **56**, 500–510 (2008)
5. Liu, M., et al.: Assessment of the fatigue transformation zone in bulk metallic glasses using positron annihilation spectroscopy. J. Appl. Phys. **105**, 093501 (2009)
6. Scudino, S., et al.: Mapping the cyclic plastic zone to elucidate the mechanisms of crack tip deformation in bulk metallic glasses. Appl. Phys. Lett. **110**, 081903 (2017)

Thermal and Mechanical Stability of Nano-Crystalline and Nano-Structured Metals

Dominic Rathmann[1], Killang Pratama[1], Andrea Bachmaier[1,2],
Michael Marx[1], and Christian Motz[1(✉)]

[1] Department of Material Science and Engineering,
Saarland University, 66123 Saarbrücken, Germany
`motz@matsci.uni-sb.de`
[2] Austrian Academy of Sciences, Erich-Schmid-Institute of Material Science,
8700 Leoben, Austria

Abstract. The aim of the current work is to study the microstructure stability under thermal and mechanical loads of a nano-structured Cu/Co alloy and nc-Nickel with different content of solute elements and nano-particles. For this annealing at different temperatures as well as fatigue loading are performed and the microstructure evolution is characterized. The effectiveness of mechanisms that impedes grain growth (solute drag, Zener pinning) are evaluated and studied for both, thermal and mechanical loads.

Keywords: Microstructure stability · Nano-crystalline metals · Fatigue
Mechanical properties

1 Materials and Experimental Set-Up

1.1 Introduction

Since the 1980th nano-crystalline materials were in the focus of materials research and become more and more relevant in technical applications [1]. A reason for the great interest in this young material class is for example the high strength gained by grain refinement to grain sizes below 100 nm. On the other hand, a big challenge for production and applications of these material is the proneness for grain growth at slightly elevated temperatures, whereby the good mechanical properties are lost. In general, grain growth is caused by the tendency of the material to reduce his surface energy. Because of the severe amount of grain boundaries in nano-crystalline materials this driving pressure is so huge, that grain growth occurs even at room temperature in some systems. A second, less studied activation for grain growth are mechanical loads, especially fatigue loads.

1.2 Materials

In the current study two types of materials are used: (i) nc-Nickel which is produced with pulsed electrodeposition (PED) and (ii) a nano-structured Cu/Co alloy which is

© Springer International Publishing AG, part of Springer Nature 2019
E. E. Gdoutos (Ed.): ICTAEM 2018, SI 5, pp. 345–346, 2019.
https://doi.org/10.1007/978-3-319-91989-8_78

synthesized by severe plastic deformation (SPD) and PED. Both materials are heat treated to obtain different microstructures, e.g. bi-modal grain size distribution, phase separation, etc.

1.3 Experimental Set-Up

The different microstructures of all materials are characterized by utilizing electron microscopy, x-ray diffraction and atom probe tomography. Furthermore, the general mechanical properties are determined by nano- and micro-hardness measurements. For selected materials fatigue crack growth experiments are performed to investigate the fatigue crack growth rate and the stability of the microstructure.

2 Results and Discussion

For the Nickel material after heat treatment nano-crystalline Ni (grain size about 30 nm) and a bi-modal microstructure (with nano-sized grains of about 40 nm and ultra-fine grains of several 100 nm) were obtained. Hardness measurements reveal a strong increase in strength for both materials compared to their coarse-grained counterparts. In the fatigue experiments the nano-crystalline Ni shows a good fatigue crack growth rate, however, a grain coarsening is visible in the plastic zone at the crack tip. Hence, the microstructure is not mechanically stable (strain induced grain growth). The bi-modal material, which should show better fatigue properties due to soft and hard regions in the microstructure (soft region should act as crack arrestor) exhibits an even worse fatigue crack growth rate. This is caused by grain boundary embrittlement due to segregation of S and P.

To avoid this grain boundary embrittlement supersaturated Cu/Co alloys are produced by SPD and PED along with different heat treatments. During annealing the supersaturated solid solution decomposes and forms a nano-crystalline microstructure of Cu and Co grains. This structure reveals to be very stable against thermal and mechanical loads. Finally, the advantages and disadvantages of both materials are discussed, where the focus lays on the fatigue properties of nano-structured materials.

References

1. Valiev, R.: Nanostructuring of metals by severe plastic deformation for advanced properties. Nat. Mater. **8**, 511–516 (2004)
2. Rathmann, D., Marx, M., Motz, C.: Crack propagation and mechanical properties of electrodeposited nickel with bimodal microstructures in the nanocrystalline and ultrafine grained regime. J. Mater. Res. (2018, in press)

Fatigue Crack-Growth Properties
of SPD-Metals

Anton Hohenwarter$^{(\boxtimes)}$ ⓘ and Thomas Leitner ⓘ

Department of Materials Physics,
Montanuniversität Leoben, Leoben 8700, Austria
anton.hohenwarter@unileoben.ac.at

Abstract. Methods of severe plastic deformation (SPD) can be principally applied to all kinds of metallic materials and are widely used to generate nanocrystalline materials that frequently exhibit improved mechanical and functional properties. The resistance against crack growth is, however, only rarely studied despite its importance for the damage tolerant design of components. In this contribution general tendencies found for the fatigue crack behavior of a large variety of SPD-processed metals and alloys, including pure metals (iron and nickel) but also alloys (NiTi, steels) is given. In general grain refinement leads to a deteriorated fatigue crack growth behavior, which can be mainly attributed to a reduction of crack closure contributions. Another significant factor seems to originate from the frequently observed transition from transgranular to intergranular fatigue fracture possessing a lower crack growth resistance than transgranular crack growth. Based on these general observations strategies to counteract these tendencies are discussed where especially the grain-aspect ratio of the microstructures may play an important role.

Keywords: Severe plastic deformation · Fatigue crack propagation
Crack closure

1 Introduction

Severe plastic deformation (SPD) is a relatively young field in materials science and deals with the structure-property relationships of metallic materials subjected to socalled SPD-processes [1]. These techniques are able to process materials to deformation strains that are much higher than the ones normally provided by classical deformation processes, such as rolling. Most of our results are based on materials processed by high pressure torsion, which is able to deform relatively hard materials without crack formation to the highest ever reported strains. In this process disks are subjected to pressures of several gigapascals and deformed by shear straining. Through this intense shear deformation the grain size can be reduced down to the nanocrystalline grain size regime depending on various technical and material parameters. Especially the presence of a hydrostatic pressure component and the fact that the specimen geometry is not changed during deformation is significant for crack-free processing and the achievement of extraordinarily high strains in the range of several thousand percent.

© Springer International Publishing AG, part of Springer Nature 2019
E. E. Gdoutos (Ed.): ICTAEM 2018, SI 5, pp. 347–349, 2019.
https://doi.org/10.1007/978-3-319-91989-8_79

2 Results

In Fig. 1 measurements performed on an austenitic steel in the microcrystalline (MC) and nanocrystalline (NC) state are presented, which largely reflect the tendencies found in other severely plastically deformed materials.

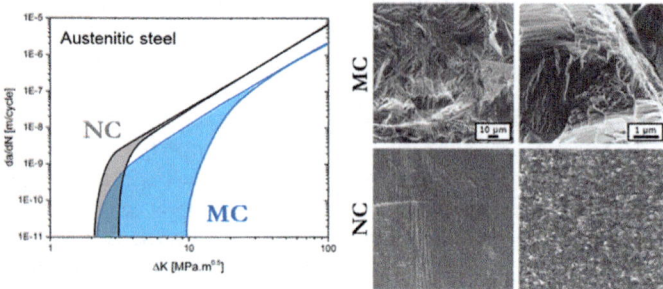

Fig. 1. Fatigue-crack growth measurements performed for the microcrystalline (MC) and nanocrystalline (NC) state of an austenitic steel. For each material the right curve represents a low stress-ratio and the left one a high stress-ratio.

By decreasing the grain size of the material the threshold for the onset of crack propagation shifts to lower values (Fig. 1 right). This is mainly related to the reduction of roughness induced crack closure, which is also reflected by typical fractographs (Fig. 1 left). In the Paris-regime the fatigue crack growth rates are higher in the NC-state, even though roughness induced crack closure should play only a minor role. The change of the crack path from transgranular in the MC-state to intergranular in the NC-state seems to be crucial (Fig. 1 right). The intrinsic fatigue crack growth resistance seems to be lower for cracks propagating along the grain boundaries compared to propagation within the grains.

In order to mitigate these tendencies there are two main strategies. First, in the case of NC-metals with very small grain size heat-treatments can be used to further increase the strength of the material leading to a reduction of the fatigue crack growth rate [2]. Second, most SPD-processes deliver aligned and elongated microstructures and introduce a pronounced anisotropy in the fracture and fatigue properties [3]. By tailoring the grain shape or by choosing an appropriate orientation of the sample or of a component with respect to the alignment of the structure the resistance against fatigue crack propagation can be optimized.

References

1. Estrin, Y., Vinogradov, A.: Extreme grain refinement by severe plastic deformation: A wealth of challenging science. Acta Mater. **61**, 782–817 (2013)
2. Leitner, T., Pillmeier, S., Kormout, K.S., Pippan, R., Hohenwarter, A.: Simultaneous enhancement of strength and fatigue crack growth behavior of nanocrystalline steels by annealing. Scripta Mater. **139**, 39–43 (2017)
3. Leitner, T., Hohenwarter, A., Ochensberger, W., Pippan, R.: Fatigue crack growth anisotropy in ultrafine-grained iron. Acta Mater. **126**, 154–165 (2017)

Factors Affecting Temperature Rise in Shear Bands in a Simulated CuZr Metallic Glass

Chunguang Tang, Wanqiang Xu, Jiaojiao Yi, and Michael Ferry$^{(\boxtimes)}$

School of Materials Science and Engineering,
The University of New South Wales, Sydney 2052, Australia
m.ferry@unsw.edu.au

Abstract. The temperature rise in shear bands is of significant importance for the mechanical behaviour of metallic glasses since it changes their atomic structure and viscosity. However, experimental measurement of any temperature rise of a shear band is difficult, due to their temporal and spatial localization within the bulk. Molecular dynamics simulations were carried out on a CuZr metallic glass under tensile loading. It is shown that the observed temperature rise in a shear band, ranging from ~ 25 K up to the melting point of the alloy, correlates linearly with the maximum sliding velocity of the shear band, which is a function of both sample size and loading rate. In response to the high energy flux into the shear band, shear band bifurcation occurs and hinders further temperature rise well above the melting point. This negative feedback mechanism imposes an upper limit of temperature rise in a shear band before the theoretical limit of shear velocity is reached.

Keywords: Metallic glass · Shear band · Plastic deformation
Temperature rise

1 Introduction

Metallic glasses are promising structural materials due to their ultrahigh strength and other desirable properties, such as light weight and high wear and corrosion resistance. Different from their ductile crystalline alloys, metallic glasses localize the plastic deformation into one or a few major shear bands that propagate in an autocatalytic manner and result in catastrophic failure shortly after plastic deformation.

Due to high elastic limit of metallic glasses, the large strain energy can be stored during the elastic deformation stage [1]. The sudden release of the strain energy could result in substantial temperature rise in the shear band, which changes the structure and viscosity of the material and, hence, is of great importance to the mechanical behavior. Nevertheless, direct experimental measurement of the temperature rise of a shear band is difficult due to its temporal and spatial localization (in the scales of ns-µs and nm), and indirect measurements of surface temperature by infrared imaging techniques reveal only minor temperature rise (<30 K) although a light spectrum analysis indicates temperature rise more than 1000 K for impact test. The low temporal and spatial resolutions (ms and µm) of these techniques, however, make necessary the estimation of temperature within the shear band based on a series of assumptions, such as

© Springer International Publishing AG, part of Springer Nature 2019
E. E. Gdoutos (Ed.): ICTAEM 2018, SI 5, pp. 350–351, 2019.
https://doi.org/10.1007/978-3-319-91989-8_80

adiabatic approximation or the thin-film solution of the heat diffusion equation, which in some cases lead to unrealistic predictions like ~ 10000 K.

A revolutionary progress was made by Lewandowski and Greer [2] who obtained temporal and spatial resolutions around 30 ps and 100 nm via a fusible coating method. The authors estimated that the temperature rise at the shear band center could reach 3100–8300 K, depending on the propagation velocity of the band. However, later these authors proposed a lower temperature range based on different interpretation of the velocity of the band (Fig. 1).

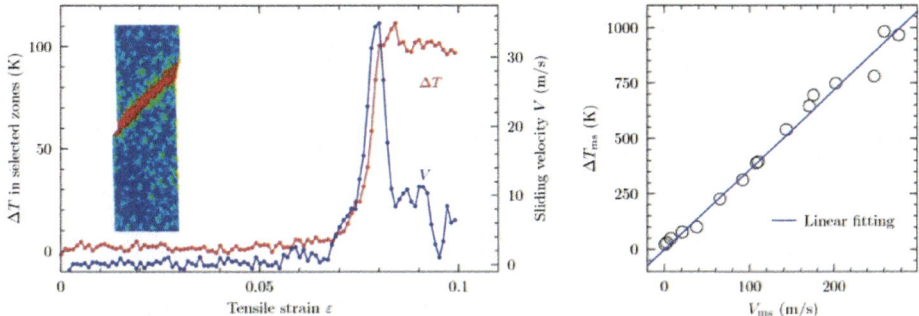

Fig. 1. Left: A shear band (in red) formed in metallic glass $Cu_{50}Zr_{50}$ of size $\sim 2 \times 20 \times 60$ nm^3 under loading rate of 10^8/s, and its temperature rise and sliding velocity as a function of tensile strain. Right: Linear relationship between temperature rise and sliding velocity in samples of various sizes and under various rates of tensile loading.

2   Results

By performing molecular dynamics studies on $Cu_{50}Zr_{50}$ metallic glass, we have examined the effect of tensile loading rate and sample size on the temperature rise in shear bands. We found that the popular adiabatic assumption to be untenable. A linear correlation between sliding velocity and the rise in temperature in the band was observed up to the melting point of the sample, and a negative feedback mechanism via shear band bifurcation preventing further temperature rise to well above the melting point of the alloy, which is different from previous conjectures.

References

1. Greer, A.L., Cheng, Y.Q., Ma, E.: Shear bands in metallic glasses. Mater. Sci. Eng., R **74**, 71–132 (2013)
2. Lewandowski, J.J., Greer, A.L.: Temperature rise at shear bands in metallic glasses. Nat. Mater. **5**, 15–18 (2006)

Structure Modulation and Nanocrystallization of Metallic Glasses: How to Tune Mechanical Properties

J. Eckert[1,2(✉)]

[1] Erich Schmid Institute of Materials Science, Austrian Academy of Sciences,
Jahnstraße 12, 8700 Leoben, Austria
juergen.eckert@unileoben.ac.at,
juergen.eckert@oeaw.ac.t
[2] Department Materials Physics, Montanuniversität Leoben,
Jahnstraße 12, 8700 Leoben, Austria

Abstract. Metallic glasses are known for their outstanding mechanical strength but limited plasticity. Significant progress has been made in recent years in how to optimize processing conditions for bulk glass formation, net-shape forming and the microscopic mechanism of failure. However, the details of the correlation between atomic structure, defects and thermo-mechanical treatments utilized for structure modification and their impact on shear band nucleation and propagation for achieving macroscopic ductility are still not well understood.

This talk attempts to shed light on structural (re)ordering, recovery and rejuvenation mechanisms, as well as nanocrystallization phenomena in different metallic glasses when they are subjected to different casting conditions, relaxation or thermoplastic net shaping. The findings will be discussed with respect to short- and medium-range order modulation, defect generation and annihilation, and precipitation of secondary phases. The structural changes will be correlated with changes in plastic deformability and failure mechanisms, and the effectiveness of composition tuning and thermo-mechanical processing for plasticity improvement will be analyzed in order to derive design aspects and processing guidelines for property optimization of metallic glasses.

Keywords: Metallic glasses · Nanocrystallization · Mechanical properties

1 Structure Modulation and Mechanical Properties

1.1 Motivation and Background

Metallic glasses (MGs) are an exciting class of materials due to their unique properties, such as high strength and good wear resistance [1]. A major drawback is their limited ductility caused by the formation of shear bands leading to catastrophic failure. While progress in the understanding of early stages of deformation localization was made using colloidal solids or molecular dynamics simulations [2], an experimental understanding is still lacking. The reason is the disordered nature of MGs inhibiting direct imaging of the fundamental deformation mechanisms. Another drawback is that the

© Springer International Publishing AG, part of Springer Nature 2019
E. E. Gdoutos (Ed.): ICTAEM 2018, SI 5, pp. 352–353, 2019.
https://doi.org/10.1007/978-3-319-91989-8_81

structure and properties of MGs strongly depend on their processing history [3] and thus the atomic structure and the resulting properties can be largely tuned through thermo-mechanical treatments utilized for structure modification [4, 5].

1.2 Influence of Thermal and Mechanical History on Properties of MGs

The influence of structural (re)ordering, recovery and rejuvenation mechanisms, as well as nanocrystallization phenomena has been tested for a variety of metallic glasses with different composition. The MGs were subjected to different casting conditions, as well relaxation treatment or thermoplastic net shaping in the supercooled liquid regime. These different treatments induce changes in the short- and medium-range order of the glass or may even lead to precipitation of secondary phases via crystallization at elevated temperatures. The structural changes are correlated with changes in plastic deformability and failure mechanisms, as revealed by detailed structure investigations and mechanical testing. Recent results on the effectiveness of composition tuning, microalloying and thermo-mechanical processing for plasticity improvement suggest that possible plasticity improvement is closely correlated with structural heterogeneities and the behavior of different structural motifs under localized stress and strain fields.

References

1. Ashby, M.F., Greer, A.L.: Metallic glasses as structural materials. Scripta Mater. **54**, 321–326 (2006)
2. Cubuk, E.D., et al.: Structure-property relationships from universal signatures of plasticity in disordered solids. Science **358**, 1033–1037 (2017)
3. Scudino, S., Shakur Shahabi, H., Stoica, M., Kaban, I., Escher, B., Kühn, U., Vaughan, G.B. M., Eckert, J.: Structural features of plastic deformation in bulk metallic glasses. Appl. Phys. Lett. **106**, 1–4 (2015). 031903
4. Sarac, B., Zhang, L., Kosiba, K., Pauly, S., Stoica, M., Eckert, J.: Towards the better: Intrinsic property amelioration in bulk metallic glasses. Sci. Rep. **6**, 1–8 (2016). 27271
5. Bian, X.L., Wang, G., Yi, J., Jia, Y.D., Bednarcik, J., Zhai, Q.J., Kaban, I., Sarac, B., Mühlbacher, M., Spieckermann, F., Keckes, J., Eckert, J.: Atomic origin for rejuvenation of a Zr-based metallic glass at cryogenic temperature. J. Alloy. Compd. **718**, 254–259 (2017)

The Size Effect of Spherical Indenter and Creep Behavior in Amorphous Alloy

Kang Sun, Gang Wang$^{(\boxtimes)}$, and Qing Wang

Shanghai University, Shanghai 200444, China
g.wang@shu.edu.cn

Abstract. We have studied the influence of different loading rate and different radius of spherical indenter on the nano-indentation creep process of {(Ce0.2La0.8)0.78Ni0.22}75Al25 at room temperature and describing the creep behavior by using elastic-viscoelastic principle. With the increase of loading rate, the strain rate sensitivity increases, due to the activation of multiple shear zones at higher loading rate, causing the materials becoming soften. Comparing the parameters between the spherical indenter of 2 μm and 5 μm, the values under 2 μm are much bigger, because large size of STZ under 2 μm enhance shear capacity of BMG, promoting the occurrence of multiple shear zone, increasing the ability of plastic deformation, leading a larger displacement. These results highlight the fact the important parameters getting from the creep function are all have different degrees of loading rate sensitivity and compared the data of different radius of spherical indenter, concluding "the smaller, the stronger".

Keywords: Amorphous alloy · Spherical indenter · Creep · Size effect

1 Experimental Procedure

The master alloy, {(Ce0.2La0.8)0.78Ni0.22}75Al25, was prepared by arc-melting a mixture of pure metal elements in a titanium-guttered high-purity argon atmosphere. Using XRD and DSC to characterize the amorphous structure and basic parameters, then doing nano-indentation experiments. Loading into 8 mN with different loading rate (80 mN/s, 8 mN/s and 0.8 mN/s), keeping 30 s at the maximum load, then unloading at the rate of 8 mN/s to 10% in the maximum, and constant 60 s in order to eliminate the influence of thermal drift, finally completely discharge. Exploring the size effect of spherical indenter on Ce-based metallic glass, different size of spherical indenter was used in nano-indentation test which were 2 μm and 5 μm.

2 Results and Discussion

From load-displacement curve, we can see, the width of the platform has obvious loading rate sensitivity, namely the greater loading rate leads to the width of the platform, in other words, the sample under lower loading rate shows better creep resistance. And the indenter did not return to the starting point after completely

© Springer International Publishing AG, part of Springer Nature 2019
E. E. Gdoutos (Ed.): ICTAEM 2018, SI 5, pp. 354–355, 2019.
https://doi.org/10.1007/978-3-319-91989-8_82

uninstall, indicating permanent plastic deformation occurred during the process of nano-indentation.

Through the data from the creep steady stage of nano-indentation, we can evaluate the strain rate sensitivity coefficient m, which is an important parameter of the material when discussing the time dependence deformation. According to Johnson's extension point mode [1] and the calculation of Bower [2], during the nano-indentation creep steady stage, the indentation hardness H and the indentation stress rate $\dot{\varepsilon}$ exist the power-law relationship:

$$H = C_1(\dot{\varepsilon})^m$$

If m = 1, corresponding to the Newtonian rheological model; if m < 1, corresponding to non-Newtonian rheological model. The value of m equal 0.079, 0.039, 0.028 correspond to the loading rate 80 mN/s, 8 mN/s and 0.8 mN/s, indicating the nano-indentation creep behavior at room temperature is inhomogeneous.

Oyen [3] sets up a nano-indentation experiment method, analyzing the mechanical behavior of spherical indentation. By using the elastic- viscoelastic principle elicit the integral solution of creep, comparing the one step creep solution, putting forward the concept of "slope correction factor" [4], using this solution can get the mechanical deformation's parameters and time constant. slope correction factor increases obviously from 1.1315 at 80 mN/s to 523.89 at 0.8 mN/s, indicating at lower loading rate, the slope correction factor should not be neglected.

When exploring the influence of size effect, from Johnson and Samwer formula [5], we can calculate the volume of STZ changes in the range of 3.7 (5 μm) to 11.9 nm^3 (2 μm) during the creep process, the number of atoms approximately increase from 170 (5 μm) to 541 (2 μm). We also get the corresponding delay spectrum, the sharp peak in the delay position indicates the material under 2 μm has more relaxation state with 5 μm indenter, and the wide peak suggests there is a distrution of free activation energy in the shear transformation process.

The nano-indentation creep behavior of {(Ce0.2La0.8)0.78Ni0.22}75Al25 under spherical indenter has obvious size effect, compared with the radius of 5 μm, 2 μm radius of the spherical head corresponding to relatively larger indentation hardness, bigger size of shear transition zone, great creep compliance and more remarkable relation state, which is called "the smaller the stronger".

References

1. Johnson, K.L.: Contact Mechanics. Cambridge University Press, Cambridge (1985)
2. Bower, A.F., Fleck, N.A., Needleman, A., et al.: Indentation of a power law creeping solid. Proc. R. Soc. Lond. Ser. A Math. Phys. Sci. **441**(1911), 97–124 (1993)
3. Oyen, M.L.: Spherical indentation creep following ramp loading. J. Mater. Res. **20**(8), 2094–2100 (2005)
4. Oyen, M.L.: Mater. Res. Soc. Proc. **841**, 211–216 (2005)
5. Mayr, S.G.: Activation energy of shear transformation zones: a key for understanding rheology of glasses and liquids. Phys. Rev. Lett. **97**(19), 195501 (2006)

Nanodiffraction Strain Mapping of Metallic Glasses During *In Situ* Deformation

Christoph Gammer[1(✉)], Thomas C. Pekin[2,3], Colin Ophus[2],
Andrew M. Minor[2,3], and Jürgen Eckert[1,4]

[1] Erich Schmid Institute of Materials Science, Austrian Academy of Sciences,
Jahnstraße 12, 8700 Leoben, Austria
christoph.gammer@oeaw.ac.at
[2] National Center for Electron Microscopy, Molecular Foundry,
Lawrence Berkeley National Laboratory, Berkeley, CA 94720, USA
[3] Department of Materials Science and Engineering,
University of California, Berkeley, CA 94720, USA
[4] Department Materials Physics, Montanuniversität Leoben,
Jahnstraße 12, 8700 Leoben, Austria

Abstract. Bulk metallic glasses are an exciting class of materials. Their mechanical properties are fundamentally different from their crystalline counterparts, due to the disordered structure. An experimental understanding of the fundamental deformation mechanisms of metallic glasses at the nanoscale is still lacking. Therefore, *in situ* deformation is carried out inside a transmission electron microscope. Scanning nanobeam electron diffraction is used to map the local elastic strain during deformation. The strain maps are determined by fitting an ellipse to the first order diffraction ring at every probe position. A direct electron detector enables acquiring the diffraction patterns at a sufficient speed to perform strain mapping during continuous *in situ* deformation.

Keywords: Metallic glass · *In situ* deformation
Transmission electron microscope (TEM)

Bulk metallic glasses are an exciting new class of materials due to their unique mechanical properties, such as high strength and good wear resistance [1]. However, potential applications are hindered by their low ductility caused by the formation of shear bands leading to catastrophic failure. While progress in the understanding of deformation localization was made using colloidal solids or molecular dynamics simulations [2], an experimental understanding of the fundamental deformation mechanisms in metallic glasses is still lacking due to their disordered nature inhibiting direct imaging. Therefore, to study the deformation of metallic glasses at the nanoscale, *in situ* deformation is carried out inside a transmission electron microscope (TEM) using a Hysitron PI-95 Picoindenter. Samples for *in situ* deformation are made from a bulk CuZrAlAg rod produced by suction-casting using focused ion beam machining. In addition to compression tests, tension tests of the specimens are carried out.

Figure 1 shows the result from an *in situ* tensile test acquired in TEM bright-field mode. The sample is 200 nm thick. The corresponding load displacement curve shows elastic deformation followed by abrupt fracture with no indication of plasticity.

© Springer International Publishing AG, part of Springer Nature 2019
E. E. Gdoutos (Ed.): ICTAEM 2018, SI 5, pp. 356–357, 2019.
https://doi.org/10.1007/978-3-319-91989-8_83

In addition, we will demonstrate that the local elastic strain can be measured during *in situ* deformation with nanometer resolution. Our method is based on scanning nanobeam electron diffraction [3]. While the electron beam scans over the sample, a full diffraction pattern is recorded for every probe position. By fitting an ellipse to the first order diffraction ring a strain map can be determined. A Gatan K2 IS direct electron detector is used to acquire the diffraction patterns at a rate of 400 frames/s, allowing to perform strain mapping during continuous deformation without stopping or pausing the experiment. The resulting strain maps reveal significant inhomogeneities showing the importance of measuring the local transient strain field during deformation.

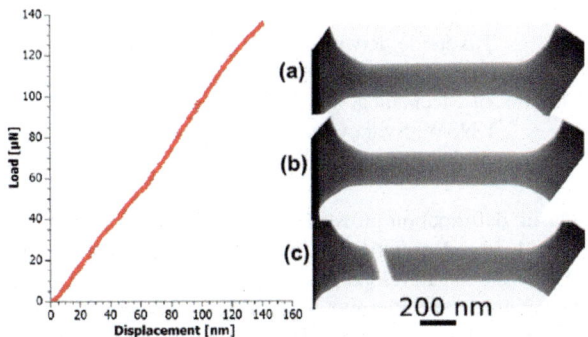

Fig. 1. The stress-strain curve recorded during *in situ* TEM deformation of a metallic glass shows linear elastic elongation of the tensile specimen flowed by fracture. Two frames corresponding to the initial state and the elongated state are shown in (a) and (b). After elongation the sample fractures abruptly and shows no ductility (c).

Acknowledgements. The authors acknowledge support of the European Research Council under the ERC Advanced Grant INTELHYB (grant ERC-2013-ADG-340025). Work at the Molecular Foundry was supported by the Office of Science, Office of Basic Energy Sciences of the U.S. Department of Energy under Contract No. DE-AC02-05CH11231.

References

1. Ashby, M.F., Greer, A.L.: Metallic glasses as structural materials. Scripta Mater. **54**, 321 (2006)
2. Cubuk, E.D., et al.: Structure-property relationships from universal signatures of plasticity in disordered solids. Science **358**, 1033–1037 (2017)
3. Gammer, C., Kacher, J., Czarnik, C., Warren, O.L., Ciston, J., Minor, A.M.: Local and transient nanoscale strain mapping during in situ deformation. Appl. Phys. Lett. **109**, 081906 (2016)

Coarse Grained Models of Heterogeneous Plasticity Using the Discrete Element Method—Bulk Metallic Glasses and Beyond

Agnieszka Truszkowska[1], P. Alex Greaney[2(⊠)], T. Matthew Evans[1],
and Jamie J. Kruzic[3]

[1] School of Civil and Construction Engineering,
Oregon State University, Corvallis, OR 97331, USA
[2] Mechanical Engineering Department,
Materials Science and Engineering Program, University of California—
Riverside, Riverside, CA 92501, USA
greaney@ucr.edu
[3] School of Mechanical and Manufacturing Engineering,
UNSW, Sydney, NSW 2052, Australia

Abstract. Plastic deformation proceeds through a sequence of stochastic local slip followed by load redistribution. With continued deformation this builds up complex stress fields and develops a heterogeneous pat- tern of local strength, leading to the emergence of microvoids and cracks. The goal of this research is to develop a coarse grained model for crystal plasticity that captures the physics emergent form stochastic heterogeneous deformation. The method based on the discrete element method (DEM), an approach developed for modeling of granular materials and recently adapted for amorphous brittle solids. DEM models the material as a collection of interacting elements. The framework naturally captures the elastic coupling due to geometric frustration in a system under heterogeneous deformation and the emergent phenomena that develop from it. This paper presents the accomplishment of intermediate steps towards modeling crystal plasticity: modeling anisotropic elasticity, and modeling iso- tropic plasticity.

Keywords: Computational mechanics · Discrete element method
Anisotropic elasticity · Coarse grained plasticity

1 Introduction

The discrete element method (DEM) is technique developed for modeling granular materials such as sand and soil [1], but it was quickly extended to model bound materials such as packstone and other rocks. In recent years it has been considerable success in applying DEM for coarse grained simulation of fracture and damage accumulation in ceramics, glasses, and composites like nacer. The long term goal of our research is to further extend DEM to be able to efficiently model crystal plasticity. Here we will report on two intermediate steps towards this goal: (1) modeling cubically anisotropic elasticity in DEM, and (2) modeling isotropic plasticity.

© Springer International Publishing AG, part of Springer Nature 2019
E. E. Gdoutos (Ed.): ICTAEM 2018, SI 5, pp. 358–359, 2019.
https://doi.org/10.1007/978-3-319-91989-8_84

2 Results

In DEM, the mechanical properties of elements are not modeled explicitly, but are captured through contact laws or "bonds" describing the interaction between elements. To generate a cubically anisotropic elastic response from an isotropic random packing of elements a set of orientation depended bond stiffness function were developed that possess cubic symmetry. Figure 1(a) shows the domain of elasticity accessible to the model revealed a random sampling of the model parameter space. The model can produce elasticity with anisotropy ratio either greater or smaller than 1. To model plasticity a set of bond deformation laws were developed which allow elements to slip twist and roll past one an- other. These were first used to model non-hardening elasticity observed in bulk metallic glasses, and extended model plasticity with work hardening as shown in Fig. 1(b). These models of plasticity exhibited emergent behavior of strain localization and necking that is seen in these materials.

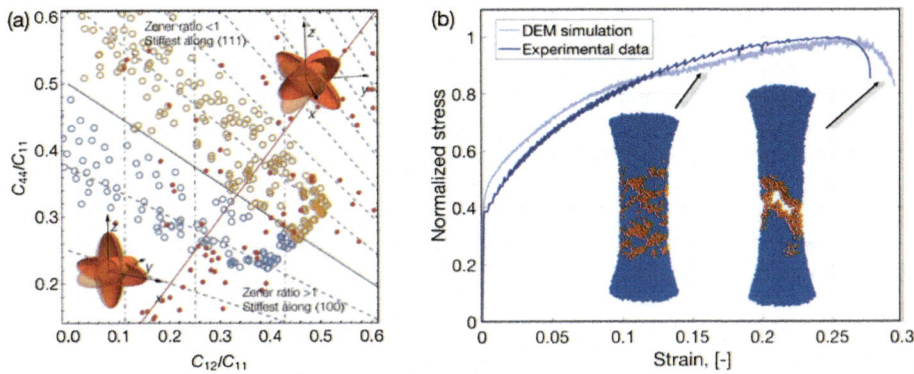

Fig. 1. (a) DEM model of cubic elasticity using angular dependent contact stiffnesses: a random sampling of model input parameters reveals the domain of elasticity space accessible to the model. The stiffness distributions are inset. The elastic moduli for known cubic materials are shown in red. (b) DEM model of plasticity with hardening—the model has been calibrated to match the tensile testing stress-strain curve for Ni20Cr; the model captures shear banding (orange elements) and fails with a correct fracture angle.

3 Conclusions

This work represent the first attempt of adapting DEM to model metallic materials. The results for both amorphous and crystalline metals are promising.

Reference

1. Cundall, P.A., Strack, O.D.L.: A discrete numerical model for granular assemblies. Geotechnique **29**(1), 47–65 (1979)

Rejuvenation of Disorder-Containing Materials

Glenn H. Balbus[1], McLean P. Echlin[1], Charlette M. Grigorian[2],
Timothy J. Rupert[2], Tresa M. Pollock[1], and Daniel S. Gianola[1(✉)]

[1] University of California Santa Barbara, Santa Barbara, CA 93106, USA
gianola@ucsb.edu
[2] University of California Irvine, Irvine, CA 93106, USA

Abstract. Here, we report on experimental studies of metallic glass and
nanocrystalline materials and novel synthesis and processing routes for controlling the structural state – and as a consequence, the mechanical properties.
A particular focus will be on strategies for rejuvenation of disorder with the goal
of suppressing shear localization and endowing damage tolerance. We also
describe a microscopic structural quantity designed by machine learning to be
maximally predictive of plastic rearrangements and further demonstrate a causal
link between this measure and both the size of rearrangements and the macroscopic yield strain. We find remarkable commonality in all of these quantities in
disordered materials with vastly different inter-particle interactions and spanning
a large range of elastic modulus and particle size.

Keywords: Rejuvenation · Disordered materials · Nanocrystalline materials
Softness

1 Background

The nonequilibrium nature of kinetically frozen solids such as metallic glasses
(MGs) is at once responsible for their unusual properties, complex and cooperative
deformation mechanisms, and their ability to explore various metastable states in the
rugged potential energy landscape. These features coupled with the presence of a glass
transition temperature, above which the solid flows like a supercooled liquid, open the
door to thermoplastic forming operations at low thermal budget as well as thermo-
mechanical treatments that can either age (structurally relax) or rejuvenate the glass.
Thus, glasses can exist in various structural states depending on their synthesis method
and thermomechanical history. Nanocrystalline (NC) metals, also considered to be
far-from-equilibrium materials owing to the large fraction of atoms residing near grain
boundaries (GBs), share many commonalities with MGs both in terms of plastic
deformation and its dependence on processing history. Despite these similarities, the
disorder intrinsic to both classes of materials has precluded the development of
structure-property relationships that can capture the multiplicity of energetic states that
glasses and GBs may possess.

We describe and use a microscopic structural quantity designed by machine
learning to be maximally predictive of plastic rearrangements and further demonstrate a

© Springer International Publishing AG, part of Springer Nature 2019
E. E. Gdoutos (Ed.): ICTAEM 2018, SI 5, pp. 360–361, 2019.
https://doi.org/10.1007/978-3-319-91989-8_85

causal link between this measure and both the size of rearrangements and the macroscopic yield strain [1]. We find remarkable commonality in all of these quantities in disordered materials with vastly different inter-particle interactions and spanning a large range of elastic modulus and particle size. We further explore the commonalities between disorder-containing NC and MG materials by using femtosecond laser processing as a unique non-equilibrium process that can generate complex stress states due to ultrafast electronic excitation and subsequent relaxation events. Experiments on NC Al-O and Cu-Zr alloys indicate that sub-ablation femtosecond laser pulses cause a dramatic reduction in hardness accompanied by negligible changes in grain size. Parallels between our results and rejuvenation processes in glassy systems will be discussed in the context of controlling metastable structural configurations through novel processing routes (Fig. 1).

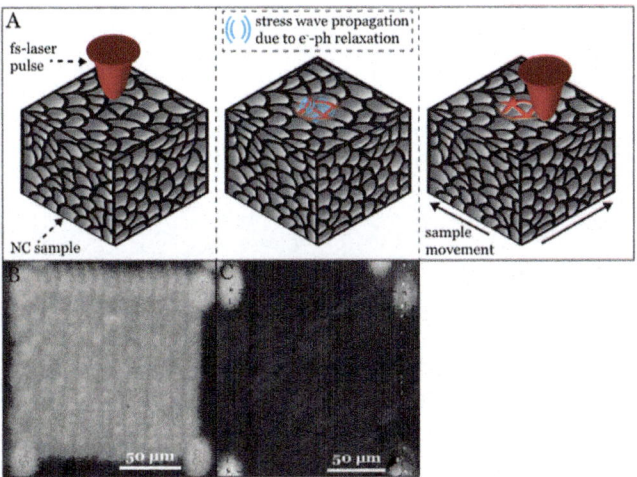

Fig. 1. (A) Schematic showing the fs-laser treatment procedure. A single fs-pulse irradiates the NC sample, which then emits elastic waves. Once this process has completed, the sample is translated and irradiated with a series of pulses until the desired size is reached. (B) SEM image of the Al-4.8 at %O sample irradiated with a fluence above the ablation threshold. The four lighter areas on the corners of the square are fiducial marks. (C) SEM image of the Al-4.8 at %O sample irradiated with a fluence below the ablation threshold.

Reference

1. Cubuk, E.D., Ivancic, R.J.S., Schoenholz, S.S., Strickland, D.J., Basu, A., Davidson, Z.S., Fontaine, J., Hor, J.L., Huang, Y.-R., Jiang, Y., Keim, N.C., Koshigan, K.D., Lefever, J.A., Liu, T., Ma, X.-G., Magagnosc, D.J., Morrow, E., Ortiz, C.P., Rieser, J.M., Shavit, A., Still, T., Xu, Y., Zhang, Y., Nordstrom, K.N., Arratia, P.E., Carpick, R.W., Durian, D.J., Fakhraai, Z., Jerolmack, D.J., Lee, D., Li, J., Riggleman, R., Turner, K.T., Yodh, A.G., Gianola, D.S., Liu, A.J.: Structure-property relationships from universal signatures of plasticity in disordered solids. Science **358**, 1033 (2017)

Symposium on: "Dynamic Response of Elastic and Viscoelastic Solids Elastostatic and Elastodynamic Problems for Thermosensitive and Nonhomogeneous Solids Dynamic Problems in Mechanics of Coupled Fields," by Roman Kushnir

Determination of Vibrations at Different Points in the Ground After the Passage of a Spherical Wave Through a Vibration-Absorbing Obstacle

N. A. Lokteva[1(✉)] and D. V. Tarlakovskii[2]

[1] Moscow Aviation Institute (National Research University),
A-80, GSP-3, Volokolamskoye shosse 4, Moscow 125993, Russia
nlok@rambler.ru
[2] Institute of Mechanics Lomonosov Moscow State University,
Michurinski prospekt 1, Moscow 119192, Russia

Abstract. The vibration-absorbing properties of the plate under the action of a spherical harmonic wave in the soil are studied. In the soil model, an elastic isotropic medium is used. The main goal is to determine the total vector field of accelerations. The mathematical formulation of the problem includes the assignment of the incident wave, the equations of motion of the soil and the plates, the boundary conditions for the slab and the soil, the conditions at infinity, and the conditions of contact of the earth with the obstacle, where we neglect the connection of the plate to the ground. The motion of the plate is described by the system of equations of Paimushin V.N. The kinematic parameters of the plate and the parameters of the disturbed stress-strain state of the soil are represented in the form of double trigonometric series satisfying the boundary conditions. After that, the constants of integration, displacement and vibration acceleration are determined.

Keywords: Vibration absorption · Three-layer vibration-damping obstacle
Elastic medium

1 Introduction

In current times buildings and facilities are exposed to various external negative effects arising from utility equipment, industrial machinery, as well as transportation means (such as shallow-depth city railroad systems, heavy trucks, railway trains, trams) that cause huge dynamic loads [1]. Vibration-absorbing barriers placed between the vibration source and the object to be protected is one of the ways of protection of foundations against ground vibrations [2, 3].

This paper is intended to study vibration-absorbing properties of a plate exposed to subsurface spherical harmonic waves. In practice, such situation may correspond to vibrations caused by a point source situated nearby the barrier.

© Springer International Publishing AG, part of Springer Nature 2019
E. E. Gdoutos (Ed.): ICTAEM 2018, SI 5, pp. 365–370, 2019.
https://doi.org/10.1007/978-3-319-91989-8_86

2 Statement of Problem

Let us consider an elastic plate surrounded from both sides by half-spaces "1" and "2" filled with the ground. The coordinate system $Oxyz$ is of Cartesian type. It is assumed that the plane Oxy for the plate is a median one and limited along axes Ox and Oy, having the length of l. The beginning of the coordinates is assumed to be situated in the upper right corner of the plate.

When undisturbed, the ground is considered to be undeformed. The obstacle is overrun by a harmonic tensile wave of an normal stress amplitude p_* at the front and frequency ω, coming from the negative Oz-axis. The normal vector towards the front of the wave is in the plane Oxy.

The main purpose of this study is finding the resultant vector field of acceleration $\mathbf{a}$ (vibration accelerations) in the second half-space as a function of the frequency ω and spatial coordinates x, y and z depending on the parameters of the plate. The stated purpose of the problem solution is refined as follows. It is necessary to find the coordinates of the vibration acceleration field

$$a_x = -\omega^2 u_1^{(2)}, \; a_y = -\omega^2 u_2^{(2)}, \; a_z = -\omega^2 w^{(2)}, \tag{1}$$

and the module

$$a = \sqrt{a_x^2 + a_y^2 + a_z^2}, \tag{2}$$

where $u_1^{(2)}, u_2^{(2)}$ are displacements of the medium "2" by coordinates x, y; $w^{(2)}$ is the normal displacement of the second medium.

The mathematical statement of the problem includes setting the ingoing wave, equations of displacements of the ground and plate, boundary conditions for the plate and ground, condition for infinity, as well as a condition of the ground-to-obstacle contact where the adhesion of the plate with the ground is ignored.

3 Equations of Displacements of Sandwich Plate

The displacement of the plate is described by a system of Paymushin V.N. equations [4]. Two bearing layers are of elastic isotropic type with the modulus of elasticity E, Poisson's ratio v, and thickness of $2t$. The filler is orthotropic of a honeycomb configuration with the elasticity modulus E_z, the Poisson's ratio v_z, and thickness of $2h$. The bearing layers are affected by normal outer stress loads p_1 and p_2. Tangential displacements are indicated by $u_1^{(k)}$ and $u_2^{(k)}$ along the axes Ox and Oy. $w^{(k)}$ is the normal displacement of the bearing layer, q^1 and q^2 are transverse tangent lines of the stress in the filler by axes Ox and Oy. The Paymushin V.N. equation system which describes the movement of the plate has the following form:

$$\rho_c \ddot{u}_1^c = L_{11}(u_1^c) + L_{12}(u_2^c), \ \rho_c \ddot{u}_2^c = L_{21}(u_1^c) + L_{22}(u_2^c),$$

$$\rho_a \ddot{u}_1^a = L_{11}(u_1^a) + L_{12}(u_2^a) + 2q^1, \ \rho_a \ddot{u}_2^a = L_{21}(u_1^a) + L_{22}(u_2^a) + 2q^2,$$

$$\rho_c \ddot{w}_c - \underline{m_c \Delta \ddot{w}_c} + \underline{\rho_{wq}\left(\ddot{q}_{,x}^1 + \ddot{q}_{,y}^1\right)} = -D\Delta^2 w_c + 2k_1\left(q_{,x}^1 + q_{,y}^1\right) + p_1 - p_2,$$

$$\rho_{aw} \ddot{w}_a - \underline{m_a \Delta \ddot{w}_a} = -D\Delta^2 w_a - 2c_3 w_a + p_1 + p_2, \tag{3}$$

$$\rho_{q1} \ddot{q}^1 - \underline{\rho_{wq1} \ddot{w}_{c,x}} = u_1^a - k_1 w_{c,x} - k_2\left(q_{,x}^1 + q_{,y}^2\right)_{,x} + k_{31}q^1,$$

$$\rho_{q2} \ddot{q}^2 - \underline{\rho_{wq2} \ddot{w}_{c,y}} = u_2^a - k_1 w_{c,y} - k_2\left(q_{,x}^1 + q_{,y}^2\right)_{,y} + k_{32}q^2;$$

where $u_i^c = u_i^{(1)} + u_i^{(2)}, u_i^a = u_i^{(1)} - u_i^{(2)} \ (i = 1, 2), \ w_c = w^{(1)} + w^{(2)}, w_a = w^{(1)} - w.$

The boundary conditions correspond to a hinge edge of the plate. All functions change harmonically.

The plate's kinematic parameters are represented in the form of two-fold trigonometric series meeting the boundary conditions. The amplitudes of ingoing and passed waves are expanded into series in a similar way. The solution of an equation system results into finding the values of the kinematic parameters depending on amplitudes of wave pressure in mediums "1" and "2".

4   Equations of Ground Displacements and Ingoing Wave

A homogeneous elastic isotropic medium is used as a model of the ground. A closed equation system describing its displacement has the following form [5]:

– motion equation:

$$\rho \ddot{u}_1 = \frac{\partial \sigma_{11}}{\partial x} + \frac{\partial \sigma_{12}}{\partial y} + \frac{\partial \sigma_{13}}{\partial z}, \ \rho \ddot{u}_2 = \frac{\partial \sigma_{21}}{\partial x} + \frac{\partial \sigma_{22}}{\partial y} + \frac{\partial \sigma_{23}}{\partial z}, \ \rho \ddot{w}$$

$$= \frac{\partial \sigma_{31}}{\partial x} + \frac{\partial \sigma_{32}}{\partial y} + \frac{\partial \sigma_{33}}{\partial z}; \tag{4}$$

– Cauchy relations

$$\varepsilon_{11} = \frac{\partial u_1}{\partial x}, \ \varepsilon_{13} = \frac{1}{2}\left(\frac{\partial u_1}{\partial z} + \frac{\partial w}{\partial x}\right), \ \varepsilon_{33} = \frac{\partial w}{\partial z}, \ \varepsilon_{22} = \frac{\partial u_2}{\partial y}, \ \varepsilon_{23} = \frac{1}{2}\left(\frac{\partial u_2}{\partial y} + \frac{\partial w}{\partial z}\right),$$

$$\varepsilon_{12} = \frac{1}{2}\left(\frac{\partial u_2}{\partial x} + \frac{\partial u_1}{\partial y}\right), \ \theta = \frac{\partial u_1}{\partial x} + \frac{\partial u_2}{\partial y} + \frac{\partial w}{\partial z}; \tag{5}$$

– physical law

$$\sigma_{11} = \lambda\theta + 2\mu\varepsilon_{11}, \ \sigma_{13} = 2\mu\varepsilon_{13}, \ \sigma_{33} = \lambda\theta + 2\mu\varepsilon_{33},$$

$$\sigma_{22} = \lambda\theta + 2\mu\varepsilon_{22}, \ \sigma_{23} = 2\mu\varepsilon_{23}, \tag{6}$$

where: σ_{ij} and ε_{ij} are components of stress and deformation tensors; θ is the coefficient of volume expansion; ρ and λ, μ are density and elastic Lame constants of the ground; dots hereinafter mark time derivatives t.

The system (3)–(5) equals the displacement equations (Lame equations):

$$\rho\ddot{u}_1 = (\lambda+\mu)\frac{\partial\theta}{\partial x}+\mu\Delta u_1, \ \rho\ddot{u}_2 = (\lambda+\mu)\frac{\partial\theta}{\partial y}+\mu\Delta u_2,$$

$$\rho\ddot{w} = (\lambda+\mu)\frac{\partial\theta}{\partial z}+\mu\Delta w, \ \Delta = \frac{\partial^2}{\partial x^2}+\frac{\partial^2}{\partial y^2}+\frac{\partial^2}{\partial z^2}. \tag{7}$$

Similarly, the system (4)–(6) equals the equations with respect to scalar potential φ and vector potential $\boldsymbol{\psi} = (\psi_1, \psi_2, \psi_3)$ of elastic displacements:

$$\ddot{\varphi} = c_1^2\Delta\psi, \ \ddot{\boldsymbol{\psi}} = c_2^2\Delta\boldsymbol{\psi}, \ \mathrm{div}\boldsymbol{\psi} = 0, \ c_1^2 = \frac{\lambda+2\mu}{\rho}, \ c_2^2 = \frac{\mu}{\rho}. \tag{8}$$

Here c_1 and c_2 are velocities of stress-strain and shear waves.

$$u_1 = \frac{\partial\varphi}{\partial x}+\frac{\partial\varphi}{\partial y}-\frac{\partial\psi}{\partial z}, \ u_2 = \frac{\partial\varphi}{\partial y}+\frac{\partial\varphi}{\partial z}-\frac{\partial\psi}{\partial x}, w = \frac{\partial\varphi}{\partial z}+\frac{\partial\psi}{\partial x}-\frac{\partial\psi}{\partial y}. \tag{9}$$

Below we will consider only harmonic waves of frequency ω.

All values are also expanded into two-fold trigonometric series.

As the area occupied by the ground is boundless, the potentials of equation solutions (7) must satisfy the Sommerfeld radiation conditions:

$$\frac{\partial\varphi}{\partial r}+ik_1\varphi = o\left(\frac{1}{r}\right), \frac{\partial\psi}{\partial r}+ik_2\psi = o\left(\frac{1}{r}\right), \ r\to\infty \tag{10}$$

where $r = \sqrt{x^2+z^2+y^2}$ is the radius vector length.

To set an ingoing flat harmonic wave, a spherical wave is considered [5] which travels along the positive direction of Oz axis. In this case in (11) we assume $\varphi = \varphi(z)$, $\psi = \psi(z)$. As a result, we arrive at the below equation with respect to the potential amplitude:

$$\varphi_a'' + k_1^2\varphi_a = 0, \psi_a'' + k_2^2\psi_a = 0. \tag{11}$$

Its solution meets the corresponding condition (9). Hence, the potential has the form of a progressing wave:

$$\varphi = A_\varphi r_1^{-1}e^{-ik_1(z-c_1t)}, \ \psi = A_\psi r_1^{-1}e^{-ik_2(z-c_2t)}. \tag{12}$$

where A_φ, A_ψ are arbitrary constants, $r_1 = \sqrt{x^2+y^2+(z+L)^2}$.

By plugging this equality consequently into (8), (5) and (4) taking into account that $\sigma_{33}|_{t=0,\,z=0} = p_*$, we will obtain the formulae of amplitudes of displacements, deformations and stresses in the ingoing wave.

5 Boundary Problem of Interaction of Harmonic Wave with Plate in Ground

Hence, the boundary conditions with respect to coefficients of the series will take the following form:

– stress and strain at the boundaries with the media "1" and "2"

$$p_1 = (\sigma_{33*} + \sigma_{33})|_{z=0},\ p_2 = -\sigma_{33}^{(2)}\Big|_{z=0},\ \sigma_{31}^{(1)}\Big|_{z=0} = \sigma_{31}|_{z=0},\ \sigma_{32}^{(1)}\Big|_{z=0} = \sigma_{32}|_{z=0};\quad (13)$$

– normal and tangential displacements

$$\left(w^{(1)} + w_*\right)\Big|_{z=0} = w_0^{(1)}\Big|_{z=0},\ u_1^{(1)}\Big|_{z=0} = u_{1*}^{(1)}\Big|_{z=0},\ u_2^{(1)}\Big|_{z=0} = u_{2*}^{(1)}\Big|_{z=0}.\quad (14)$$

The values having asterisk * correspond to the values of functions in the ingoing wave.

To find the coefficients of the series corresponding to a disturbed stress-strained condition in the media, let us plug the potentials expanded into two-fold trigonometrical series into the Eq. (10):

$$\frac{\partial^2 \varphi_{nm}^{(l)}}{\partial z^2} + \text{sign}\big(k_1 - (\lambda_{1n}^2 + \lambda_{2m}^2)\big)\kappa_{1nm}^2(\omega^2)\varphi_{nm}^{(l)} = 0\,,$$

$$\frac{\partial^2 \psi_{inm}^{(l)}}{\partial z^2} + \text{sign}\big(k_2 - (\lambda_{1n}^2 + \lambda_{2m}^2)\big)\kappa_{2nm}^2(\omega^2)\psi_{inm}^{(l)} = 0\,,\quad (15)$$

$$\kappa_{lnm}(\omega^2) = \kappa_l\big(\lambda_{1n}^2, \lambda_{2m}^2, \omega^2\big) = \sqrt{\left|k_l^2 - \left(\lambda_{1n}^2 + \lambda_{2m}^2\right)^2\right|}\,.$$

Their common solutions satisfying the Sommerfeld conditions are written in details as follows:

$$\varphi_{nm}^{(l)}(z,\omega) = C_1^{(l)}(\omega)\Big[e^{i\kappa_{1nm}(\omega^2)z}H(k_1 - (\lambda_{1n} + \lambda_{2m})) + e^{\kappa_{1nm}(\omega^2)z}H((\lambda_{1n} + \lambda_{2m}) - k_1)\Big],$$

$$\psi_{qnm}^{(l)}(z,\omega) = C_{2q}^{(l)}(\omega)\Big[e^{i\kappa_{2nm}(\omega^2)z}H(k_2 - (\lambda_{1n} + \lambda_{2m})) + e^{\kappa_{2nm}(\omega^2)z}H((\lambda_{1n} + \lambda_{2m}) - k_2)\Big].$$

$$(16)$$

Constant integrating $C_1^{(l)}(\omega), C_{2q}^{(l)}(\omega)$ where l is the number of the medium, q is numbers of axes Ox, Ox and Oz are found from boundary conditions (12, 13). By plugging the resultant constant values into expressions of displacements, we will obtain coefficients of expansions into displacements series in the medium "2":

Then, based on the formulae (1) and (2) it became possible to find the module of vibration acceleration and its components.

Example. Let us assume that $l_1 = 1м, l_2 = 1м$, with thickness of the filler $h = 0.1м$. The bearing layers are made of DIN17100-type steel with density of $\rho_0 = 7859кг / м^3$, Young's modulus of $E_0 = 2 \cdot 10^{11} MПa$, Poisson's ratio of $v_0 = 0,28$. Filler material is aluminum of Al-Mn type with density of $\rho_0 = 2730кг / м^3$, Young's modulus of $E_0 = 0,71 \cdot 10^{-5} MПa$, Poisson's ratio of $v_0 = 0,3$. The materials of the media "1" and "2" are assumed to be fill-up ground compacted under the humidity degree of 0.5, Young's modulus of $E = 10^8 MПa$, density of $\rho = 1600кг / м^3$ [6] (Fig. 1).

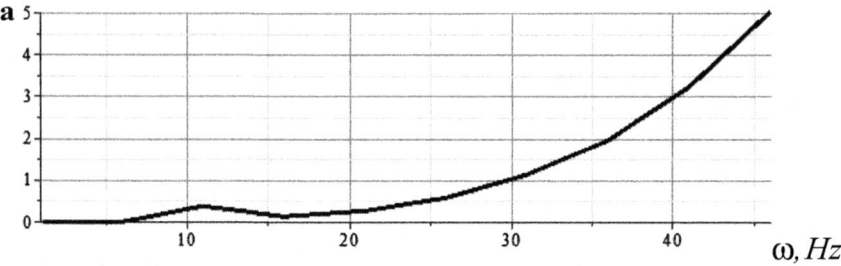

Fig. 1. Dependence of the acceleration field on the frequency of the incoming waves

References

1. Umek, A.: Dynamic responses of building foundations to incident elastic waves. Ph.D. thesis Illinois Institute of Technology, December 1973
2. Kostrov, B.V.: Movement of rigid solid plate sealed into elastic medium affected by flat wave. PMM **28**(1) (1964). (in Russian)
3. Rylko, M.Y.: On movement of rigid rectangular inclusions in elastic medium affected by flat wave. MTT No I (1977). (in Russian)
4. Ivanov, V.A., Paymushin, V.N.: Refined statement of dynamic problems of sandwich shells with transversal soft fillings and numerical analytical method of their solution. Prikladnaya mekhanika i tekhnicheskaya phisika (Appl. Mech. Phys.) **36**(4), 147–151 (1995). (in Russian)
5. Gorshkov, A.G., Medvedsky, A.L., Rabinsky, L.N., Tarlakovsky, D.V.: Waves in continuum media: textbook for Universities. FIAMATLIT, 472 p., Moscow (2004). (in Russian)
6. Code of design and construction SP 23-105-2004 Assessment of vibration for design and construction of underground railway systems. In: GOSSTROY ROSSII, Moscow (2014). (in Russian)

The Direct-Integration Method for 3D Elastic Analysis of Transversally-Isotropic Nonhomogeneous Solids

Yuriy Tokovyy[(⊠)]

Pidstryhach Institute for Applied Problems of Mechanics and Mathematics,
National Academy of Sciences of Ukraine, Lviv 79060, Ukraine
tokovyy@gmail.com

Abstract. A technique for the analysis of nonhomogeneous transversally-isotropic materials is suggested for the case of 3D elasticity formulation. By making use of the direct integration method, the formulated problems are reduced to a set of governing equations for the stress-tensor components. In order to construct the solutions to the latter equations in an explicit form, an advanced solution technique is developed on the basis of the resolvent-kernel method.

Keywords: Direct integration method · Nonhomogeneous solid
3D elasticity problem · Transversally-isotropic material

1 Introduction

1.1 Objectives and Motivation

The 3D elastic analysis of transversally-isotropic materials (TIM) presents a challenge due to the complexity of the governing system of partial-differential equations involving dissimilar elastic moduli [1]. Therefore, the considerable amount of literature has been published on particular classes of the elasticity problems implying certain simplifying assumptions, e.g., the theories of pates and shells, plane (stress/strain) problems, anti-plane problems, problems on the pure torsion, bending etc. [2, 3].

The implementation of methods involving potential functions for the TIM is often associated with both analytical and numerical complications. First of all, the analysis involves the derivatives of higher order in comparison to the original equations either in terms of stresses or displacements. Second, as a matter of fact, the coefficients or eigenvalues of the relevant auxiliary or spectral problems are expressed through the combinations of dissimilar material constants [4, 5]. The problem becomes even more involved when assuming the TIM solids to be spatially nonhomogeneous so that the material properties exhibit arbitrary variation from point to point within the solid.

1.2 Direct Integration Method

Herein, we employ the direct integration method (DIM) [6] in order to construct closed form analytical solutions to 3D elasticity problems for TIM nonhomogeneous

© Springer International Publishing AG, part of Springer Nature 2019
E. E. Gdoutos (Ed.): ICTAEM 2018, SI 5, pp. 371–376, 2019.
https://doi.org/10.1007/978-3-319-91989-8_87

composite solids subjected to external force loadings. In our previous works, this method has been efficiently implemented for the solution of a 3D elasticity problem for an exponentially inhomogeneous layer [7], and a 3D thermoelasticity problem [8].

The application of this method implies the following solution scheme. First, the so called key functions (key stresses) are selected out of the six stress-tensor components. Next, as a result of the integration of the equilibrium equations, the relations are set up which enable one to determine the remaining stress-tensor components in terms of the key functions. The set of boundary conditions imposed for the different stresses is equivalently reduced to the set of boundary and integral conditions for the key functions. On the basis of both the relations obtained and the compatibility equations in terms of strains along with the constitutive equations, the set of governing compatibility equations can be derived for the key stresses. By making use of an appropriate technique, the separation of variables in the derived governing equations can be performed and thereby the latter equations are solved in an explicit form by making use of the previously derived boundary and integral conditions for the key functions. After the key functions are found, the remaining stress-tensor components can be computed by making use of the relations derived from the equilibrium equations.

The advantages of our method are the following. The method is applied to the problem stated in terms of stresses. This does not cause the order increment in the governing differential equations, what is the case if one solves the problem in terms of displacements. Such a point is vital for numerical implementation. The solution has the form of an explicit functional dependence on the applied loadings, which is important for the analysis and further implementation of the solutions. The method deals with the feasible functions; it agrees with the fundamental principles of mechanics. The method can be applied for the analysis of TIM nonhomogeneous materials with arbitrary dependences of the material properties on the spatial coordinates.

2 Application of the DIM to the Solution of 3D Elasticity Problems for Nonhomogeneous TIM Solids

2.1 Formulation of the Problem

Consider a TIM solid related to the dimensionless Cartesian coordinate system $Oxyz$ so that the isotropy plane is parallel to xOy. Assume the solid to be in the state of elastic equilibrium 3D, which, in the absence of body forces, is governed by [1]:

– the equilibrium equations

$$\frac{\partial \sigma_{xx}}{\partial x} + \frac{\partial \sigma_{xy}}{\partial y} + \frac{\partial \sigma_{xz}}{\partial z} = 0, \quad \text{etc.,} \tag{1}$$

– the compatibility equations

$$\frac{\partial^2 \varepsilon_{xy}}{\partial x \partial y} = \frac{\partial^2 \varepsilon_{xx}}{\partial x^2} + \frac{\partial^2 \varepsilon_{yy}}{\partial y^2}, \quad \text{etc.,} \tag{2}$$

$$2\frac{\partial^2 \varepsilon_{xx}}{\partial y \partial z} = \frac{\partial}{\partial x}\left(\frac{\partial \varepsilon_{xy}}{\partial z} + \frac{\partial \varepsilon_{xz}}{\partial y} - \frac{\partial \varepsilon_{yz}}{\partial x}\right), \quad \text{etc.,} \tag{3}$$

– the constitutive equations

$$
\begin{aligned}
\varepsilon_{xx} &= \frac{\sigma_{xx} - \nu\sigma_{yy}}{E} - \frac{\nu'\sigma_{zz}}{E'}, & \varepsilon_{xz} &= \frac{\sigma_{xz}}{G'}, \\
\varepsilon_{yy} &= \frac{\sigma_{yy} - \nu\sigma_{xx}}{E} - \frac{\nu'\sigma_{zz}}{E'}, & \varepsilon_{yz} &= \frac{\sigma_{yz}}{G'}, \\
\varepsilon_{zz} &= \frac{\sigma_{zz}}{E'} - \nu'\frac{\sigma_{xx} + \sigma_{yy}}{E'}, & \varepsilon_{xy} &= \frac{\sigma_{xy}}{G}.
\end{aligned} \tag{4}
$$

Here, σ_{ij} and ε_{ij} are components of the stress- and strain-tensor $\hat{\sigma}$ and $\hat{\varepsilon}$, respectively, ν and ν' are the Poisson ratios characterizing the transverse contraction in response to the tension applied in the plane of isotropy and transversely to this plane, respectively, E and E' are the Young moduli in the plane of isotropy and in the transversal direction, $G = E/(2+2\nu)$ and G' are the shearing moduli in the plane of isotropy and in the transversal direction. Note that here and, if used, in formulae below, "etc." means two more equation of the same kind obtained by the cyclic permutation of indices i and j within the range $\{x \to y \to z \to x\}$. Assume the material properties to be the arbitrary functions of the coordinate z.

On the limiting boundary S, the solid is exposed to the static external loadings

$$\hat{\sigma} \cdot \vec{n} = \vec{f}, \tag{5}$$

where $\vec{n}$ is the outward normal vector of the surface S and $\vec{f}$ is the given vector of external forces.

2.2   Reduction to the Governing Equations

In order to solve the formulated problem (1)–(5), we employ the DIM as presented in [6–8]. Making use of Eq. (1), we can express the shearing stresses as

$$2\frac{\partial^2 \sigma_{xy}}{\partial x \partial y} = \frac{\partial^2 \sigma_{zz}}{\partial z^2} - \frac{\partial^2 \sigma_{xx}}{\partial x^2} - \frac{\partial^2 \sigma_{yy}}{\partial y^2}, \quad \text{etc.} \tag{6}$$

Putting (6) into (2) in view of Eqs. (1) and (4) yields, after some algebra, the governing equation

$$(1 + \nu(z))\Delta\sigma_{zz} + E(z)\beta_1(z)\Delta_{xy}\sigma_{zz} = \Delta_{xy}\sigma, \tag{7}$$

for the stresses σ_{zz} and

$$\sigma = \sigma_{xx} + \sigma_{yy} + \sigma_{zz}. \tag{8}$$

Here, $\Delta_{xy} = \frac{\partial^2}{\partial x^2} + \frac{\partial^2}{\partial y^2}$, $\Delta = \Delta_{xy} + \frac{\partial^2}{\partial z^2}$, and $\beta_1(z) = \frac{\nu'(z)E(z) - \nu(z)E'(z)}{E(z)E'(z)}$.

Similarly we obtain two more equations,

$$\Delta\left(\frac{\sigma_{xx}}{2G(z)}\right) + \beta_2(z)\Delta_{xy}\sigma_{xx}$$

$$= \frac{d}{dz}\left(\frac{1}{G'(z)}\right)\frac{\partial\sigma_{xz}}{\partial x} + \Delta(\beta_1(z)\sigma_{zz}) - \beta_2(z)\Delta\sigma_{zz} \qquad (9)$$

$$- \beta_3(z)\frac{\partial^2\sigma_{zz}}{\partial x^2} + \beta_5(z)\frac{\partial^2\sigma}{\partial y^2} + \beta_4(z)\frac{\partial^2\sigma}{\partial x^2} + \frac{\partial^2}{\partial z^2}\left(\frac{v(z)\sigma}{E(z)}\right)$$

and

$$\Delta\left(\frac{\sigma_{yy}}{2G(z)}\right) + \beta_2(z)\Delta_{xy}\sigma_{yy}$$

$$= \frac{d}{dz}\left(\frac{1}{G'(z)}\right)\frac{\partial\sigma_{yz}}{\partial y} + \Delta(\beta_1(z)\sigma_{zz}) - \beta_2(z)\Delta\sigma_{zz} \qquad (10)$$

$$- \beta_3(z)\frac{\partial^2\sigma_{zz}}{\partial y^2} + \beta_5(z)\frac{\partial^2\sigma}{\partial x^2} + \beta_4(z)\frac{\partial^2\sigma}{\partial y^2} + \frac{\partial^2}{\partial z^2}\left(\frac{v(z)\sigma}{E(z)}\right),$$

expressing the in-plane normal stresses σ_{xx} and σ_{yy} in terms of the stresses σ, σ_{zz}, σ_{xz} and σ_{yz}, respectively. Here,

$$\beta_2(z) = \frac{G(z) - G'(z)}{2G(z)G'(z)}, \qquad \beta_3(z) = \frac{2(1 + v'(z))G'(z) - E'(z)}{2E'(z)G'(z)},$$

$$\beta_4(z) = \frac{v'(z)E(z) - E'(z)}{E(z)E'(z)}, \qquad \beta_5(z) = \frac{E(z) - 2G'(z)}{2E(z)G'(z)}.$$

Combining Eqs. (7) and (9), (10) with Eq. (8) in mind, we obtain the following equation

$$\Delta\left(\frac{1 - v(z)}{E(z)}\sigma\right) - \beta_1(z)\Delta_{xy}\sigma = \frac{\sigma_{zz}}{2}\frac{d^2}{dz^2}\left(\frac{1}{G'(z)}\right) + \beta_6(z)\Delta_{xy}\sigma_{zz}$$

$$- \beta_2(z)\frac{\partial^2\sigma_{zz}}{\partial z^2} + \frac{\partial^2}{\partial z^2}((2\beta_1(z) - \beta_2(z))\sigma_{zz}) \qquad (11)$$

for the total stress σ and the stress σ_{zz} only. Here, $\beta_6(z) = \frac{E'(z) - E(z)}{E(z)E'(z)}$.

2.3   Solution of the Governing Equations

For the solution of the derived set of equations, we employ an appropriate method for the separation of variables, which relies on the geometry of the section of body K by a plane parallel to xOy and on the boundary conditions, given on the limiting surface. If, for example, the shape of the mentioned section is an infinite plane and the boundary conditions impose the stresses to vanish at the points on infinity, the Fourier double-integral transform can be employed [7, 8]. If the section is finite in certain direction, then, e.g., the infinite series by the appropriate eigenfunctions can be in use [6].

By solving then Eqs. (7) and (11), we can determine the stresses σ and σ_{zz}. Then, making use of Eqs. (6) and (9), the in-plane normal stresses can be determined.

After the normal stresses are found, the shearing stresses can be found from Eq. (6). For each case, the corresponding boundary and integral conditions [6–8] can be obtained by making use of Eq. (5) and the appropriate integration of Eq. (1).

3   Numerical Examples and Discussion

As an example, consider the computation of the stress-tensor components in a TIM layer ($|z| \leq 1$) subjected to the force loadings

$$\sigma_{zz}(x, y, \pm 1) = -p \exp(-c_x x^2 - c_y y^2), \quad \sigma_{xz}(x, y, \pm 1) = \sigma_{yz}(x, y, \pm 1) = 0. \quad (12)$$

where p is an arbitrary constant in the dimension of stresses, $c_x, c_y = \text{const}$. In this case, the Fourier double-integral transform with respect to x and y has been employed for the separation of variables in Eqs. (7)–(11).

The effect of the transversal isotropy is demonstrated in Fig. 1. The full-field distribution of the transversal stress shown in this figure demonstrates the normal stresses attaining the maximum values in the vicinity of the loading zones of the boundary planes $z = \pm 1$; they decrease when moving away from the boundaries in the z-direction and vanish when $x, y \to \pm\infty$. The stress σ_z is compressive within the affected area. Comparing Figs. 1a and 1b, we can observe how significant is the effect of transversal isotropy in the distribution of the stress-tensor components for the considered 3D problem.

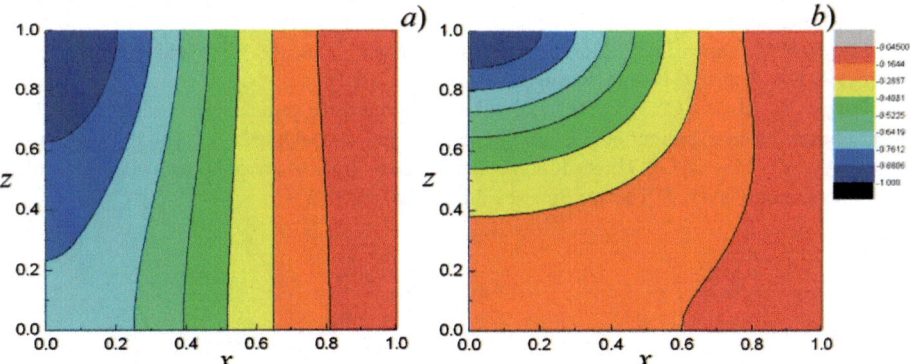

Fig. 1. The dimensionless transversal stress σ_{zz}/p in the cross-section $y = 0$ of a TIM composite layer due to the force loading (12) with $c_x = 3$ and $c_y = 5$ in the case of (a) the carbon fiber ($E = 15\,\text{GPa}$, $E' = 232\,\text{GPa}$, $v = 0.49$, $v' = 0.28$, $G = 5.03\,\text{GPa}$, $G' = 24\,\text{GPa}$) and (b) the ceramic PZT–4 ($E = 81.28\,\text{GPa}$, $E' = 64.53\,\text{GPa}$, $v = 0.33$, $v' = 0.34$, $G = 30.56\,\text{GPa}$, $G' = 25.6\,\text{GPa}$)

4 Conclusions

An approach for the construction of analytical solutions to the general 3D elasticity problems for nonhomogeneous TIM composite solids is presented for the case when the solids are subjected to external force loadings on the limiting surface. By making use of the direct integration method, the problems are reduced to the solution of governing equations for the separate stress-tensor components, which were obtained on the basis of the compatibility equations in terms of strains and solved by making use of the separation of variables.

The solution is convenient for the analysis as well as the practical computation of the stress state. It is also potentially attractive for the verification of the analytical or numerical methods developed for the analysis of nonhomogeneous TIM.

References

1. Ding, H., Chen, W., Zhang, L.: Elasticity of Transversely Isotropic Materials. Springer, Dordrecht (2006)
2. Wang, M.Z., Xu, B.X., Gao, C.F.: Recent general solutions in linear elasticity and their applications. Appl. Mech. Rev. **61**(3), 030803-1–030803-20 (2008)
3. Podil'chuk, Y.N.: Exact analytic solutions of three-dimensional boundary-value problems of the statics of a transversely isotropic body of canonical form (survey). Int. Appl. Mech. **33**(10), 763–787 (1997)
4. Wang, W., Shi, M.X.: On the general solutions of transversely isotropic elasticity. Int. J. Solids Struct. **35**(25), 3283–3297 (1998)
5. Eubanks, R.A., Sternberg, E.: On the axisymmetric problems elasticity theory for a medium with transverse isotropy. J. Ration. Mech. Anal. **3**, 89–101 (1954)
6. Tokovyy, Y.V.: Direct Integration Method. In: Hetnarski, R.B. (ed.) Encyclopedia of Thermal Stresses, vol. 2, pp. 951–960. Springer, Dordrecht (2014)
7. Tokovyy, Y., Ma, C.-C.: An analytical solution to the three-dimensional problem on elastic equilibrium of an exponentially-inhomogeneous layer. J. Mech. **31**(5), 545–555 (2015)
8. Tokovyy, Y., Ma, C.-C.: Three-dimensional temperature and thermal stress analysis of an inhomogeneous layer. J. Therm. Stresses **36**(8), 790–808 (2013)

Thermal and Thermoelastic State of Thermosensitive Structures Subject to Complex Heat Exchange

Roman Kushnir[(✉)]

Pidstryhach Institute for Applied Problems of Mechanics and Mathematics,
National Academy of Sciences of Ukraine, Lviv 79060, Ukraine
dyrector@iapmm.lviv.ua

Abstract. The key features of the thermal- and thermo-stressed analyses for the thermosensitive structures are considered for the case of complex heat-exchange with the surroundings. A nonlinear boundary-value problem of heat conduction is formulated and the analytical-numerical solution methods are presented. For the determination of the thermoelastic state in the bodies of such kind, the thermoelasticity equations with variable coefficients are implemented. The solutions to the later ones are constructed by making use of a technique that is similar to the perturbation method.

Keywords: Thermosensitive structure · Nonlinear heat conductivity problem
Complex heat exchange · Analytical-Numerical methods · Thermoelastic state
Perturbation method

1 Introduction

In order to perform the efficient and feasible analysis of engineering problems related to the heating/cooling process in solids, the effect of thermosensitivity (the material characteristics depend on the temperature) is to be taken into consideration when solving the relevant heat conductivity problems [1–4]. It is also important to construct the solutions to the aforementioned heat conduction problems in analytical form. This requirement is motivated, for instance, by the need of solving the thermoelasticity problems for thermosensitive bodies, for those the determined temperature is a kind of the input data, and thus, is desired to be in the analytical form.

In general, the model of a thermosensitive body emanates a nonlinear heat conductivity problem. The exact solutions to such problems can be determined when the temperature or heat flux are given on the surface by assuming the material to be "simply nonlinear" (thermal conductivity λ_t and volumetric heat capacity c_v depend on the temperature, but the relation, called thermal diffusivity $a = \lambda_t/c_v$, is assumed to be constant). For solution construction in this case, it is sufficient to use the Kirchhoff transformation in order to obtain the corresponding linear problem for Kirchhoff's variable. In the case of complex heat exchange, however, the Kirchhoff transformation linearizes the heat conductivity problem only partially. This is due to the fact that, in the heat conductivity problem for the Kirchhoff's variable, the heat conduction

© Springer International Publishing AG, part of Springer Nature 2019
E. E. Gdoutos (Ed.): ICTAEM 2018, SI 5, pp. 377–382, 2019.
https://doi.org/10.1007/978-3-319-91989-8_88

equation is nonlinear (due to dependence of the thermal diffusivity on the Kirchhoff's variable). The boundary condition of the complex heat exchange is also nonlinear due to a nonlinear expression of the temperature on the surface. Herein, we discuss several methods, developed by Kushnir and Popovych [4–6] for determination of the temperature distributions in thermosensitive bodies of canonical shape under complex (convective, radiation or convective-radiation) heat exchange through the surface. The thermoelastic state of such structures is determined from a boundary-value problem for the equations of mechanics with variable coefficients. For solving this problem a modification of the perturbation method was proposed by Kushnir and Popovych in [4].

2 Analytical-Numerical Methods for Solution of the Nonlinear Heat Conductivity Problems

First, let us consider a step-by-step linearization method for the determination of the one-dimensional transient temperature field $t(x, \tau)$ in a thermosensitive structure of simple nonlinearity under the complex heat exchange. The temperature $t(x, \tau)$ can be found from the following non-linear heat conduction equation

$$\frac{1}{x^m} \frac{\partial}{\partial x} \left(x^m \lambda_t(t) \frac{\partial t}{\partial x} \right) = c_v(t) \frac{\partial t}{\partial \tau} - W, \tag{1}$$

where $m = 0; \ 1; \ 2$ correspond to Cartesian, cylindrical and spherical coordinate systems, respectively, $a \leq x \leq b$, $a \geq 0$, $a < b \leq \infty$, the density of the heat sources W is a function of coordinate x and time τ. The surface $x = a$, for instance, is exposed to the convective-radiation heat exchange with the environment of a constant temperature t_a

$$\left[\lambda_t(t) \frac{\partial t}{\partial x} - \alpha_a(t)(t - t_a) - \sigma \varepsilon_a(t)(t^4 - t_a^4) \right]_{x=a} = 0, \tag{2}$$

while the surface $x = b$ is heated with a constant temperature t_b or a constant heat flux q_b

$$t|_{x=b} = t_b \ \text{ or } \ \lambda_t(t) \frac{\partial t}{\partial x} \Big|_{x=b} = q_b. \tag{3}$$

At the initial moment of time, the temperature is uniformly distributed within the body

$$t|_{\tau=0} = t_p. \tag{4}$$

Here, $\alpha_a(t)$ is the temperature-dependent coefficient of heat exchange between the surface and the environment; $\varepsilon_a(t)$ is the temperature-dependent emittance; σ is the Stefan-Boltzmann constant.

The method provides:

- reduction of the heat conduction problem (1)–(4) to the corresponding dimensionless problem (we can present the functional parameters $\lambda_t(t)$, $c_v(t)$, $\alpha_a(t)$, and $\varepsilon_a(t)$ in the form $\chi(t) = \chi_0 \chi^*(T)$, where $T = t/t_0$ is the dimensionless temperature, χ_0 is a reference value and $\chi^*(T)$ stands for the dimensionless function, t_0 is a reference temperature, $T_p = t_p/t_0$);
- partial linearization of the obtained problem by making use of the Kirchhoff's transform $\theta = \int\limits_{T_p}^{T} \lambda_t^*(t)dt$;
- full linearization of the nonlinear condition on the Kirchhoff's variable θ, that has been obtained from the condition of complex heat exchange due to approximation of the nonlinear term by specially constructed spline of zero or first order;
- construction of the solution to the linearized boundary value problem for θ by means of the appropriate analytical method;
- determination of the temperature in question by means of the inverse Kirchhoff transform;
- determination of the unknown parameters of spline-approximation, those remain in the expression for the temperature, by means of the collocation method.

The method of a step-by-step linearization is applicable for the determination of temperature fields in thermo-sensitive plates, half-space, solid and hollow cylinders or spheres, space with cylindrical or spherical cavities, whose surfaces are exposed to the conditions of convective, radiation or convective-radiation heat exchange [4]. This method has been efficiently used for the solution of the two-dimensional steady problem in thermosensitive bodies.

Next, the efficient method of the linearizing parameters [4–6] can be employed for the determination of temperature fields in structures with simple nonlinearity under the convective heat exchange through the limiting surfaces for an arbitrary dependence of the heat conduction coefficient on the temperature. The main feature of this method lies in the fact that the full linearization of the nonlinear condition for the Kirchhoff variable θ (obtained from the condition of convective heat exchange) is achieved by substitution of the nonlinear term $T(\theta)$ by $(1+\kappa)\theta + T_p$ with unknown parameter κ. This parameter can be found by satisfaction of the nonlinear condition for θ with required accuracy.

The method of the linearizing parameters is adapted to solution of the nonlinear steady-state and transient heat conduction problems for contacting thermosensitive bodies of a simple geometrical shape under conditions of the perfect thermal contact at the interfaces and complex heat exchange on the limiting surfaces.

3   Determination of the Stress-Strain State

The determination of thermoelastic state of the thermosensitive body is reduced to a boundary-value problem for a system of partial differential equations with variable coefficients. For solving this problem, some modifications of the perturbation method were used by Noda [3] and Kushnir and Popovych [4]. Let us illustrate the application of the latter one for the determination of the quasi-static stress-strain state in a thermosensitive structure subjected to a given centrally-symmetric temperature field $T(\rho, Fo)$ and constant pressure p on the limiting surface. Under the foregoing conditions, the stress-strain state is governed by the non-zero dimensionless radial displacement $\bar{u}$, in terms of which the dimensionless radial $\sigma_\rho = \sigma_r/(2G_0\alpha_{t0}t_0)$ and circumferential $\sigma_\phi = \sigma_\varphi/2G_0\alpha_{t0}t_0$ stresses are expressed

$$\sigma_\rho = \bar{G}(T)\left[(1-v(T))\frac{\partial\bar{u}}{\partial\rho} + 2v(T)\frac{\bar{u}}{\rho} - (1-v(T))\Phi^*(T)\right],$$

$$\sigma_\phi = \bar{G}(T)\left[\frac{v(T)\partial\bar{u}}{\partial\rho} + \frac{\bar{u}}{\rho} - (1-v(T))\Phi^*(T)\right], \tag{5}$$

where $\Phi^*(T) = \frac{1+v(T)}{1-v(T)}\int\limits_{T_p}^{T}\alpha_t^*(T)dT, \bar{G}(T) = \frac{G^*(T)}{(1-2v(T))}$ and the temperature-dependent share modulus $G(T)$, the coefficient of linear thermal expansion α and the Poisson's ratio v are given in the form $\chi(t) = \chi_0\chi^*(T)$.

By substituting the relation (5) into the equilibrium equation, we can obtain the differential equation with variable coefficients for determination $\bar{u}$ [4, 6]

$$\frac{\partial}{\partial\rho}\left(\frac{1}{\rho^2}\frac{\partial}{\partial\rho}(\rho^2\bar{u})\right) = \frac{\partial\Phi^*}{\partial\rho} - \psi(T)\left(\frac{\partial\bar{u}}{\partial\rho} + 2m(T)\frac{\bar{u}}{\rho} - \Phi^*(T)\right), \tag{6}$$

where

$$\psi(T) = \frac{\partial}{\partial\rho}\left(\ln\left(G^*(T)\frac{1-v(T)}{1-2v(T)}\right)\right),$$

$$m(T) = \left(\frac{\partial}{\partial\rho}\left(G^*(T)\frac{1-v(T)}{1-2v(T)}\right)\right)^{-1}\frac{\partial}{\partial\rho}\left(\frac{G^*(T)v(T)}{1-2v(T)}\right).$$

The solution to Eq. (6) is to be constructed by making use of the perturbation method. In the same time, we shall consider the differential equation with variable coefficients

$$\frac{\partial}{\partial\rho}\left(\frac{1}{\rho^2}\frac{\partial}{\partial\rho}(\rho^2\bar{u})\right) = \frac{\partial\Phi^*(T)}{\partial\rho} + \psi(T)\Phi^*(T) - \varepsilon\psi(T)\left(\frac{\partial\bar{u}}{\partial\rho} + 2m(T)\frac{\bar{u}}{\rho}\right), \tag{7}$$

which coincides with (6) when $\varepsilon = 1$.

The solution to Eq. (7) is given in the form of expansion in powers of the parameter ε

$$\bar{u} = \sum_{k=0}^{\infty}\varepsilon^k\bar{u}_k(\rho, Fo). \tag{8}$$

Substituting (8) into Eq. (7) and equating the terms at equal powers ε, we obtain the differential equation relative to the zero component $\bar{u}_0$

$$\frac{\partial}{\partial \rho}\left(\frac{1}{\rho^2}\frac{\partial}{\partial \rho}(\rho^2 \bar{u}_0)\right) = \frac{\partial \Phi^*(T)}{\partial \rho} + \psi(T)\Phi^*(T) \tag{9}$$

and the recursion sequence of differential equation relative to the component $\bar{u}_k$ $(k \geq 1)$

$$\frac{\partial}{\partial \rho}\left(\frac{1}{\rho^2}\frac{\partial}{\partial \rho}(\rho^2 \bar{u}_k)\right) = -f_{k-1}(\rho, Fo) \tag{10}$$

of displacement $\bar{u}$, where $f_{k-1}(\rho, Fo) = \psi(T)\left(\frac{\partial \bar{u}_{k-1}}{\partial \rho} + 2m(T)\frac{\bar{u}_{k-1}}{\rho}\right)$.

Taking into account the above, the solution to Eq. (6)

$$\bar{u} = \sum_{k=0}^{\infty} \bar{u}_k(\rho, Fo), \tag{11}$$

where $\bar{u}_0$ and $\bar{u}_k$ present the solutions to Eqs. (9) and (10), which, respectively, are of the form

$$\bar{u}_0 = c_{10}\rho + \rho^{-2}(c_{20} + H(\rho) - \frac{H_3(\rho)}{3}) + \frac{\rho H_0(\rho)}{3},$$

$$\bar{u}_k = c_{1k}\rho + \rho^{-2}(c_{2k} + \frac{H_3^{k-1}(\rho)}{3}) - \frac{\rho H_0^{k-1}(\rho)}{3}. \tag{12}$$

Here, c_{ik} $(i = 1, 2)$ are the constants of integration, which are determined from the boundary conditions; $H(\rho) = \int_a^\rho \xi^2 \Phi^*(\xi, Fo)d\xi$, $H_m(\rho) = \int_a^\rho \xi^m \psi(T)\Phi^*(\xi, Fo)d\xi$, $H_m^{(k-1)}(\rho) = \int_a^\rho \xi^m f_{k-1}(\xi, Fo)d\xi$.

Taking into account the presentation (11), the thermal stresses are calculated by formulas $\{\sigma_\rho, \sigma_\phi\} = \sum_{k=0}^{\infty} \{\sigma_{\rho k}, \sigma_{\phi k}\}$, where the terms of corresponding series are given in [4, 6].

4 Conclusions

The analysis of the thermo-stressed state for a thermosensitive structure has shown the following results [4–6]: the thermal stresses occur in the load-free thermosensitive solids even for a linear distribution of the temperature (whereas they are absent in non-thermosensitive solids in the same situation); on the basis of the simplified (due to neglecting the thermal dependences for some of the material characteristics) mathematical models, one can obtain the distributions of the temperature and thermal stresses, which differs substantially from the actual ones; the unsubstantial linearization of the heat conduction problems can bring to quantitative discrepancy, as well as to the unfeasible results; for the structure members made of steel, the discrepancy between the temperatures computed for thermosensitive and non-thermosensitive models within the temperature range 273–673 $°K$ falls within 10%, for the displacements – 20 – 40%, for the stresses – 25 – 30%. These discrepancies appear to be even greater for the cases of the radiative heating or piece-wisely homogeneous solids.

References

1. Nowinski, J.: Transient thermoelastic problem for an infinite medium with a spherical cavity exhibiting temperature-dependent properties. J. Appl. Mech. **29**(2), 399–407 (1962)
2. Podstrihach, Y.S., Kolyano, Y.M.: Nonstationary Temperature Fields and Stresses in Thin Plates. Naukova Dumka, Kyiv (1972)
3. Noda, N.: Thermal stresses in materials with temperature-dependent properties. In: Hetnarski, R.B. (ed.) Thermal Stresses I, pp. 391–483. Elsevier, North-Holland (1986)
4. Kushnir, R.M., Popovych, V.S.: Thermoelasticity of Thermosensitive Solids. Spolom, L'viv (2009)
5. Kushnir, R.M., Popovych, V.S.: Heat conduction problems of thermosensitive solids under complex heat exchange. In: Vikhrenko, V.S. (ed.) Heat Conduction – Basic Research, pp. 131–154. InTech, Rijeka (2011)
6. Popovych, V.: Methods for determination of the thermo-stressed state of thermosensitive solids under complex heat exchange conditions. In: Hetnarski, R.B. (ed.) Encyclopedia of Thermal Stresses, vol. 6, pp. 2997–3008. Springer, Dordrecht (2014)

Analytical Solutions of Some Nonstationary Contact Problems with Moving Boundaries

Grigory V. Fedotenkov[1,2(✉)] and Dmitry V. Tarlakovskii[1,2]

[1] Moscow Aviation Institute (National Research University),
Volokolamskoye shosse 4, Moscow 119192, Russian Federation
greghome@mail.ru
[2] Dynamic Testing Laboratory, Institute of Mechanics, Lomonosov Moscow
State University, Michurinsky prospect 1, Moscow 119192, Russian Federation
https://mai.ru

Abstract. The paper considers plain nonstationary contact problems for perfectly rigid die translating into an elastic half-space. As a general case, the boundary of the contact area is assumed to be moving one. The method of functionally invariant solutions is used as a tool of analytic solution and study of distribution of contact stresses. By making use this method, the formulations of the problems are reduced to Riemann-Hilbert problem. It is shown that the distribution of stresses near the boundaries of the region differs for four ranges of contact speed: superseismic, first transseismic, second transseismic and subseismic. It is also shown that in the case when the given stresses or displacements are arbitrary continuous functions, the problem reduces to the considered way with help of approximating the given functions by homogeneous polynomials. The authors found and studied the analytic solution of the problems of indentation with conical and parabolic dies.

Keywords: Plain nonstationary contact problems · Perfectly rigid die
Elastic half-space · Method of functionally invariant solutions
Riemann-Hilbert problem · Keldysh-Sedov mixed boundary problem

1 Statement of the Problem

At the initial time a perfectly rigid die starts to be pressed into an elastic isotropic half-space. The die is moving vertically according to a predetermined motion law. The coordinate system $Oxyz$ is of Cartesian type. The Oz-axis points inward the half-space and the Ox-axis runs along the undisturbed free surface of the half-space. As the die is assumed to extend infinitely along the Oy axis, we will consider plane problems. The contact between the die and the half-space is assumed to be non-adhesive. The motion of the half-space is described by the equations of motion in the φ and ψ elastic potentials [1, 2].

The relationships between displacements and stresses are given by the Cauchy relations end the Hook's low. The half-space has zero initial conditions. The boundary conditions has complicated mixed type with moving boundaries.

© Springer International Publishing AG, part of Springer Nature 2019
E. E. Gdoutos (Ed.): ICTAEM 2018, SI 5, pp. 383–384, 2019.
https://doi.org/10.1007/978-3-319-91989-8_89

2   Analytical Solution Method

Let us assume that the elastic potentials are homogeneous function of n degree. Therefore, by employing the method of functionally invariant solutions [2] the problem can be reduced to finding one unknown function $F(\theta)$ of a complex variable θ from the boundary conditions on the real axis Im $\theta = 0$ [2].

In simplest cases this problem reduces to Dirichlet problem [3], and in more complicated cases, to Keldysh-Sedov mixed boundary problem [3] or to Riemann-Hilbert problem [4].

3   Results

By making use of the method of functionally invariant solutions we obtained analytic results of the solution of nonstationary contact problems of interaction of a cone and parabolic dies with an elastic half-space. We also obtained and studied the specificities of contact strains around the moving boundary of the contact area.

4   Conclusions

We obtained the analytic solutions of a number of plane nonstationary contact problems for perfectly rigid dies and elastic half-space. The paper offers a generalization of a method of solution of more sophisticated contact problems for dies of an undefined cross section. The method obtained can also be applied when solving problems of crack propagation in an elastic body.

Acknowledgements. The authors would like to acknowledge the financial support of the Russian Foundation for Basic Research (project 16-08-00260 A).

References

1. Tarlakovskii, D.V., Fedotenkov, G.V.: Non-stationary problems for elastic half-plane with moving point of changing boundary conditions. PNRPU Mech. Bull. **3**, 188–206 (2016). https://doi.org/10.15593/perm.mech/2016.3.13
2. Poruchikov, V.B.: Methods of the Classical Theory of Elastodynamics. Springer, Heidelberg (1993)
3. Lavrentiev, M.A., Shabat, B.V.: Methods of The Theory of Functions of Complex Variable. Nauka, Moscow (1973)
4. Gakhov, F.D.: Boundary Value Problems. Dover, New York (1990)

Influence of Viscoelasticity Properties on Propagation of Two-Dimensional Non-stationary Waves in a Half-Plane

Ekaterina Korovaytseva[(⊠)] and Dmitry Tarlakovskii

Institute of Mechanics, Lomonosov Moscow State University,
Michurinsky prospect, 1, Moscow 119192, Russia
katrell@mail.ru

Abstract. This paper studies a linear two-dimensional non-stationary problem of a viscoelastic half-plane with normal displacements applied on its boundary.

The solution is presented in the form of generalized convolution of the corresponding solutions of two-dimensional problem for elastic half-plane and one-dimensional problem for viscoelastic half-plane. Both solutions are written as convolutions of boundary conditions with surface Green functions as kernels. To build Green functions Laplace time transform is used. Two-dimensional elastic half-plane problem is written in potentials. Inversion of the transforms is carried out analytically. For plane elasticity theory problem analytic representations of the transforms are used, for one-dimensional viscoelastic problem the transform is expanded in series in powers of relaxation kernel which is taken in the form of two-parameter exponential function. As a result formulas for normal and tangential stresses at the boundaries of viscoelastic half-plane are obtained. This allows analyzing the distribution of stresses in time and coordinate and comparing it with the solution of the corresponding problem for elastic half-plane.

Keywords: Viscoelasticity · Half-space · Green function
Generalized convolution · Laplace transform

Viscoelastic half-plane movement with coinciding volume and shear relaxation kernels is considered. At the initial time medium is at rest, on the boundary of the half-space the displacements are given, while at infinity perturbations don't exist. We assume that hereditary properties are described by two-dimensional exponential relaxation kernel. Using the confirmation proved in [1], we write down the constrained solution of the problem as generalized convolution of corresponding plane elasticity theory problem solution and one-dimensional viscoelasticity theory problem solution. The convolution is carried out in parameter which has the meaning of time for elastic problem and coordinate for one-dimensional viscoelasticity theory problem. Boundary conditions for these auxiliary problems are determined as follows: for plane elastic problem we take the part of original problem boundary conditions which depends only on coordinate, while for one-dimensional viscoelasticity theory problem we take the part which depends only on time. Stresses in viscoelastic half-space can also be written using the representation mentioned above as generalized convolution of stresses in elastic

© Springer International Publishing AG, part of Springer Nature 2019
E. E. Gdoutos (Ed.): ICTAEM 2018, SI 5, pp. 385–386, 2019.
https://doi.org/10.1007/978-3-319-91989-8_90

half-space and a function of time and coordinates constructed on the basis of one-dimensional viscoelastic problem solution.

The solution of two-dimensional problem for elastic half-plane is represented following [2, 3] in integral form as convolution of Green functions and boundary conditions. Green functions are determined writing the problem in constrained in the half-plane scalar potential and non-zero component of vector potential of the displacements. Then we restrict ourselves to determination of stresses on the boundary of the half-plane. Applying integral coordinate Fourier transform and time Laplace transform to the relations of the elastic problem, we can obtain transforms of surface Green functions. For the originals of these functions calculation we use an algorithm of sequential inversion of Laplace and Fourier transforms described in [2].

The solution of one-dimensional viscoelastic problem is determined analytically, which on the one hand allows constructing analytical form of the original problem solution, but on the other hand is valid only for limited time and coordinate intervals and certain values of relaxation kernel parameters.

Stresses in two-dimensional viscoelastic problem can be calculated only numerically as the expressions for them constructed using solutions of the auxiliary problems mentioned above contain an integral which can't be taken analytically. Calculations show that contrary to elastic medium the wave front observed doesn't have discontinuities of the first kind.

Acknowledgement. The reported study was funded by RFBR, according to the research project No. 16-38-60074 mol_a_dk.

References

1. Ilyasov, M.Kh.: Non-Stationary Viscoelastic Waves. Azerbaijan Hava Yollary, Baku (2011)
2. Gorshkov, A.G., Medvedskii, A.L., Rabinskii, L.N., Tarlakovskii, D.V.: Waves in Continuous Media. Physmatlit, Moscow (2004)
3. Gorshkov, A.G., Tarlakovskii, D.V.: Dynamic Contact Problems with Moving Borders. Physmatlit, Moscow (1995)

Non-stationary Dynamic Problems of Linear Viscoelasticity with a Constant Poisson's Ratio

Sergey Pshenichnov[(✉)]

Institute of Mechanics, Lomonosov Moscow State University,
Michurinsky prospect, 1, Moscow 119192, Russia
serp56@yandex.ru

Abstract. The problems of propagation of non-stationary waves in linear viscoelastic bodies on condition that the Poisson's ratio of the material does not change through the time are considered. The issues of finding of the solutions of such problems by the method of Laplace transform in time are discussed. The general form of the solution in transforms is presented. The case when a hereditary kernel is an exponential two-parametrical one is considered. We have demonstrated that in such case the singular points of the Laplace transforms are connected by a simple relation with singular points of the Laplace transforms for the corresponding elastic body. There have been conditions established under which the poles of transform have the first order and the original is simpler. As an example, the analytical solution of the problem of one-dimensional non-stationary longitudinal wave propagation in a viscoelastic cylinder is presented. This solution is valid within the whole range of time without the assumption of smallness of viscosity.

Keywords: Viscoelastic bodies · Non-stationary waves · Laplace transform

1 Introduction

In studies of transient wave processes in viscoelastic bodies an important role is played by the analytical methods of construction the solutions of non-stationary dynamic problems of linear viscoelasticity. One of such methods developed by Ilyasov involves a special type of convolution [1]. There is a method of modal expansion developed by Zheltkov [2] and method of biorthogonal spectral expansion, developed by Lychev [3]. The Laplace transformation in time followed by inversion is the most common procedure of construction the solutions of the considered problems (for instance, [4, 5]). Let us remark here that the mathematical complexity significantly limits the class of the studied problems. The most results were received either in a limited time range or with low viscosity or they were represented in a hardly analyzable form.

The purpose of this work is to consider the issues related to constructing the solutions of problems of the above-noted types with a time-independent Poisson's ratio.

© Springer International Publishing AG, part of Springer Nature 2019
E. E. Gdoutos (Ed.): ICTAEM 2018, SI 5, pp. 387–388, 2019.
https://doi.org/10.1007/978-3-319-91989-8_91

2   The Issues Discussed and the Main Results

Dynamic initial-boundary value problems describing transient wave processes in linear viscoelastic bodies in the framework of the Boltzmann-Volterra model for small deformations are considered. It is assumed that the following conditions are met: the area of disturbance is limited; creep of material is limited; the rigid body displacements are excluded.

The integral Laplace transform in time is applied. In the work [6] some important general properties of the solution in transforms are established. In this paper, a special case when the Poisson's ratio of the material does not depend on time is considered. Then the solution of the problem in transforms with allowance for the above conditions can be represented as a sum of two terms, one of which is a series of eigenfunctions of the problem of free vibrations of the corresponding linearly elastic body. The sufficient conditions for the absence of branch points for solutions in transforms are formulated. Questions concerning the order of its poles are considered.

The case when hereditary kernel is an exponential two-parametrical one is considered. It has been established that in such case the singular points of the Laplace transforms are connected by a simple relation with singular points of the Laplace transforms for the corresponding elastic body. The sufficient conditions under which the poles of transform have the first order and the original is simpler are formulated.

As an example, a problem of propagation of a non-stationary longitudinal wave in a cross section of a viscoelastic infinitely long cylinder being initially in a non-perturbed state is considered. The interior surface of the cylinder is rigidly fixed, while the outer surface is exposed to an axisymmetric radial load evenly distributed along the cylinder's element since the initial moment. The solution was presented in a rather simple form and remains true within the whole range of time changing.

Acknowledgements. The reported study was funded by Russian Foundation for Basic Research, according to the research projects No. 16-08-00260 a, 18-08-00471 a.

References

1. Ilyasov, M.K.: Non-stationary Viscoelastic Waves. Azerbaijan Hava Yollary, Baku (2011). (in Russian)
2. Jeltkov, V.I., Tolokonnikov, L.A., Khromova, N.G.: The transient functions in the dynamics of viscoelastic bodies. Acad. Sci. USSR Rep. **329**(6), 718–719 (1993). (in Russian)
3. Lycheva, T.N., Lychev, S.A.: Spectral decompositions in dynamical viscoelastic problems. PNRPU Mech. Bull. **4**, 120–150 (2016). (in Russian)
4. Filippov, I.G., Cheban, V.G.: The Mathematical Theory of Vibrations of Elastic and Viscoelastic Plates and Rods. Stiintsa, Chisinau (1988). (in Russian)
5. Colombaro, I., Giusti, A., Mainardi, F.: On the propagation of transient waves in a viscoelastic Bessel medium. Z. Angew. Math. Phys. **68**, 62 (2017). https://doi.org/10.1007/s00033-017-0808-6
6. Pshenichnov, S.G.: Nonstationary dynamic problems of linear viscoelasticity. Mech. Solids **48**(1), 68–78 (2013)

Transient Contact Problem for Liquid Filled Concentric Spherical Shells and a Rigid Barrier

Elena Yu. Mikhailova[1]([⊠]), Dmitry V. Tarlakovskii[1,2], and Grigory V. Fedotenkov[1,2]

[1] Moscow Aviation Institute (National Research University), Volokolamskoye shosse 4, Moscow 119192, Russian Federation
mihel6@yandex.ru
[2] Dynamic Testing Laboratory, Institute of Mechanics, Lomonosov Moscow State University, Michurinsky prospect 1, Moscow 119192, Russian Federation
https://mai.ru

Abstract. A vertical impact of fluid-filled spherical shells (indenter) with a rigid barrier (foundation) is investigated. The contact between the shells and the filling and between the shell and the foundation is adhesive-free. The basic resolving integral equation resulting from the boundary condition caused by the impact between the shells system and the foundation is obtained. The kernel of this equation is the transient function for a system of shells with a filler. The transient function for the indenter is constructed analytically with help of Laplace integral transform and Fourier series. The system of governing equations is derived, and the numerical-analytical algorithm of solution of the formulated problem are described.

Keywords: Transient contact problems · Fluid-filled spherical shell
Transient functions · Integral equations · Laplace integral transform
Fourier series

1 Statement of the Problem

This paper investigates a vertical impact of a system of fluid-filled elastic shells (indenter) on a rigid barrier (foundation). The contact between the shell and the filling and between the shell and the foundation is adhesive-free. The mathematical model of this dynamical process includes equations of motions of shells of Timoshenko model in displacements [1–3]; equation of liquid motion, liquid's physical relation and a relation between the liquid velocity at the point and the potential of velocity [1]. The zone of contact between the shells system and the rigid barrier is defined by the condition of intersection of the surface of the foundation with the undeformed surface of the shell. The statement of the problem is finished by the initial conditions.

© Springer International Publishing AG, part of Springer Nature 2019
E. E. Gdoutos (Ed.): ICTAEM 2018, SI 5, pp. 389–391, 2019.
https://doi.org/10.1007/978-3-319-91989-8_92

2 Solution Method

For the purpose of investigation of the stress-strain condition we have obtained the basic resolving equation resulting from the boundary condition caused by the impact between the shells system and the foundation. This equation includes normal displacements of the external shell which are connected with contact pressure through the integral relation [4]. It is completed to a closed system of resolving equations with the equations of mathematical model of the process of contact interaction. Such system involves a numerical analytical algorithm based on the quadrature method. Taking account of hyperbolic type of the equations of motion of shell and half-space, an explicit scheme of integrating is used.

The indenter transient function represents normal displacements which are resulted from the solution of a system of equations with homogeneous initial value conditions and instant normal load in form of multiplication of Dirac delta functions [3, 4]. The system of equations includes shell motion equations, liquid motion equation, its physical relation and a relation which binds the velocity of the liquid at the point with the potential of velocities, as well as the initial value condition determined by the condition of no adhesion between the shell and liquid.

A method of separation of variables is used for solution of this system. All the required functions depending on radial and time variables are expanded into series in terms of Legendre polynomials and their derivatives. As a result we will obtain an infinite system of differential equations with respect to unknown coefficients depending on time and angle coordinate. The solution of the system is sought with a Laplace transform and constitutes a transient function of indenter in form of a series.

3 Conclusions

The transient function for the system of two concentric spherical shells with liquid layer between them was found analytically with help of Laplace integral transform and Fourier series. The numerical-analytical algorithm of solution of the transient contact problem for liquid filled concentric spherical shells and a rigid barrier is constructed.

Acknowledgements. The authors would like to acknowledge the financial support of the Russian Foundation for Basic Research (project 16-08-00260 A).

References

1. Gorshkov, A.G., Medvedsky, A.L., Rabinsky, L.N., Tarlakovsky, D.V.: Waves in Continuum Media. FIAMATLIT, Moscow (2004)
2. Mikhailova, E.Yu., Fedotenkov, G.V.: Nonstationary axisymmetric problem of the impact of a spherical shell on an elastic half-space (initial stage of interaction). Mech. Solids **46**(2), 239–247 (2011). https://doi.org/10.3103/S0025654411020129

3. Tarlakovskii, D.V., Fedotenkov, G.V.: Two-dimensional nonstationary contact of elastic cylindrical or spherical shells. J. Mach. Manuf. Reliabil. **43**(2), 145–152 (2014). https://doi.org/10.3103/S1052618814010178
4. Tarlakovskii, D.V., Fedotenkov, G.V.: Nonstationary 3D motion of an elastic spherical shell. Mech. Solids **50**(2), 208–217 (2015). https://doi.org/10.3103/S0025654415020107

Arbitrary Oriented Defects in Anisotropic Quarter Plane

Kostyantyn Arkhypenko⬭ and Oleksandr Kryvyi$^{(\boxtimes)}$⬭

National University "Odessa Maritime Academy", Odessa, Ukraine
kryvyi-od@math.onma.edu.ua, krivoy-odessa@ukr.net

Abstract. To solve the problems about arbitrary oriented defects in anisotropic quarter plane, a method based on the use of the space of generalized functions with slow growth properties was developed. Two-dimensional integral Fourier transform was used to construct the system of fundamental solutions for anisotropic quarter plane with the concentrated jumps of stresses and displacements in this space. The latter allows the problems about the defects to be reduced to systems of singular integral equations (SIE) with fixed singularities. As an example the problem about the crack exiting to the boundary of the quarter plane has considered, for which the system of SIE with fixed singularities is obtained. To determine the singularities' indexes of the stresses and displacements at the vertices of the quarter plane and of the crack the transcendental equations are obtained. The asymptotic behavior of the solutions is investigated. The efficient numerical-analytical method for solving systems of SIE with fixed singularities is proposed.

Keywords: Arbitrary oriented defects · Anisotropic quarter plane
System of singular integral equations · Exiting crack · Fixed singularities

The stress-strain state of the anisotropic quarter plane containing the defect L_0 ($L_0 = \{y = g(x), x \in [a;b]\}$) is described by a vector $\mathbf{v} = \{v_p(x,y)\}_{p=\overline{1,5}} = \{\sigma_x, \sigma_y, \tau_{xy}, u, v\}$, $(x,y) \in \mathbb{R}^2_+ = L_+ \times L_+$, $L_+ = (0, +\infty)$, whose boundary values on the faces of the quarter plane and the defect are denoted by

$$\chi_k(x) = v_{k+1}(x, +0), k = \overline{1,4}, \ \mu_1(y) = v_1(+0, y), \mu_k(y) = v_{k+1}(+0, y), k = \overline{2,5}, x \in L_+, \quad (1)$$

$$p_k^{\pm}(t) = u_k^+(t) \pm u_k^-(t), \ k = \overline{1,4}, t \in L_0,$$

where $u_k^{\pm}(t)$– the limiting values respectively from the side of the normal $\mathbf{n}$ and from the opposite side to the contour L_0 for the components of the vector of the stresses and displacements $\mathbf{u} = \{u_k(x,y)\}_{k=\overline{1,4}} = \{\tau_s, \sigma_n, u_s, u_n\}$ in the coordinate system associated with the contour L_0.

The components of the vector $\mathbf{v}$ satisfy the equilibrium equations and the generalized Hooke's law. When setting the problem of anisotropic elasticity, two functions of the representations (1) must be known on the each faces of the quarter plane and the

© Springer International Publishing AG, part of Springer Nature 2019
E. E. Gdoutos (Ed.): ICTAEM 2018, SI 5, pp. 392–393, 2019.
https://doi.org/10.1007/978-3-319-91989-8_93

defect. The unknown functions are to be determined by solving the problem. The stress tensor's components and the displacement's vector are represented in the form:

$$v_p(x, y) = v_p^1(x, y) + v_p^2(x, y), p = \overline{1, 5}, \tag{2}$$

where $\{v_p^1(x, y)\}_{p=\overline{1,5}}$ – the solution of the boundary-value problem for the quarter plane without defect, $\{v_p^2(x, y)\}_{p=\overline{1,5}}$ – the solution of the problem about the defect in the anisotropic quarter plane with free (zero) boundary conditions at the quarter-plane boundary.

Following the papers [1–8], we will construct systems of fundamental solutions for each solution from (2) and establish the integral relations connecting functions (1) on the faces of the quarter plane and of the defect. These systems of fundamental solutions allow the problem about an arbitrarily oriented defect to reduce to system of singular integral equations (SIE) with fixed singularities.

In this paper, as an example, the problem about the crack exiting to the boundary of the quarter plane has considered, for which the system of SIE with fixed singularities is constructed. To determine the singularities' indexes of the stresses and displacements at the vertices of the quarter plane and of the crack the transcendental equations are obtained. The asymptotic behavior of the solutions is investigated. The efficient numerical-analytical method for solving systems of SIE with fixed singularities is proposed.

References

1. Krivoi, A.F., Radiolo M.V.: Features of the stress field near inclusions in composite anisotropic plane. In: Proceedings of the AS USSR. Mechanics of Solid, vol. 3, pp. 84–92 (1984)
2. Krivoi, A.F.: Arbitrarily oriented defects in composite anisotropic plane. Vest. Odessa State Univ. Ser. Phys. Math. Sci. 6(3), pp. 108–115 (2001)
3. Krivoi, A.F.: Fundamental solution for four-component anisotropic plane. Vest. Odessa State Univ. Ser. Phys. Math. Sci. 8(2), 140–149 (2003)
4. Kryvyi, O.F., Arkhypenko, K.M.: Crack going on the conjunction's line of two different anisotropic halfplanes. Math. Methods Phys.-Mech. Fields 48(3), 110–116 (2005)
5. Krivoi, A.F., Popov, G.Y.: Interface tunnel cracks in composite anisotropic space. J. Appl. Math. Mech. 72(3), 499–507 (2008)
6. Krivoi, A.F., Popov, G.Y.: Features of the stress field near tunnel inclusions in an inhomogeneous anisotropic space. Int. Appl. Mech. 44(6), 626–634 (2008)
7. Kryvyy, O.F.: Tunnel internal crack in piecewise homogeneous anisotropic space. J. Math. Sci. 198(1), 62–74 (2014)
8. Kryvyi, O.F.: Mutual influence of interface tunnel crack and interface tunnel inclusion in piecewise homogeneous anisotropic space. J. Math. Sci. 208(4), 409–416 (2015)

Interphase Circular Inclusion in a Piecewise-Homogeneous Transversely Isotropic Space Under the Action of a Heat Flux

Oleksandr Kryvyi[1]([⊠]) [iD] and Yurii Morozov[2] [iD]

[1] National University "Odessa Maritime Academy", Odessa, Ukraine
krivoy-odessa@ukr.net, kryvyi-od@math.onma.edu.ua
[2] Institute of Mechanical Engineering, Odessa National Polytechnic University,
Odessa, Ukraine

Abstract. In the space of generalized functions of slow growth $\Im'(\mathbb{R}^3)$ a discontinuous solution of the stationary thermoelasticity problem for a piecewise homogeneous transversally isotropic space is constructed in the case of arbitrary loading. Using the constructed discontinuous solution and the properties of the functions from $\Im'(\mathbb{R}^3)$, two-dimensional singular integral relations are obtained which allow the problem of interphase defects in an inhomogeneous transversely isotropic space allow to reduce to systems of two-dimensional singular integral equations (SIE) with kernels that are expressed in terms of elementary functions. An exact solution of the thermoelasticity problem for interphase circular inclusion is constructed, which is in complete coupling with different transversely isotropic half-spaces. Expressions for the jumps of the normal and tangential stresses are obtained and the dependences of the translational displacements of the inclusion on the temperature, the resultant load, the main moment, and the thermomechanical characteristics of transversely isotropic materials are obtained.

Keywords: Thermoelasticity problem · Interphase circular inclusion
Singular integral equations
Piecewise-homogeneous transversely isotropic space

1 Formulation of the Problem

The problem of stationary thermoelasticity for a piecewise homogeneous transversally isotropic space under the influence of a heat flux of a given intensity $q_0(x_1, x_2, x_3)$. In the plane $x_3 = 0$ of connection of two transversally-isotropic spaces, contains the insulated absolutely rigid inclusion which occupies area: $\Omega : \{0 \le r \le a, \, 0 \le \varphi \le 2\pi\}$. The inclusion is subject to an arbitrary load, the action of which results in a resultant force $\mathbf{P} = (P_1, P_2, P_3)$ and the main moment $\mathbf{M} = (M_1, M_2, M_3)$. The location of the inclusion faces after deformation describes the functions:

© Springer International Publishing AG, part of Springer Nature 2019
E. E. Gdoutos (Ed.): ICTAEM 2018, SI 5, pp. 394–396, 2019.
https://doi.org/10.1007/978-3-319-91989-8_94

$$\zeta_6^{\pm} = \zeta_6^0 + \vartheta_0^{\pm}(x_1, x_2), \ \zeta_k^{\pm} = \zeta_k^0, \ k = 4, 5, \ (x_1, x_2) \in \Omega$$

$$\zeta_4^0 = \delta_1 - \varphi_3 x_3, \ \zeta_5^0 = \delta_2 + \varphi_3 x_1, \ \zeta_6^0 = \delta_3 + \varphi_2 x_2 + \varphi_1 x_2, \tag{1}$$

$$\{\zeta_k^{\pm}\}_{k=1}^{8} = \{\sigma_3(\mathbf{x}), \sigma_4(\mathbf{x}), \sigma_5(\mathbf{x}), u_1(\mathbf{x}), u_2(\mathbf{x}), u_3(\mathbf{x}), T(\mathbf{x}), Q(\mathbf{x})\}|_{x_3 = \pm 0}, \ \mathbf{x} = (x_1, x_2, x_3)$$

$$\sigma = \{\sigma_k\}_{k=1}^{6} = \{\sigma_x, \sigma_y, \sigma_z, \tau_{yz}, \tau_{xz}, \tau_{xy}\}, \mathbf{u} = \{u_k\}_{k=1}^{3} = \{u, v, w\}$$

Is considered the case when the heat flow is specified on the inclusion and the inclusion is completely linked to the half-spaces, in this case the boundary conditions have the form [1, 2]:

$$\chi_4^{\pm}(x_1, x_2) = (1 \pm 1)\zeta_4^0, \ \chi_5^{\pm}(x_1, x_2) = (1 \pm 1)\zeta_5^0, \ \chi_6^{\pm}(x_1, x_2) = \vartheta^{\pm}(x_1, x_2) + (1 \pm 1)\zeta_6^0,$$

$$\vartheta^{\pm} = \vartheta_0^+ \pm \vartheta_0^-, \ \zeta_8^{\pm}(x_1, x_2, \pm 0) = q(x_1, x_2), \ (x_1, x_2) \in \Omega \tag{2}$$

Considering the conditions

$$\chi_k^-(x_1, x_2) = 0, k = \overline{1, 6}, \ \zeta_7(x_1, x_2, +0) = \zeta_7(x_1, x_2, -0)$$

$$\lambda_3^+ \partial_2 \zeta_7(x_1, x_2, +0) = \lambda_3^- \partial_2 \zeta_7(x_1, x_2, -0), \ (x_1, x_2) \notin \Omega, \tag{3}$$

which reflect the fact of complete coupling of the half-spaces beyond the limits of inclusion and using the technique described in [1–7], relatively unknown stress and temperature jumps, following system of two-dimensional singular integral equations (SIE). Applying, to the solution of the system SIE, the approach outlined in the works [3–6], we get explicit expressions for jumps of stresses on the inclusion. In particular for the jump in the normal stresses, we have

$$\langle \sigma_z \rangle^- = v_1^-(\rho, \varphi) = -\frac{1}{\pi \rho^2} \partial_\rho \int_\rho^a \frac{t \, \eta_{10}^-(t) dt}{\sqrt{t^2 - \rho^2}} - \frac{1}{\pi} \mathrm{Re} \left[e^{i\varphi} \partial_\rho \int_\rho^a \frac{\eta_{11}^-(t) dt}{\sqrt{t^2 - \rho^2}} \right], \tag{4}$$

References

1. Kryvyi, O.: The discontinuous solution for the piece-homogeneous transversal isotropic medium. Oper. Theory Adv. Appl. **191**, 387–398 (2009)
2. Kryvyi, O.: Singular integral relations and equations for a piecewise homogeneous transversally isotropic space with interphase defects. J. Math. Sci. **176**(4), 515–531 (2011)
3. Kryvyi, O.F.: Mizhfazni krugovi vkljuchennja v kuskovo-odnoridnomu transversal'no-izotropnomu prostori. Prikl. Problemi meh. i mat.iss **8**, 173–183 (2010)
4. Kryvyi, O.F.: Interface crack in the inhomogeneous transversely isotropic space. Mater. Sci. **47**(6), 726–736 (2012)
5. Kryvyi, O.F.: Interface circular inclusion under mixed conditions of interaction with a piecewise homogeneous transversally isotropic space. J. Math. Sci. **184**(1), 101–119 (2012)

6. Kryvyi, O.F.: Delaminated interface inclusion in a piecewise homogeneous transversely isotropic space. Mater. Sci. **3**(2), 245–253 (2014)
7. Krivoi, A.F., Morozov, Y.: Reshenie zadachi teploprovodnosti dlja kusochno-odnorodnogo ortotropnogo prostranstva s mezhfaznymi defektami. Visnik Odes'k.nac.un-tu. Matemat. i meh.iss. 17, **3**(15), 107–119 (2012)

Electromagnetoelastic Diffusion
in Anisotropic Continuum

Olga Afanasieva[1], Dmitry Tarlakovskii[2], and Andrei Zemskov[1,2(✉)]

[1] Moscow Aviation Institute (National Research University), Moscow, Russia
azemskov1975@mail.ru
[2] Institute of Mechanics, Lomonosov Moscow State University, Moscow, Russia

Abstract. The coupled unsteady electromagnetic processes in a multicomponent elastic solid taking into account the diffusion are investigated. A general model of thermoelectromagnetoelastic diffusion is presented for an arbitrary anisotropic continuum in a curvilinear coordinate system. The transition from the general formulation to the one-dimensional problem of electromagnetoelastic diffusion in the Cartesian coordinate system is done. To solve the problem, the Laplace transform and the Fourier expansion are used.

Keywords: Electromagnetoelastic diffusion · Green's functions
Unsteady problems

1 General Formulation of the Problem

The related unsteady electromagnetoelastic processes in a multicomponent medium taking into account mass transfer phenomena are considered. The mathematical model of the problem includes:

- the Cauchy momentum equation and components of Cauchy strain tensor (inirial state is undeformed) [1, 2];
- the linearized equation of the entropy balance and the Fick's law of mass transfer [1–5];
- the Maxwell's equations and the generalized Ohm's law (the current density as a function of the velocity of medium motion and the diffusion flow is added to the classical variant) [1, 2];
- an expression for the Lorentz force and the inflow of electromagnetic energy [1, 2].

Physical relations are constructed using the thermodynamic potential method. In addition, quadratic approximations are used for the free energy function, which depends on the deformations, the changes in concentration of medium components, the components of the electric and magnetic induction vectors.

© Springer International Publishing AG, part of Springer Nature 2019
E. E. Gdoutos (Ed.): ICTAEM 2018, SI 5, pp. 397–398, 2019.
https://doi.org/10.1007/978-3-319-91989-8_95

2 Method of Solution

From a general multidimensional formulation of the problem for an arbitrary aniso-tropic medium in a curvilinear coordinate system a transition to a one-dimensional problem in a Cartesian coordinate system is performed. Additionally it is assumed that:

- the elastic medium is an ideal solid solution;
- direct piezoelectric, piezomagnetic and diffusion-electric effects are weakly expressed.

This allows us to consider the electrodynamics problem separately. The electro-magnetic fields found in this way are included in the volume perturbations of the elastic diffusion problem. In turn, the elastic-diffusion fields found here are included in the volume perturbations of the electrodynamic problem.

The solutions of one-dimensional elastic diffusion problems with volume pertur-bations were constructed in [6–8]. Analogous problems of electrodynamics are solved by methods of operational calculus.

Acknowledgments. This work was funded by the subsidy from Russian Foundation for Basic Research (RFBR) (Project 17-08-00663 A).

References

1. Sedov, L.I.: Continuum Mechanics, vol. 1, 2. Nauka, Moscow (1973). [in Russian]
2. Tarlakovskii, D.V., Vestyak, V.A., Zemskov, A.V.: Dynamic Processes in Thermoelectro-magnetoelastic and Thermoelastodiffusive Media (2014)
3. Eremeev, V.S.: Diffusion and Stresses. Energoatomizdat, Moscow (1984). [in Russian]
4. Knyazeva, A.G.: Introduction to Locally-Equilibrium Thermodynamics of Physical and Chemical Transformations in Deformable Media. Ivan Fedorov, Tomsk (2014). [in Russian]
5. Zemskov, A.V., Tarlakovskii, D.V.: Approximate solution of three-dimensional problem for elastic diffusion in orthotropic layer. J. Math. Sci. **203**(2), 221–238 (2014)
6. Davydov, S.A., Zemskov, A.V., Tarlakovskii, D.V.: An elastic half-space under the action of one-dimensional time-dependent diffusion perturbations. Lobachevskii J. Math. **36**(4), 503–509 (2015)
7. Zemskov, A.V., Tarlakovskii, D.V.: Two-dimensional nonstationary problem elastic for diffusion an isotropic one-component layer. J. Appl. Mech. Techn. Phys. **56**(6), 1023–1030 (2015)
8. Igumnov, L.A., Tarlakovskii, D.V., Zemskov, A.V.: A two-dimensional nonstationary problem of elastic diffusion for an orthotropic one-component layer. Lobachevskii J. Math. **38**(5), 808–817 (2017)

An Analysis of Electrically Conducting and Magnetically Permeable Interface Crack in a Piezoelectromagnetic Bimaterial

V. Loboda$^{(\boxtimes)}$ and A. Grynevych

Oles Honchar Dnipro National University,
72 Gagarin Av., Dnipro 49010, Ukraine
loboda@dnu.dp.ua

Abstract. Exact analytical approach for plane stain investigation of piezoelectric/piezomagnetic bimaterial with magnetically permeable and electrically conducting crack located at the interface between its components is suggested. It is assumed that a normal and shear stresses field and also the electric and magnetic fields paralleled to the crack are given at infinity. First, an open crack is considered, and then it is assumed that the crack faces may have a frictionless contact at a certain section of unknown length, which is adjacent to one of the crack tip. In the last case the combined Dirichlet-Riemann boundary value problem and Hilbert problem have been formulated and the exact analytical solution has been presented. For both cases a numerical illustrations of the obtained solutions were done for different values of mechanical and electric fields at infinity. The dependence of magnetic, electric and mechanical characteristics on the mechanical load and electric field were investigated.

Keywords: Magnetically permeable interface crack
Piezoelectric/piezomagnetic bimaterial · Contact zone

1 Formulation of the Problem

Two bonded semi-infinite spaces $x_3 > 0$ and $x_3 < 0$ having an interface crack in the segment $x_1 \in (c, b)$ of the interface are considered. The materials are assumed to be poled in the direction x_3 and the crack is considered as free of mechanical loading, electrically conducting, magnetically permeable and having the total electric charge D_0. The frictionless contact zone can arise in the zone $x_1 \in (a, b)$ at the right crack tip. It means that the boundary conditions at the crack faces are the following:

$$\sigma_{i3}^{(m)}(x_1, 0) = 0, \ E_1^{(m)}(x_1, 0) = 0,$$
$$H_1^{(1)}(x_1, 0) = H_1^{(2)}(x_1, 0), \ B_3^{(1)}(x_1, 0) = B_3^{(2)}(x_1, 0) \quad \text{for} \quad x_1 \in (c, a), \tag{1}$$

$$\langle u_3(x_1) \rangle = 0, \ \sigma_{13}^{(m)}(x_1, 0) = 0, \ E_1^{(m)}(x_1, 0) = 0, \ \langle \sigma_{33}(x_1) \rangle = 0 \quad \text{for} \quad x_1 \in (a, b) \tag{2}$$

where the superscripts (1) and (2) are related to the upper and lower semi-infinite spaces and the brackets $\langle \ \rangle$ mean the jump of the associated function over the interface.

© Springer International Publishing AG, part of Springer Nature 2019
E. E. Gdoutos (Ed.): ICTAEM 2018, SI 5, pp. 399–400, 2019.
https://doi.org/10.1007/978-3-319-91989-8_96

It is also assumed that uniformly distributed $\sigma_{13}^{\infty} = \tau^{\infty}$, $\sigma_{33}^{\infty} = \sigma^{\infty}$, $E_1^{\infty} = e^{\infty}$, $H_1^{\infty} = h^{\infty}$ are prescribed at infinity.

2 Solution of the Problem and the Results

The presentations of electro-magneto-mechanical fields at the material interface via the functions $F_j(z)$ which are analytic in the whole plane except the crack region are obtained.

At the beginning the open crack model, which takes place for $a = b$, was considered. In this case the following Hilbert problems

$$F_1^+ (x_1) + \gamma_1 F_1^- (x_1) = -\theta_1, F_4^+ (x_1) + F_4^- (x_1) = -\theta_4 \quad \text{for} \quad x_1 \in (c, b), \quad (3)$$

are formulated and solved analytically by the method of [1]. This solution leads to the physically unrealistic oscillating singularity at the crack tips, therefore the case of $a < b$ was also considered. This case corresponds to the existing of the contact zone and leads to the additional equation

$$\text{Im} \, F_1^{\pm}(x_1) = 0 \quad \text{for} \quad x_1 \in (a, b), \quad (4)$$

for the function $F_1(z)$. First Eqs. (5) and (6) constitute the combined Dirichlet-Riemann boundary value problem. The solutions of these problems under the conditions at infinity and additional conditions, which reflect displacement uniqueness and total electric and magnetic charges, are found in an analytical form. Due to this solution the analytical expressions for all required mechanical, electric and magnetic quantities along the interface are presented in a close form for any position of the point a.

The conditions of nonpositiveness of the normal stress in the contact zone and non-penetration of the crack faces on $x_1 \in (a, b)$ are used for the determination of the real contact zone length. Satisfaction of the mentioned conditions leads to the transcendental equation with respect to $\lambda = (b - a)/(b - c)$. The solution of this equation can easily be found numerically and its maximum root from the interval $(0, 1)$ must be used.

Due to the numerical analysis of the obtained analytical solutions the influence of the external mechanical, electrical and magnetic loading as well as the total electric crack charge on the stresses, electric and magnetic fields, contact zone length and the electrical, magnetic and mechanical intensity factors were studied.

Reference

1. Muskhelishvili, N.I.: Some Basic Problems of Mathematical Theory of Elasticity, 707 p. Noordhoff International Publishing, Leyden (1977)

Interaction of Antiplane Shear Waves with Elastic Fiber in the Presence of a Thin Interphase Piezoceramic Layer

Yaroslav Kunets[1], Roman Kushnir[1(✉)], Valeriy Matus[1],
and Oleksandr Trofymchuk[2]

[1] Pidstryhach Institute for Applied Problems of Mechanics and Mathematics,
Ukrainian National Academy of Sciences, Lviv, Ukraine
dyrector@iapmm.lviv.ua
[2] Institute of Telecommunications and Global Information Space,
Ukrainian National Academy of Sciences, Kyiv, Ukraine

Abstract. An analytical-numerical method for the investigation of steady-state wave fields in an unbounded elastic medium scattered by a non-canonical form of elastic fiber in the presence of a thin piezoceramic interphase of variable thickness and low stiffness is proposed. The components of the electroelastic system are in perfect mechanical contact; the electric induction is zero on the surface of the interphase. The interphase material belongs to the 6 mm crystallographic class. The elastic system is in the conditions of antiplane motion. The research algorithm is based on the modified null field method. The influence of mechanical and geometric parameters of the composite on the amplitude-frequency characteristics of SH-waves scattered by a fiber of non-canonical form into a far-field zone is analyzed.

Keywords: Scattering of SH-waves
Elastic fiber of non-classical cross-section · Thin piezoceramic interphase
Null-field approach

1 Introduction

The influence of thin interlayers on wave phenomena in the matrix-filler composites reinforced with elastic fibers has received a considerable attention in the scientific literature. The topicality of these problems is mainly emanated by the need to simulate the interphase damages in the matrix composites and to evaluate their subsequent influence on the effective properties of elastic structures [1–4]. In the vast majority of papers, structures with elastic fillers of canonical shape and interphase layers of constant thickness are considered. The method of null-field, developed in this paper, can be regarded as one of the effective tools for solving the problems on scattering waves of various physical nature in the objects with complex geometry.

The authors have already studied the cases of SH-waves scattering by fibers of non-canonical shape, when a thin interphase layer is elastic. In this paper, the case of a thin piezoceramic interphase layer of variable thickness and low stiffness is investigated.

© Springer International Publishing AG, part of Springer Nature 2019
E. E. Gdoutos (Ed.): ICTAEM 2018, SI 5, pp. 401–403, 2019.
https://doi.org/10.1007/978-3-319-91989-8_97

2 Introduction

Consider a homogeneous elastic unbounded medium with shear modulus μ_1 and density ρ_1. Assume the medium to contain a fiber with parameters μ_2 and ρ_2, respectively. The fiber is surrounded by a thin piezoceramic layer with shear modulus μ_0, density ρ_0 and electromechanical coupling factor η. The material of the layer belongs to the crystallographic class 6 mm, and the axis of symmetry of the sixth order is perpendicular to the plane $x_1 x_2$ [4, 5]. Here, $\mathbf{x} = (x_1, x_2)$ are the Cartesian coordinates. The thickness of the layer $h(\mathbf{x})$ is much smaller than the characteristic cross-sectional dimension of the fiber.

For the case of antiplane steady-state motion, the displacements in both matrix and fiber meet the Helmholtz equation [3]. In the interphase layer, the appropriate equation of motion and the electrostatic equation for the electric potential are satisfied [4, 5].

The components of the composite are assumed to be in the perfect mechanical contact. On the surface of the layer, the electric induction is zero. The radiation condition at infinity is also satisfied [3, 5].

We consider the case when the stiffness of the layer is much lesser than the stiffness of the matrix and fiber, and the lengths of the transverse waves in the composite are much larger than the thickness of the layer. Under such restrictions, it is advisable to use the effective asymptotically exact conditions prescribing the case when the mechanical contact between the matrix and the fiber occurs through a thin piezoceramic layer [3, 5, 6]

$$\frac{\mu_0}{\mu_1}\left(1+\eta^2\right)\left[u^1(\mathbf{x}) - u^2(\mathbf{x})\right] = h(\mathbf{x})\frac{\partial u^1(\mathbf{x})}{\partial n}, \quad \mu_1\frac{\partial u^1(\mathbf{x})}{\partial n} = \mu_2\frac{\partial u^2(\mathbf{x})}{\partial n}, \quad \mathbf{x} \in \partial W.$$

Here, $u^1(\mathbf{x})$ and $u^2(\mathbf{x})$ are the displacement in the matrix and fiber; ∂W there is a fiber contour.

Assume a plane harmonic SH-wave to travel towards the fiber. The amplitude of the scattered wave is found by means of the method of the null-field [1, 2]. In this case, the integral representations are used for displacements in the matrix and fiber. Our goal is to determine the unknown displacements and stresses on the contour of the fiber in the form of trigonometric Fourier series. Substituting these series into the equations obtained from the method of the null field [1, 2] and taking into account the effective boundary conditions on the fiber contour, we obtain a system of linear algebraic equations of infinite order with respect to the coefficients of the series. The solution of the system is constructed by the method of reduction.

3 Conclusions

Numerical analysis was carried out for various electro-elastic and geometric parameters of the composite. It is established that the spectrum of scattering amplitudes is not resonant for the values of the contrast parameter of the material of a thin interphase layer $\mu_0/(\varepsilon\mu_1) = O(1)$ in the ranges of low and medium frequencies. When the waveform size of the scaterer increases, the difference between the levels of these amplitudes is negligible. For very small rigidities of the interphase layers and various mechanical properties of matrices and fibers, the spectrum have a pronounced resonant character. Their background practically does not depend on the electro-elastic parameters of the composite. The locations of resonances approximate the natural frequencies for bodies defining the cross-sections of fibers. Moreover, the increment of the material stiffness in the layer leads increases the width of the resonances and their displacement to the right.

References

1. Kunets, Y.I., Matus, V.V., Mykhas'kiv, V.V., Boström, A., Zhang, Ch.: Scattering of a SH-wave by an elastic fiber of nonclassical cross section with an interface crack. Mech. Compos. Mater. **44**, 165–172 (2008)
2. Matus, V., Kunets, Y., Mykhas'kiv, V., Boström, A., Zhang, C.: Wave propagation in 2-D elastic composites with partially debondedfibers by the null field approach. Waves in Random and Complex Media **19**, 654–669 (2009)
3. Kunets, J., Matus, V.: Scattering of SH-waves by elastic fibre of noncanonicalcross-section given a thin interphase layer. Physico-Math. Model. Inf. Technol. **20**, 132–139 (2014). [in Ukrainian]
4. Auld, B.A.: Acoustic Fields and Waves in Solids. Krieger, Malabar (1990)
5. Sulym, G.T., Kunets, Y.I., Rabosh, R.V.: Assymptotic analysis of the dynamic interaction between a thin rectilinear piezoelectric inclusion and an elastic medium under the antiplanesheat. Bull. Donetsk Univ. **1**, 137–141 (2008). [in Ukrainian]
6. Kit, H.S., Kunets, Y.I., Yemets, V.F.: Elastodynamic scattering from a thin–walled inclusion of low rigidity. Int. J. Eng. Sci. **37**, 331–345 (1999)

The Averaged Dynamic Models of Bending and Longitudinal Vibrations for Elastic Sandwich Plates

Marina Yu. Ryazantseva[(✉)]

Institute of Mechanics, Lomonosov Moscow State University,
Michurinsky prospect 1, Moscow 119192, Russian Federation
common@imec.msu.ru

Abstract. The focus of this work is to derive the averaged 2D dynamic theory for elastic sandwich plates of symmetric structure on thickness which can describe the stress-strain state asymptotically correct on long waves and qualitative correct on sufficiently short waves. The governing two-dimensional dynamic equations for sandwich plates are derived with the help of variational asymptotic method worked out by Berdichevsky [1] and which is used to rigorously reduce three-dimensional problem into two-dimensional problem.

Keywords: Sandwich composites · 2D dynamic theory · Stress-strain state
Long waves · Variational asymptotic method · 3D energy functional
Asymptotic analysis

1 Introduction

Although sandwich composites have found increasing applications in modern engineering due to their superior mechanical properties and enhanced manufacturing technology, their application is not as extensive as one might expect. The main reason is that many new analytic models and computational schemes are constructed for specific problems without generalization. The models based on the system of kinematical hypotheses are often too complicated and computationally inefficient to be used in practice (see, for example [2]). In static the problem of derivation of 2D averaged bending theory for elastic sandwich plates from 3D dynamic theory of elasticity was solved by Berdichevsky [3] by an asymptotic analysis of 3D energy functional. It was shown when the sandwich composite plates can be described by classical theory based on Kirchhoff's hypotheses and when may need a modified theory. In dynamic the picture of deformations is more complicated. Unfortunately the averaged dynamic theory cannot be constructed as asymptotic theory and therefore it is necessary to combine an asymptotic analysis with heuristic approach.

2 Statement of the Problem

Consider a three-layered elastic plate of symmetric structure on thickness h. Each layer is made of homogeneous isotropic material, the contact between the layers is ideal. Denote by x^α ($\alpha = 1, 2$) the Cartesian coordinates in the mid-plane of the plate, and by x the normal coordinate to mid-plane. In dynamic problem the displacements of a point of the plate with coordinates x, x^α are denoted by $w^\alpha(x^\alpha, x, t)$, $w(x^\alpha, x, t)$, t is time. For symmetric elastic plates the solution of any three-dimensional problems can be represented as a superposition of two solutions: one with even function of x, $w(x^\alpha, x, t)$ and odd functions x, $w(x^\alpha, x, t)$ is bending or flexural vibrations; contrary situation is connected with longitudinal vibrations. Therefore the constructing two-dimensional dynamic model will consist of two systems of equations. The problem is to construct 2D dynamic equations as a theory of leading approximation which is asymptotically correct on the long waves and qualitative correct on the sufficiently short waves.

3 The Refined Dynamic Theory of Sandwich Elastic Plates

The main idea assumed as a basis of deriving the refined dynamic theory of sandwich elastic plates is to take into consideration the interaction of low-frequency modes of vibrations with first high-frequency forms.

Flexural Vibrations. The main hypotheses of the bending theory considering the classical flexural mode and first high-frequency mode describing the transverse shear can be represented in the following expressions for the displacements: $w = u + \Delta u y(\zeta)$ $h^2/4$, $w_\alpha = \chi_\alpha q(\zeta)h/2 - u_{,\alpha}\zeta h/2$. Function $u(x^\alpha, t)$ describes the bending of mid-surface of the sandwich plate. Functions $\chi_\beta(x^\alpha, t)$ describe the effect of transverse shear. The displacement distributions on the thickness are described by the functions $y(\zeta)$, $q(\zeta)$, the respective analytic forms were obtained in [4].

The development starts with variation formulation of three-dimensional elasticity dynamic problem with expressions for displacements represented above. The governing two-dimensional equations are constructed applying the asymptotic analysis and averaging method to three-dimensional energy functional. The averaged equations obtained in that way contain all leading effects.

Longitudinal Vibrations. The similar procedure was used to derive two-dimensional theory of longitudinal vibrations sandwich plates based on an interaction classical mode of vibrations with two first high-frequency forms.

4 Conclusions

The dynamic behavior of elastic sandwich plates of symmetric structure on thickness (hard-skin composite) is proposed to study on the basis of refined averaged theory considering the high-frequency vibration modes. The governing hyperbolic equations allow describing such effects as transverse shear, transverse compressions and geometric dispersion of waves.

Acknowledgements. The authors would like to acknowledge the financial support of the Russian Foundation for Basic Research (project 16-08-00260 A).

References

1. Berdichevski, V.L.: Variational Principles of Continuum Mechanics. Springer, Heidelberg (2009)
2. Heng, H., Salim, B., Michel, P.-F., El Mustafa, D.: Review and assessment of various theories for modeling sandwich composites. Compos. Struct. **84**, 282–292 (2008)
3. Berdichevski, V.L.: An asymptotic theory of sandwich plates. Int. J. Eng. Sci. **48**, 383–404 (2010)
4. Ryazantseva, M.Y., Antonov, F.K.: Harmonic running waves in sandwich plates. Int. J. Eng. Sci. **59**, 184–193 (2012)

Propagation of the Coupled Waves in the Electromagnetoelastic Thick-Walled Sphere

Vladimir Vestyak[1($\boxtimes$)] and Dmitry Tarlakovskii[2]

[1] Moscow Aviation Institute (National Research University), Moscow, Russia
v.a.vestyak@mail.ru
[2] Institute of Mechanics, Lomonosov Moscow State University, Moscow, Russia

Abstract. Unsteady oscillations of a thick-walled sphere under the action of a mechanic and electric fields on its surfaces are considered. The expansion in serieses on Legendre polynomials and the Laplace time transform are used. The inverse Laplace transform is performed using a specially developed algorithm.

Keywords: Electromagnetoelasticity · Coupled problems
Time-dependent axisymmetric problems · Thick-walled sphere
Green's functions

The process of propagation of nonstationary axisymmetric waves are considered in an electromagnetoelastic thick-walled sphere under the action of nonstationary surface perturbations which given on the boundary in the form of radial and tangential displacements or the tangential component of the electric field strength vector. Using the Cauchy momentum equation, the Maxwell electrodynamics equations, the generalized Ohm's law, and the expression for the Lorentz force, this process is described by a closed system of dimensionless equations with respect to the magnetic field strength H and to the displacements u and v linearized with respect to some initial state [1]. At the initial time moment the sphere is in the unperturbed state.

To solve the initial-boundary value problem the unknown radial components of the displacement vector, the electric field strength, the charge density, the Lorentz force, the current density, the right-hand sides of the boundary conditions are expanding into series on Legendre polynomials $P_n(\cos\theta)$ and similar tangential components - into series on Gegenbauer polynomials $C_{n-1}^{3/2}(\cos\theta)$. In the space of Laplace transforms on time-domain the equations and boundary conditions are written in terms of coefficients of the above series and form boundary value problems for each n. It is necessary to expand all the unknown coefficients into power series in terms of the dimensionless parameter, connecting the mechanical and electromagnetic components of the problem. As a result of this expansion, we obtain a sequence of boundary value problems recurrent in the m (m - term number of power series). If $m = 0$ then problem is purely elastic and its solution is known [2].

The solution of the formulated boundary value problems for $m \geq 1$ is represented in the integral form. The kernels in the integral representations are Green's functions of boundary value problems, the finding of which is described in detail in [1]. The Green's

© Springer International Publishing AG, part of Springer Nature 2019
E. E. Gdoutos (Ed.): ICTAEM 2018, SI 5, pp. 407–408, 2019.
https://doi.org/10.1007/978-3-319-91989-8_99

functions of the electromagnetic part of the problem are finding in the quasistatic approximation. Then finding the original of non-zero component of the magnetic field strength vector is greatly simplified by the fact that when conversion to the originals there is no need for time integration.

Under condition $n \geq 1$ the recurrent system includes originals of previous representations. The integrand in representation of the nonzero component of the coefficients of the expansion of the stress vector includes the radial derivative of the displacement. To exclude this, the formula is modified by the integration by parts. Then the integral representations for coefficients corresponding to the components of the electric field strength can be found from it.

The described algorithm makes it possible to analyze problems on the radial and axisymmetric oscillations of an electromagnetoelastic thick-walled sphere with the mutual influence of mechanical and electromagnetic fields.

Acknowledgments. This work was funded by the subsidy from RFBR 18-08-00493 A.

References

1. Vestyak, V.A., Tarlakovskii, D.V.: Unsteady axisymmetric deformation of an elastic thick-walled sphere under the action of volume forces. J. Appl. Mech. Tech. Phys. **56**(6), 984–994 (2015)
2. Gorshkov, A.G., Tarlakovskiy, D.V.: Transient Aerohydroelasticity of Spherical Bodies. Springer, Heidelberg (2001). 289 p.

Influence of Residual Stress Caused by Cladding of Anticorrosion Layer in Nozzle Zone of RPV WWER-1000 on Results of Brittle Strength Evaluation

Oleh Makhnenko[✉] and Elena Kostenevich

The E.O. Paton Electric Welding Institute,
11 Kazimir Malevich Street, Kyiv 03680, Ukraine
makhnenko@paton.kiev.ua

Abstract. Structural integrity and lifetime assessment of nuclear power plant components and RPV requires a responsible data on residual stresses. A significant complexity of experimental measurement of residual stresses in existing RPV prevents taking them into account in evaluation of safe operation. Significant progress in computer simulation of welding stresses and their interaction with temperature and operational loads using modern mathematical methods creates preconditions for more precise evaluation of non-relaxed residual stresses in RPV taking into account the real technological parameters of multi-pass welding, cladding of corrosion protective layer and PWHT. Using exist and calculated precise data the influence of residual stresses in the nozzle zone of RPV WWER-1000 on brittle strength during PTS event was evaluated.

Keywords: RPV · Residual stresses · Structural integrity

The nozzle zone of the reactor pressure vessel (RPV) WWER-1000 is a critical area considering the resistance to brittle fracture during the pressurized thermal shock (PTS) events of pouring through the nozzles of cold borated water. Residual stresses induced by welding and cladding may influence on brittle strength of RPV structure [1]. Now for these purposes, rather approximate data contained in the recommendations of the various procedures such as MPK-CXP-2000 [2], VERLIFE [3] and others [4] are used.

The distribution of residual stresses in the base material of the RPV nozzle zone, which is the bainitic 2.5Cr-Mo-V steel (15H2NMFA), depends on the microstructure phase content and respectively on the mechanical properties in the melting zone (MZ) and heat affected zone (HAZ). Arc cladding under flux of austenitic material was performed by the strip with width of 35–60 mm in two layers of a total thickness 9 mm with preheating and interpass temperature up to 250 °C. For the computer simulation of residual stresses microstructural phase transformations were taken into account by the method, based on the use of Avrami equation [5, 6]. After the 10 h of PWHT at 650 °C the numerical calculated residual stress (hoop component) is presented on Fig. 1.

The application of the residual stresses obtained numerically with rather higher level in comparison with the recommendations of VERLIFE (the hoop component is higher by about 30%) for estimating the structural integrity of the WWER-1000 RPV

© Springer International Publishing AG, part of Springer Nature 2019
E. E. Gdoutos (Ed.): ICTAEM 2018, SI 5, pp. 409–411, 2019.
https://doi.org/10.1007/978-3-319-91989-8_100

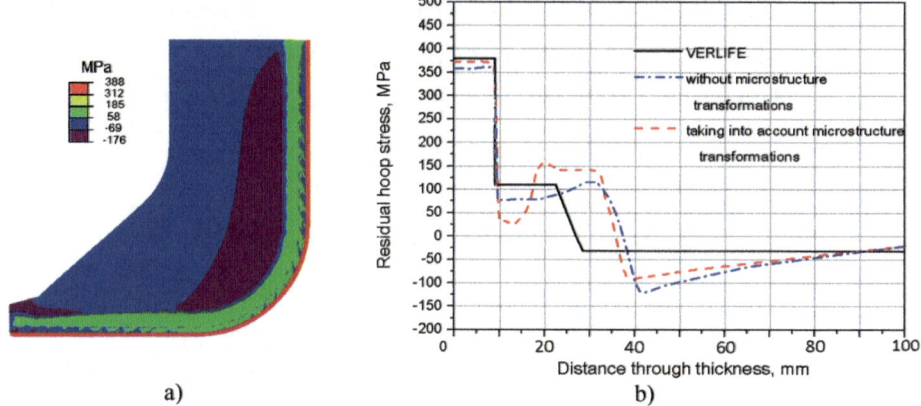

a)   b)

Fig. 1. Distribution of residual hoop stress $\sigma_{\beta\beta}$ in crosssection (a) and through thickness (b) of nozzle zone.

during PTS event increases the conservativeness of linear fracture mechanic analysis according to the SIF approximately by 2% (Fig. 2). No taking into account the residual stresses reduces the conservatism by more than 10%. This confirms the requirement to take into account the residual stresses for the calculation assessment of service life of the reactor components. Moreover, the existing data on residual stresses distributions in RPV may be not sufficiently conservative.

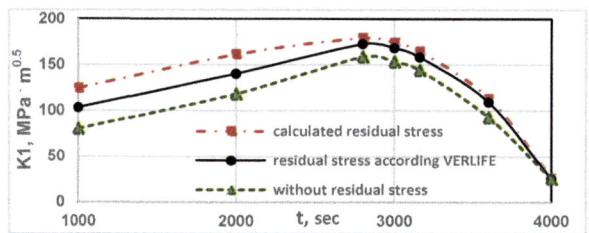

Fig. 2. SIF in the tip of underclad axes crack $(a = 50\,°\text{mm}; a/c = 0,3; c = 167\,°\text{mm})$ during PTS event.

References

1. Guidance on the Reactor Pressure Vessel PTS Assessment for WWER Nuclear Power Plants. International Atomic Energy Agency, WWER-SC-157 (1996)
2. Procedure for Lifetime Assessment of RPV in WWER during operation MPK-CXP-2000. S-Petersburg–Moscow, 52p. (2000). (in Russian)
3. Unified Procedure for Lifetime Assessment of Components and Piping in WWER NPPs "VERLIFE", ver. 5th Framework Programme of EU (2003)

4. Kostylev, V.I., Margolin, B.Z.: Determination of residual stress and strain fields caused by cladding and tempering of reactor pressure vessels. Int. J. Pressure Vessels Pip. **77**(12), 723–735 (2000)
5. Deng, D., Tong, Y., Ma, N., Murakawa, H.: Prediction of the residual welding stress in 2.25Cr-1Mo steel by taking into account the effect of the solid-state phase transformations. Acta Metall. Sin. **26**(3), 333–339 (2013)
6. Makhnenko, V.I., Velikoivanenko, E.A., Pochinok, V.E., Makhnenko, O.V., Rozynka, G.Ph., Pivtorak, N.I.: Numerical methods for the prediction of welding stress and distortions. In: Welding and Surfacing Reviews, vol. 13, Part 1, 146 p. (1999)

Development of the Theory of Multicomponent Media for Describing Dynamic Processes in Materials of Complex Rheology

N. F. Morozov[2,3], D. A. Indeitsev[2(✉)], K. L. Muratikov[1],
A. L. Glazov[1], and D. S. Vavilov[2]

[1] Ioffe Physical-Technical Institute, Russian Academy of Sciences,
Politekhnicheskaya ul. 26, St. Petersburg 194021, Russia
[2] Institute of Problems of Mechanical Engineering,
Russian Academy of Sciences, Bolshoi pr. 61, St. Petersburg 199178, Russia
Dmitry.Indeitsev@gmail.com
[3] St. Petersburg State University,
Universitetskaya nab. 7/9, St. Petersburg 199034, Russia

Abstract. In the present paper we consider the problem of describing the dynamics of material with complex rheology, when it is necessary to take into account that it consists of several subsystems with different relaxation times. Such approach, in which the deformable solid is considered as a multicomponent medium, greatly expands the possibilities of continuum mechanics allowing to describe the processes, occurring at different scale levels. Here this problem is demonstrated by using the example of the interaction between the crystalline lattice and the electron gas. The experiments on different conductors demonstrate that their dynamic response on a laser pulse of nanosecond duration is different from the form predicted by classical theory of thermoelasticity. In this case it should be supplemented by equations describing kinematics of the electron gas, which may exert a serious influence on behavior of the lattice, resulting in significant increase of the stretching phase in comparison with the classical solution.

Keywords: Thermoelasticity · Dynamic response · Electron gas

Dynamic problems describing the mechanical processes in one of the components in materials with complex rheology are of great interest. In such materials, processes in one of the components are accompanied by relaxation processes, leading to significant changes in the structures of the other components [1, 2]. In an experimental study, such transformations are usually accompanied by a subsequent slow relaxation, characterized by responses with lower-frequency spectra.

As is well known, the main difference between multicomponent models of a continuous medium and the theory of mixtures is the possibility of considering the kinematics of each component without reducing the element of a continuous medium to a single velocity of the center of mass while maintaining the connection between the degrees of freedom through internal interaction forces and source terms contained in the balance equations masses for each component.

This problem occurs when the laser is exposed to the surface of multicomponent materials. Among such materials, an important case is represented by conductive materials (metals, semiconductors). In addition to the dynamical processes in the lattice, kinematic processes in the electron gas are awakened in parallel. The interaction of electrons and lattice leads to the need to consider the processes in both subsystems of conductors simultaneously.

This approach becomes especially important in the study of dynamic processes under the action of laser radiation. In this case, the laser radiation energy, as a rule, is absorbed by the electronic subsystem in the first stage of the interaction and is transferred to the lattice only as a result of the subsequent interaction. When considering fast processes (femto-picosecond range), this situation leads to the need to use a two-temperature model in which the electron gas and lattice temperatures have different values [3].

In this case, the classical, coupled equations of thermoelasticity, taking into account the effect of laser irradiation only through thermal and thermoelastic phenomena in the crystal lattice, do not completely describe the formation of the acoustic pulse. Due to this fact, in this paper we study the problem of the influence of dynamic processes in the electronic subsystem on the deformation processes in the lattice. It is shown that accounting for the electronic processes can lead to a qualitative modification of deformation processes in conductors. In particular, they are significantly lengthen relaxation time in comparison with dynamic processes that take into account only the lattice component.

The development of modern methods for the diagnosis of defects in materials (first of all in metals) essentially depends on the physics of deformation processes in lattice. Their development is of paramount importance for acoustic methods, which are based on recording the deformation processes of the material lattice. Thus, the problem of describing the formation of a dynamic response in a lattice of conducting materials with allowance for the excitation of an electron gas is a fundamentally new fundamental problem.

The physical model developed for the first time which takes into account the impact of the heated electron gas on the dynamic of the crystal lattice not only due to the thermal expansion of the lattice, but also due to an additional reaction caused by the dynamic motion of the electrons themselves. In this case, the effects of excitation of the electron gas and electron-drag effects due to the motion of the thermal fronts are considered taking into account the interaction with the lattice component of the material.

In its essence, classical thermoelasticity is expanded by additional equations that take into account the dynamics of electrons, as well as their variable composition [4]. Within the framework of the proposed model, the features of deformation processes near certain model defects in conducting materials action are considered under laser irradiation.

The proposed model is implemented in the software product and tested in accordance with the experimental data. In particular, the experimental data obtained by laser ultrasonic methods for metals indented by Vickers and Rockwell test methods are given [5–7].

References

1. Indeitsev, D.A., Meshcheryakov, Y.I., Kuchmin, A.Y., Vavilov, D.S.: Multi-scale model of steady-wave shock in medium with relaxation. Acta Mech. **226**(3), 917–930 (2015)
2. Indeitsev, D.A., Skubov, D.Y., Vavilov, D.S.: Problems of describing phase transitions in solids. In: Belyaev, A.K., Irschik, H., Krommer, M. (eds.) Mechanics and Model-Based Control of Advanced Engineering Systems, pp. 181–188. Springer, Vienna (2014)
3. Jiang, L., Tsai, H.L.: Improved two-temperature model and its application in ultrashort laser heating of metal films. J. Heat Transf. **127**(10), 1167–1173 (2005)
4. Tzou, D.Y., Chen, J.K., Beram, J.E.: Recent development of ultrafast thermoelasticity. J. Thermal Stress. **28**(6–7), 563–594 (2005)
5. Glazov, A.L., Morozov, N.F., Muratikov, K.L.: The effect of external stresses on the behavior of photoacoustic signals inside Vickers indenter marks on a steel surface. Tech. Phys. Lett. **42**(1), 67–70 (2016)
6. Glazov, A.L., Morozov, N.F., Muratikov, K.L.: Variations of photoacoustic signals within the Vickers indent in metals under external stresses by the examples of steel and nanocopper. Phys. Solid State **58**(9), 1735–1743 (2016)
7. Glazov, A.L., Morozov, N.F., Muratikov, K.L.: Photoacoustic microscopy of Vickers indentations in metals with piezoelectric detection. Int. J. Thermophys. **38**(7), 113-1–113-13 (2017)

Author Index

Printed by Printforce, the Netherlands